A·N·N·U·A·L EDITIONS

D0127693

Global Issues
05/06
Twenty-first Edition

EDITOR

Robert M. Jackson

California State University, Chico

Robert M. Jackson is a professor of political science and dean of the School of Graduate, International, and Sponsored Programs at California State University, Chico. In addition to teaching, he has published articles on the international political economy, international relations simulations, and political behavior. His special research interest is in the way northern California is becoming increasingly linked to the Pacific Basin. His travels include China, Japan, Hong Kong, Taiwan, Singapore, Malaysia, Portugal, Spain, Morocco, Costa Rica, El Salvador, Honduras, Guatemala, Mexico, Germany, Belgium, the Netherlands, Russia, and Czechoslovakia.

McGraw-Hill/Dushkin

2460 Kerper Blvd., Dubuque, IA 52001

Visit us on the Internet
http://www.dushkin.com

Credits

1. **Global Issues in the Twenty First Century: An Overview**
   Unit photo—© by Photo Disc, Inc.
2. **Population and Food Production**
   Unit photo—© CORBIS/Royalty-Free
3. **The Global Environment and Natural Resources Utilization**
   Unit photo—© Getty Images/PhotoLink
4. **Political Economy**
   Unit photo—© Getty Images/PhotoLink/D. Falconer
5. **Conflict**
   Unit photo—© Getty Images/StockTrek
6. **Cooperation**
   Unit photo—© Photograph by Master Sgt. Lono Kollars, U.S. Air Force,
   courtesy of Department of Defense
7. **Values and Visions**
   Unit photo—© Getty Images/Doug Menuez

Copyright

Cataloging in Publication Data
Main entry under title: Annual Editions: Global Issues. 2005/2006.
1. Global Issues—Periodicals. I. Jackson, Robert M., *comp.* II. Title: Global Issues.
ISBN 0–07–3112178   658'.05   ISSN 1093–278X

Twenty-first Edition

Cover image © PhotoDisc, Inc.
Printed in the United States of America   1234567890QPDQPD98765   Printed on Recycled Paper

Preface

In publishing ANNUAL EDITIONS we recognize the enormous role played by the magazines, newspapers, and journals of the public press in providing current, first-rate educational information in a broad spectrum of interest areas. Many of these articles are appropriate for students, researchers, and professionals seeking accurate, current material to help bridge the gap between principles and theories and the real world. These articles, however, become more useful for study when those of lasting value are carefully collected, organized, indexed, and reproduced in a low-cost format, which provides easy and permanent access when the material is needed. That is the role played by ANNUAL EDITIONS.

The beginning of the new millennium was celebrated with considerable fanfare. The prevailing mood in much of the world was that there was a great deal for which we could congratulate ourselves. The very act of sequentially watching on television live celebrations from one time zone to the next was proclaimed as a testimonial to globalization and the benefits of modern technology. The tragic events of September 11, 2001, however, were a stark reminder of the intense emotions and methods of destruction available to those determined to challenge the status quo. The subsequent wars in Afghanistan and Iraq along with continuing acts of terror have dampened the optimism that was expressed at the outset of the twenty-first century.

While the mass media may focus on the latest crisis for a few weeks or months, the broad forces that are shaping the world are seldom given the in-depth analysis that they warrant. Scholarly research about these historic forces of change can be found in a wide variety of publications, but these are not readily accessible. In addition, students just beginning to study global issues can be discouraged by the terminology and abstract concepts that characterize much of the scholarly literature. In selecting and organizing the materials for this book, we have been mindful of the needs of beginning students and have, thus, selected articles that invite the student into the subject matter.

Each unit begins with an introductory article(s) providing a broad overview of the subject area to be studied. The following articles examine in more detail specific case studies which often identify the positive steps being taken to remedy problems. Recent events are a continual reminder that the world faces many serious challenges, the magnitude of which would discourage even the most stouthearted individual. While identifying problems is easier than solving them, it is encouraging to know that many are being addressed.

Perhaps the most striking feature of the study of contemporary global issues is the absence of any single, widely held theory that explains what is taking place. As a result, we have made a conscious effort to present a wide variety of points of view. The most important consideration has been to present global issues from an international perspective, rather than from a purely American or Western point of view. By encompassing materials originally published in different countries and written by authors of various nationalities, the anthology represents the great diversity of opinions that people hold. Two writers examining the same phenomenon may reach very different conclusions. It is not just a question of who is right or wrong, but rather understanding that people from different vantage points can have differing perspectives on an issue.

Another major consideration when organizing these materials was to explore the complex interrelationship of factors that produce social problems such as poverty. Too often, discussions of this problem (and others like it) are reduced to arguments about the fallacies of not following the correct economic policy or not having the correct form of government. As a result, many people overlook the interplay of historic, cultural, environmental, economic, and political factors that form complex webs that bring about many different problems. Every effort has been made to select materials that illustrate this complex interaction of factors, stimulating the beginning student to consider realistic rather than overly simplistic approaches to the pressing problems that threaten the existence of civilization.

In addition to an annotated *table of contents* and a *topic guide,* included in this edition of *Annual Editions: Global Issues* are *World Wide Web* sites that can be used to further explore topics addressed in the articles.

This is the twentieth-first edition of *Annual Editions: Global Issues.* When looking back over more than two decades of work, a great deal has taken place in world affairs, and the contents and organization of the book reflect these changes. Nonetheless there is one underlying constant. It is my continuing goal to work with the editors and staff at McGraw-Hill/Dushkin to provide materials that encourage the readers of this book to develop a life-long appreciation of the complex and rapidly changing world in which we live. This collection of articles is an invitation to further explore the global issues of the twenty-first century and become personally involved in the great issues of our time.

Finally, materials in this book were selected for both their intellectual insights and readability. Timely and well-written materials should stimulate good classroom lectures and discussions. I hope that students and teachers will enjoy using this book. Readers can have input into the next edition by completing and returning the postage-paid article rating form in the back of the book.

Robert M. Jackson Editor

Editor

Contents

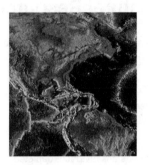

UNIT 1
Global Issues in the Twenty First Century: An Overview

UNIT 2
Population and Food Production

The concepts in bold italics are developed in the article. For further expansion, please refer to the Topic Guide and the Index.

UNIT 3
The Global Environment and Natural Resources Utilization

UNIT 4
Political Economy

The concepts in bold italics are developed in the article. For further expansion, please refer to the Topic Guide and the Index.

The concepts in bold italics are developed in the article. For further expansion, please refer to the Topic Guide and the Index.

UNIT 5
Conflict

The concepts in bold italics are developed in the article. For further expansion, please refer to the Topic Guide and the Index.

UNIT 6
Cooperation

The concepts in bold italics are developed in the article. For further expansion, please refer to the Topic Guide and the Index.

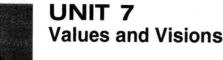

UNIT 7
Values and Visions

The concepts in bold italics are developed in the article. For further expansion, please refer to the Topic Guide and the Index.

Topic Guide

This topic guide suggests how the selections in this book relate to the subjects covered in your course. You may want to use the topics listed on these pages to search the Web more easily.

On the following pages a number of Web sites have been gathered specifically for this book. They are arranged to reflect the units of this *Annual Edition*. You can link to these sites by going to the DUSHKIN ONLINE support site at *http://www.dushkin.com/online/*.

ALL THE ARTICLES THAT RELATE TO EACH TOPIC ARE LISTED BELOW THE BOLD-FACED TERM.

Agriculture
1. A Special Moment in History
7. Bittersweet Harvest: The Debate Over Genetically Modified Crops
8. Deflating the World's Bubble Economy
10. Water Scarcity Could Overwhelm the Next Generation
20. Is Chile a Neoliberal Success?
34. The Ultimate Crop Insurance

Communication
3. Five Meta-Trends Changing the World
12. The Complexities and Contradictions of Globalization
15. Will Globalization Go Bankrupt?
25. What's Wrong With This Picture?

Conservation
1. A Special Moment in History
7. Bittersweet Harvest: The Debate Over Genetically Modified Crops
8. Deflating the World's Bubble Economy
9. Shifting the Pain: World's Resources Feed California's Growing Appetite
10. Water Scarcity Could Overwhelm the Next Generation
11. Vanishing Alaska
22. Thirty Years of Petro-Politics
34. The Ultimate Crop Insurance
43. Why Environmental Ethics Matters to International Relations

Cultural Customs and Values
2. America's Sticky Power
3. Five Meta-Trends Changing the World
4. Holy Orders: Religious Opposition to Modern States
11. Vanishing Alaska
12. The Complexities and Contradictions of Globalization
14. The Five Wars of Globalization
16. Soccer vs. McWorld
21. The Fall of the House of Saud
24. How Nike Figured Out China
25. What's Wrong With This Picture?
28. Lifting the Veil: Understanding the Roots of Islamic Militancy
29. The Great War on Militant Islam
34. The Ultimate Crop Insurance
35. Medicine Without Doctors
38. Are Human Rights Universal?
41. Women Waging Peace
43. Why Environmental Ethics Matters to International Relations

Demographics
1. A Special Moment in History
5. The Big Crunch
6. Scary Strains
11. Vanishing Alaska
35. Medicine Without Doctors

Dependencies, International
2. America's Sticky Power
7. Bittersweet Harvest: The Debate Over Genetically Modified Crops
9. Shifting the Pain: World's Resources Feed California's Growing Appetite

13. Three Cheers for Global Capitalism
17. Croesus and Caesar
20. Is Chile a Neoliberal Success?

Development, economic
2. America's Sticky Power
3. Five Meta-Trends Changing the World
4. Holy Orders: Religious Opposition to Modern States
4. Holy Orders: Religious Opposition to Modern States
5. The Big Crunch
7. Bittersweet Harvest: The Debate Over Genetically Modified Crops
9. Shifting the Pain: World's Resources Feed California's Growing Appetite
10. Water Scarcity Could Overwhelm the Next Generation
12. The Complexities and Contradictions of Globalization
13. Three Cheers for Global Capitalism
14. The Five Wars of Globalization
15. Will Globalization Go Bankrupt?
18. Where the Money Went
19. Render Unto Caesar: Putin and the Oligarchs
20. Is Chile a Neoliberal Success?
23. India's Hype, Hope, and Hazards
24. How Nike Figured Out China
35. Medicine Without Doctors
36. Countdown to Eradication
38. Are Human Rights Universal?
39. The Grameen Bank

Development, social
3. Five Meta-Trends Changing the World
5. The Big Crunch
7. Bittersweet Harvest: The Debate Over Genetically Modified Crops
9. Shifting the Pain: World's Resources Feed California's Growing Appetite
10. Water Scarcity Could Overwhelm the Next Generation
11. Vanishing Alaska
12. The Complexities and Contradictions of Globalization
14. The Five Wars of Globalization
15. Will Globalization Go Bankrupt?
19. Render Unto Caesar: Putin and the Oligarchs
23. India's Hype, Hope, and Hazards
34. The Ultimate Crop Insurance
35. Medicine Without Doctors
36. Countdown to Eradication
38. Are Human Rights Universal?

Energy
8. Deflating the World's Bubble Economy
21. The Fall of the House of Saud
22. Thirty Years of Petro-Politics

Environment
1. A Special Moment in History
7. Bittersweet Harvest: The Debate Over Genetically Modified Crops
8. Deflating the World's Bubble Economy
9. Shifting the Pain: World's Resources Feed California's Growing Appetite
11. Vanishing Alaska

xiii

World Wide Web Sites

The following World Wide Web sites have been carefully researched and selected to support the articles found in this reader. The easiest way to access these selected sites is to go to our DUSHKIN ONLINE support site at *http://www.dushkin.com/online/*.

AE: Global Issues 05/06

The following sites were available at the time of publication. Visit our Web site—we update DUSHKIN ONLINE regularly to reflect any changes.

General Sources

U.S. Information Agency (USIA)
http://usinfo.state.gov

USIA's home page provides definitions, related documentation, and discussions of topics of concern to students of global issues. The site addresses today's Hot Topics as well as ongoing issues that form the foundation of the field.

World Wide Web Virtual Library: International Affairs Resources
http://www.etown.edu/vl/

Surf this site and its extensive links to learn about specific countries and regions, to research various think tanks and international organizations, and to study such vital topics as international law, development, the international economy, human rights, and peacekeeping.

UNIT 1: Global Issues in the Twenty First Century: An Overview

The Henry L. Stimson Center
http://www.stimson.org

The Stimson Center, a nonpartisan organization, focuses on issues where policy, technology, and politics intersect. Use this site to find varying assessments of U.S. foreign policy in the post–cold war world and to research other topics.

The Heritage Foundation
http://www.heritage.org

This page offers discussion about and links to many sites having to do with foreign policy and foreign affairs, including news and commentary, policy review, events, and a resource bank.

IISDnet
http://www.iisd.org/default.asp

The International Institute for Sustainable Development presents information through links to business, sustainable development, and developing ideas. "Linkages" is its multimedia resource for policymakers.

The North-South Institute
http://www.nsi-ins.ca/ensi/index.html

Searching this site of the North-South Institute, which works to strengthen international development cooperation and enhance gender and social equity, will help you find information and debates on a variety of global issues.

UNIT 2: Population and Food Production

The Hunger Project
http://www.thp.org

Browse through this nonprofit organization's site, whose goal is the sustainable end to global hunger through leadership at all levels of society. The Hunger Project contends that the persistence of hunger is at the heart of the major security issues threatening our planet.

Penn Library: Resources by Subject
http://www.library.upenn.edu/cgi-bin/res/sr.cgi

This vast site is rich in links to information about subjects of interest to students of global issues. Its extensive population and demography resources address such concerns as migration, family planning, and health and nutrition in various world regions.

World Health Organization
http://www.who.int

This home page of the World Health Organization will provide you with links to a wealth of statistical and analytical information about health and the environment in the developing world.

WWW Virtual Library: Demography & Population Studies
http://demography.anu.edu.au/VirtualLibrary/

A definitive guide to demography and population studies can be found at this site. It contains a multitude of important links to information about global poverty and hunger.

UNIT 3: The Global Environment And Natural Resources Utilization

National Geographic Society
http://www.nationalgeographic.com

This site provides links to material related to the atmosphere, the oceans, and other environmental topics.

National Oceanic and Atmospheric Administration (NOAA)
http://www.noaa.gov

Through this home page of NOAA, part of the U.S. Department of Commerce, you can find information about coastal issues, fisheries, climate, and more. The site provides many links to research materials and to other Web resources.

SocioSite: Sociological Subject Areas
http://www.pscw.uva.nl/sociosite/TOPICS/

This huge site provides many references of interest to those interested in global issues, such as links to information on ecology and the impact of consumerism.

United Nations Environment Programme (UNEP)
http://www.unep.ch

Consult this home page of UNEP for links to critical topics of concern to students of global issues, including desertification, migratory species, and the impact of trade on the environment.

UNIT 4: Political Economy

Belfer Center for Science and International Affairs (BCSIA)
http://ksgwww.harvard.edu/csia/

BCSIA is the hub of Harvard University's John F. Kennedy School of Government's research, teaching, and training in international affairs related to security, environment, and technology.

U.S. Agency for International Development
http://www.info.usaid.gov

Broad and overlapping issues such as democracy, population and health, economic growth, and development are covered on this

Web site. It provides specific information about different regions and countries.

The World Bank Group

http://www.worldbank.org

News, press releases, summaries of new projects, speeches, publications, and coverage of numerous topics regarding development, countries, and regions are provided at this World Bank site. It also contains links to other important global financial organizations.

UNIT 5: Conflict

DefenseLINK

http://www.defenselink.mil

Learn about security news and research-related publications at this U.S. Department of Defense site. Links to related sites of interest are provided. The information systems BosniaLINK and GulfLINK can also be found here. Use the search function to investigate such issues as land mines.

Federation of American Scientists (FAS)

http://www.fas.org

FAS, a nonprofit policy organization, maintains this site to provide coverage of and links to such topics as global security, peace, and governance in the post–cold war world. It notes a variety of resources of value to students of global issues.

ISN International Relations and Security Network

http://www.isn.ethz.ch

This site, maintained by the Center for Security Studies and Conflict Research, is a clearinghouse for information on international relations and security policy. Topics are listed by category (Traditional Dimensions of Security, New Dimensions of Security, and Related Fields) and by major world region.

The NATO Integrated Data Service (NIDS)

http://www.nato.int/structur/nids/nids.htm

NIDS was created to bring information on security-related matters to within easy reach of the widest possible audience. Check out this Web site to review North Atlantic Treaty Organization documentation of all kinds, to read *NATO Review,* and to explore key issues in the field of European security and transatlantic cooperation.

UNIT 6: Cooperation

Carnegie Endowment for International Peace

http://www.ceip.org

An important goal of this organization is to stimulate discussion and learning among both experts and the public at large on a wide range of international issues. The site provides links to *Foreign Policy,* to the Moscow Center, to descriptions of various programs, and much more.

Commission on Global Governance

http://www.sovereignty.net/p/gov/gganalysis.htm

This site provides access to *The Report of the Commission on Global Governance,* produced by an international group of leaders who want to find ways in which the global community can better manage its affairs.

OECD/FDI Statistics

http://www.oecd.org/statistics/

Explore world trade and investment trends and statistics on this site from the Organization for Economic Cooperation and Development. It provides links to many related topics and addresses the issues on a country-by-country basis.

U.S. Institute of Peace

http://www.usip.org

USIP, which was created by the U.S. Congress to promote peaceful resolution of international conflicts, seeks to educate people and to disseminate information on how to achieve peace. Click on Highlights, Publications, Events, Research Areas, and Library and Links.

UNIT 7: Values And Visions

Human Rights Web

http://www.hrweb.org

The history of the human rights movement, text on seminal figures, landmark legal and political documents, and ideas on how individuals can get involved in helping to protect human rights around the world can be found in this valuable site.

InterAction

http://www.interaction.org

InterAction encourages grassroots action and engages government policymakers on advocacy issues. The organization's Advocacy Committee provides this site to inform people on its initiatives to expand international humanitarian relief, refugee, and development-assistance programs.

We highly recommend that you review our Web site for expanded information and our other product lines. We are continually updating and adding links to our Web site in order to offer you the most usable and useful information that will support and expand the value of your Annual Editions. You can reach us at: *http://www.dushkin.com/annualeditions/.*

World Map

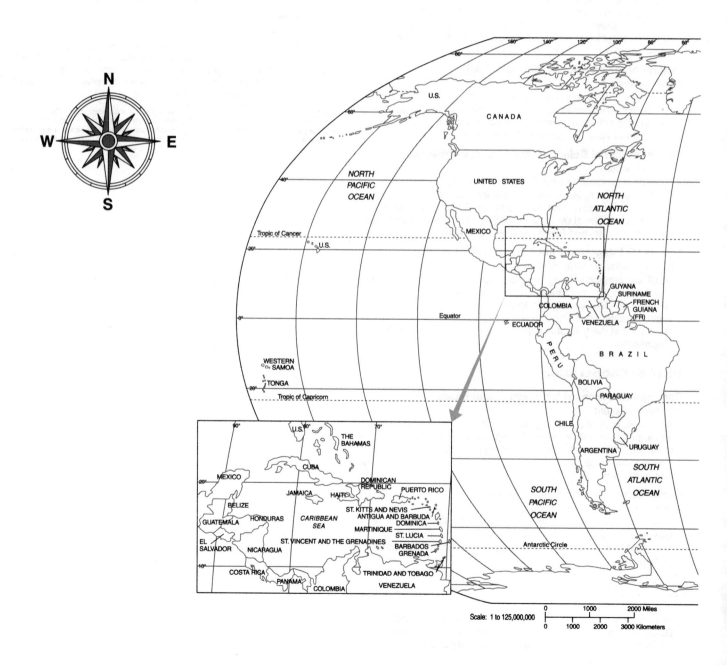

Scale: 1 to 125,000,000

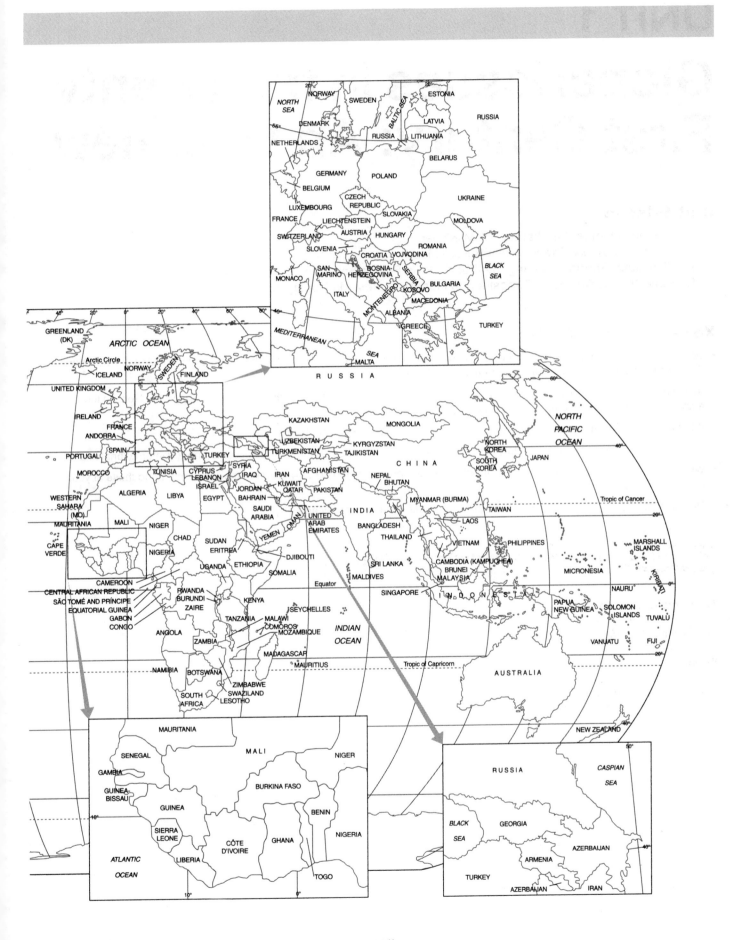

UNIT 1

Global Issues in the Twenty First Century: An Overview

Unit Selections

1. **A Special Moment in History**, Bill McKibben
2. **America's Sticky Power**, Walter Russell Mead
3. **Five Meta-Trends Changing the World**, David Pearce Snyder
4. **Holy Orders: Religious Opposition to Modern States**, Mark Juergensmeyer

Key Points to Consider

- Do the analyses of any of the authors in this section employ the assumptions implicit in the allegory of the balloon? If so, how? If not, how are the assumptions of the authors different?

- All the authors point to interactions among different factors. What are some of the relationships that they cite? How do the authors differ in terms of the relationships they emphasize?

- What assets that did not exist 100 years ago do people now have to solve problems?

- What events during the twentieth century had the greatest impact on shaping the realities of contemporary international affairs?

- What do you consider to be the five most pressing global problems of today? How do your answers compare to those of your family, friends, and classmates?

- Describe international affairs in the year 2050. How is it different from today and why?

 **Links: www.dushkin.com/online/**
These sites are annotated in the World Wide Web pages.

The Henry L. Stimson Center
http://www.stimson.org
The Heritage Foundation
http://www.heritage.org
IISDnet
http://www.iisd.org/default.asp
The North-South Institute
http://www.nsi-ins.ca/ensi/index.html

Imagine yellow paint being brushed onto an inflated, clear balloon. The yellow color, for purposes of this allegory, represents *people*. In many ways the study of global issues is first and foremost the study of people. Today, there are more human beings occupying Earth than ever before. In addition, we are in the midst of a period of unprecedented population growth. Not only are there many countries where the majority of people are under age 16, but also due to improved health care, there are more older people alive than ever before. The effect of a growing global population, however, goes beyond sheer numbers, for this trend has unprecedented impacts on natural resources and social services. An examination of population trends and the related topic of food production is a good place to begin an in-depth study of global issues.

Imagine that our fictional artist next dips the brush into a container of blue paint to represent *nature*. The natural world plays an important role in setting the international agenda. Shortages of raw materials, drought, and pollution of waterways are just a few examples of how natural resources can have global implications.

Adding blue paint to the balloon reveals one of the most important underlying concepts found in this book. Although the balloon originally was covered by both yellow and blue paint (people and nature as separate conceptual entities), the two combined produce an entirely different color: green. Talking about nature as a separate entity or people as though they were somehow removed from the forces of the natural world is a serious intellectual error. The people-nature relationship is one of the keys to understanding many of today's most important global issues.

The third color to be added to the balloon is red. This color represents *social structures*. Factors falling into this category include whether a society is urban or rural, industrial or agrarian, consumer-oriented, or dedicated to the needs of the state. The relationship between this component and the others is extremely important. The impact of political decisions on the environment, for example, is one of the most significant features of the contemporary world. Will the whales or bald eagles survive? Historically, the forces of nature determined which species survived or perished. Today, survival depends on political decisions—or indecision. Understanding the complex relationship between social structure and nature (known as "ecopolitics") is central to the study of global issues.

Added to the three primary colors is the fourth and final color of white. It represents the *meta* component (i.e., those qualities that make human beings different from other life forms). These include new ideas and inventions, culture and values, religion and spirituality, and art and literature. The addition of the white paint immediately changes the intensity and shade of the mixture of colors, again emphasizing the relationship among all four factors.

If the painter continues to ply the paintbrush over the miniature globe, a marbling effect becomes evident. From one area to the next, the shading varies because one element is more dominant than another. Further, the miniature system appears dynamic. Nothing is static; relationships are continually changing. This leads to a number of important insights: (1) there are no such things as separate elements, only connections or relationships; (2) changes in one area (such as the weather) will result in changes in all other areas; and (3) complex and dynamic relationships make it difficult to predict events accurately, so observers and policy makers are often surprised by unexpected events.

This book is organized along the basic lines of the balloon allegory. The first unit provides a broad overview of a variety of perspectives on the major forces that are shaping the world of the twenty-first century. From this "big picture" perspective more in-depth analyses follow. Unit 2, for example, focuses on population and food production. Unit 3 examines the environment and related natural resource issues. The next three units look at different aspects of the world's social structures. They explore issues of economics, national security, conflict, and international cooperation. In the final unit, a number of "meta" factors are presented.

The reader should keep in mind that, just as it was impossible to keep the individual colors from blending into new colors on the balloon, it is also impossible to separate global issues into discrete chapters in a book. Any discussion of agriculture, for example, must take into account the impact of a growing population on soil and water resources, as well as new scientific breakthroughs in food production. Therefore, the organization of this book focuses attention on issue areas; it does not mean to imply that these factors are somehow separate.

With the collapse of the Soviet empire and the end of the cold war, the outlines of a new global agenda have emerged. Rather than being based on the ideology and interests of the two superpowers, new political, economic, environmental, cultural and security issues are interacting in an unprecedented fashion. Rapid population growth, environmental decline, uneven economic progress, and global terrorist networks are all parts of a complex state of affairs for which there is no historic parallel. As we begin the twenty-first century, signs abound that we are entering a new era. In the words of Abraham Lincoln, "As our case is new, so we must think anew." Compounding this situation, however, is a whole series of old problems such as ethnic and religious rivalries.

The authors in this first unit provide a variety of perspectives on the trends that they believe are the most important to understanding the historic changes at work at the global level. This discussion is then pursued in greater detail in the following units.

It is important for the reader to note that although the authors look at the same world, they often come to different conclusions. This raises an important issue of values and beliefs, for it can be argued that there really is no objective reality, only differing perspectives. In short, the study of global issues will challenge each thoughtful reader to examine her or his own values and beliefs.

A Special Moment in History

Bill McKibben

We may live in the strangest, most thoroughly different moment since human beings took up farming, 10,000 years ago, and time more or less commenced. Since then time has flowed in one direction—toward *more*, which we have taken to be progress. At first the momentum was gradual, almost imperceptible, checked by wars and the Dark Ages and plagues and taboos; but in recent centuries it has accelerated, the curve of every graph steepening like the Himalayas rising from the Asian steppe....

But now—now may be the special time. So special that in the Western world we might each of us consider, among many other things, having only one child—that is, reproducing at a rate as low as that at which human beings have ever voluntarily reproduced. Is this really necessary? Are we finally running up against some limits?

To try to answer this question, we need to ask another: *How many of us will there be in the near future?* Here is a piece of news that may alter the way we see the planet— an indication that we live at a special moment. At least at first blush the news is hopeful. *New demographic evidence shows that it is at least possible that a child born today will live long enough to see the peak of human population.*

Around the world people are choosing to have fewer and fewer children—not just in China, where the government forces it on them, but in almost every nation outside the poorest parts of Africa.... If this keeps up, the population of the world will not quite double again; United Nations analysts offer as their mid-range projection that it will top out at 10 to 11 billion, up from just under six billion at the moment....

The good news is that we won't grow forever. The bad news is that there are six billion of us already, a number the world strains to support. One more near-doubling— four or five billion more people—will nearly double that strain. Will these be the five billion straws that break the camel's back?...

LOOKING AT LIMITS

The case that the next doubling, the one we're now experiencing, might be the difficult one can begin as readily with the Stanford biologist Peter Vitousek as with anyone else. In 1986 Vitousek decided to calculate how much of the earth's "primary productivity" went to support human beings. He added together the grain we ate, the corn we fed our cows, and the forests we cut for timber and paper; he added the losses in food as we overgrazed grassland and turned it into desert. And when he was finished adding, the number he came up with was 38.8 percent. We use 38.8 percent of everything the world's plants don't need to keep themselves alive; directly or indirectly, we consume 38.8 percent of what it is possible to eat. "That's a relatively large number," Vitousek says. "It should give pause to people who think we are far from any limits." Though he never drops the measured tone of an academic, Vitousek speaks with considerable emphasis: "There's a sense among some economists that we're *so* far from any biophysical limits. I think that's not supported by the evidence."

For another antidote to the good cheer of someone like Julian Simon, sit down with the Cornell biologist David Pimentel. He believes that we're in big trouble. Odd facts stud his conversation—for example, a nice head of iceberg lettuce is 95 percent water and contains just fifty calories of energy, but it takes 400 calories of energy to grow that head of lettuce in California's Central Valley, and another 1,800 to ship it east. ("There's practically no nutrition in the damn stuff anyway," Pimentel says. "Cabbage is a lot better, and we can grow it in upstate New York.") Pimentel has devoted the past three decades to tracking the planet's capacity, and he believes that we're already too crowded—that the earth can support only two billion people over the long run at a middle-class standard of living, and that trying to support more is doing damage. He has spent considerable time studying soil erosion, for instance. Every raindrop that hits exposed ground is like a small explosion, launching soil particles into the air. On a slope, more than half of the soil contained in those splashes is carried downhill. If crop residue—cornstalks, say—is left in the field after harvest, it helps to shield the soil: the raindrop doesn't hit hard. But in the developing world, where firewood is scarce, peasants burn those cornstalks for cooking fuel. About 60 percent of crop residues in China and 90 percent in Bangladesh are removed and burned, Pimentel says. When planting season comes, dry soils simply blow away. "Our measuring stations pick up African soils in the wind when they start to plough."

The very things that made the Green Revolution so stunning—that made the last doubling possible—now cause trouble. Irrigation ditches, for instance, water 27 percent of all arable land and help to produce a third of all crops. But when flooded soils are baked by the sun, the water evaporates and the minerals in the irrigation water are deposited on the land. A hectare (2.47 acres) can accumulate two to five tons of salt annually, and eventually plants won't grow there. Maybe 10 percent of all irrigated land is affected.

… [F]ood production grew even faster than population after the Second World War. Year after year the yield of wheat and corn and rice rocketed up about three percent annually. It's a favorite statistic of the eternal optimists. In Julian Simon's book *The Ultimate Resource* (1981) charts show just how fast the growth was, and how it continually cut the cost of food. Simon wrote, "The obvious implication of this historical trend toward cheaper food—a trend that probably extends back to the beginning of agriculture—is that real prices for food will continue to drop…. It is a fact that portends more drops in price and even less scarcity in the future."

A few years after Simon's book was published, however, the data curve began to change. That rocketing growth in grain production ceased; now the gains were coming in tiny increments, too small to keep pace with population growth. The world reaped its largest harvest of grain per capita in 1984; since then the amount of corn and wheat and rice per person has fallen by six percent. Grain stockpiles have shrunk to less than two months' supply.

No one knows quite why. The collapse of the Soviet Union contributed to the trend—cooperative farms suddenly found the fertilizer supply shut off and spare parts for the tractor hard to come by. But there were other causes, too, all around the world—the salinization of irrigated fields, the erosion of topsoil, and all the other things that environmentalists had been warning about for years. It's possible that we'll still turn production around and start it rocketing again. Charles C. Mann, writing in *Science*, quotes experts who believe that in the future a "gigantic, multi-year, multi-billion-dollar scientific effort, a

kind of agricultural 'person-on the-moon project,'" might do the trick. The next great hope of the optimists is genetic engineering, and scientists have indeed managed to induce resistance to pests and disease in some plants. To get more yield, though, a cornstalk must be made to put out another ear, and conventional breeding may have exhausted the possibilities. There's a sense that we're running into walls.

… What we are running out of is what the scientists call "sinks"—places to put the by-products of our large appetites. Not garbage dumps (we could go on using Pampers till the end of time and still have empty space left to toss them away) but the atmospheric equivalent of garbage dumps.

It wasn't hard to figure out that there were limits on how much coal smoke we could pour into the air of a single city. It took a while longer to figure out that building ever higher smokestacks merely lofted the haze farther afield, raining down acid on whatever mountain range lay to the east. Even that, however, we are slowly fixing, with scrubbers and different mixtures of fuel. We can't so easily repair the new kinds of pollution. These do not come from something going wrong—some engine without a catalytic converter, some waste-water pipe without a filter, some smokestack without a scrubber. New kinds of pollution come instead from things going as they're supposed to go—but at such a high volume that they overwhelm the planet. They come from normal human life—but there are so many of us living those normal lives that something abnormal is happening. And that something is different from the old forms of pollution that it confuses the issue even to use the word.

Consider nitrogen, for instance. But before plants can absorb it, it must become "fixed"—bonded with carbon, hydrogen, or oxygen. Nature does this trick with certain kinds of algae and soil bacteria, and with lightning. Before human beings began to alter the nitrogen cycle, these mechanisms provided 90–150 million metric tons of nitrogen a year. Now human activity adds 130–150 million more tons. Nitrogen isn't pollution—it's essential. And we are using more of it all the time. Half the industrial nitrogen fertilizer used in human history has been applied since 1984. As a result, coastal waters and estuaries bloom with toxic algae while oxygen concentrations dwindle, killing fish; as a result, nitrous oxide traps solar heat. And once the gas is in the air, it stays there for a century or more.

Or consider methane, which comes out of the back of a cow or the top of a termite mound or the bottom of a rice paddy. As a result of our determination to raise more cattle, cut down more tropical forest (thereby causing termite populations to explode), and grow more rice, methane concentrations in the atmosphere are more than twice as high as they have been for most of the past 160,000 years. And methane traps heat—very efficiently.

Or consider carbon dioxide. In fact, concentrate on carbon dioxide. If we had to pick one problem to obsess about over the next fifty years, we'd do well to make it CO_2—which is not pollution either. Carbon *mon*oxide is pollution: it kills you if you breathe enough of it. But carbon *dioxide*, carbon with two oxygen atoms, can't do a blessed thing to you. If you're reading this indoors, you're breathing more CO_2 than you'll ever get outside. For generations, in fact, engineers said that an engine burned clean if it produced only water vapor and carbon dioxide.

Here's the catch: that engine produces a *lot* of CO_2. A gallon of gas weighs about eight pounds. When it's burned in a car, about five and a half pounds of carbon, in the form of carbon dioxide, come spewing out the back. It doesn't matter if the car is a 1958 Chevy or a 1998 Saab. And no filter can reduce that flow—it's an inevitable by-product of fossil-fuel combustion, which is why CO_2 has been piling up in the atmosphere ever since the Industrial Revolution. Before we started burning oil and coal and gas, the atmosphere contained about 280 parts CO_2 per million. Now the figure is about 360. Unless we do everything we can think of to eliminate fossil fuels from our diet, the air will test out at more than 500 parts per million fifty or sixty years from now, whether it's sampled in the South Bronx or at the South Pole.

This matters because, as we all know by now, the molecular structure of this clean, natural, common element that we are adding to every cubic foot of the atmosphere surrounding us traps heat that would otherwise radiate back out to space. Far more than even methane and nitrous oxide, CO_2 causes global warming—the greenhouse effect—and climate change. Far more than any other single factor, it is turning the earth we were born on into a new planet.

… For ten years, with heavy funding from governments around the world, scientists launched satellites, monitored weather balloons, studied clouds. Their work culminated in a long-awaited report from the UN's Intergovernmental Panel on Climate Change, released in the fall of 1995. The panel's 2,000 scientists, from every corner of the globe, summed up their findings in this dry but historic bit of understatement: "The balance of evidence suggests that there is a discernible human influence on global climate." That is to say, we are heating up the planet—substantially. If we don't reduce emissions of carbon dioxide and other gases, the panel warned, temperatures will probably rise 3.6° Fahrenheit by 2100, and perhaps as much as 6.3°.

You may think you've already heard a lot about global warming. But most of our sense of the problem is behind the curve. Here's the current news: the changes are already well under way. When politicians and businessmen talk about "future risks," their rhetoric is outdated. This is not a problem for the distant future, or even for the near future. The planet has already heated up by a degree or more. We are perhaps a quarter of the way into the greenhouse era, and the effects are already being felt. From a new heaven, filled with nitrogen, methane, and carbon, a new earth is being born. If some alien astronomer is watching us, she's doubtless puzzled. This is the most obvious effect of our numbers and our appetites, and the key to understanding why the size of our population suddenly poses such a risk.

STORMY AND WARM

What does this new world feel like? For one thing, it's stormier than the old one. Data analyzed last year by Thomas Karl, of the National Oceanic and Atmospheric Administration, showed that total winter precipitation in the United States has increased by 10 percent since 1900 and that "extreme precipitation events"—rainstorms that dumped more than two inches of water in twenty-four hours and blizzards—had increased by 20 percent. That's because warmer air holds more water vapor than the colder atmosphere of the old earth; more water evaporates from the ocean, meaning more clouds, more rain, more snow. Engineers designing storm sewers, bridges, and culverts used to plan for what they called the "hundred-year storm." That is, they built to withstand the worst flooding or wind that history led them to expect in the course of a century. Since that history no longer applies, Karl says, "there isn't really a hundred-year event anymore… we seem to be getting these storms of the century every couple of years." When Grand Forks, North Dakota, disappeared beneath the Red River in the spring of last year, some meteorologists referred to it as "a 500-year flood"—meaning, essentially, that all bets are off. Meaning that these aren't acts of God. "If you look out your window, part of what you see in terms of weather is produced by ourselves," Karl says. "If you look out the window fifty years from now, we're going to be responsible for more of it."

Twenty percent more bad storms, 10 percent more winter precipitation—these are enormous numbers. It's like opening the newspaper to read that the average American is smarter by 30 IQ points. And the same data showed increases in drought, too. With more water in the atmosphere, there's less in the soil, according to Kevin Trenberth, of the National Center for Atmospheric Research. Those parts of the continent that are normally dry—the eastern sides of mountains, the plains and deserts—are even drier, as the higher average temperatures evaporate more of what rain does fall. "You get wilting plants and eventually drought faster than you would otherwise," Trenberth says. And when the rain does come, it's often so intense that much of it runs off before it can soak into the soil.

So—wetter and drier. *Different....*

The effects of… warming can be found in the largest phenomena. The oceans that cover most of the planet's surface are clearly rising, both because of melting glaciers and because water expands as it warms. As a result, low-lying Pacific islands already report surges of water washing across the atolls. "It's nice weather and all of a sudden water is pouring into your living room," one Marshall Islands resident told a newspaper reporter. "It's very clear that something is happening in the Pacific, and these islands are feeling it." Global warming will be like a much more powerful version of El Niño that covers the entire globe and lasts forever, or at least until the next big asteroid strikes.

If you want to scare yourself with guesses about what might happen in the near future, there's no shortage of possibilities. Scientists have already observed large-scale shifts in the duration of the El Niño ocean warming, for instance. The Arctic tundra has warmed so much that in some places it now gives off more carbon dioxide than it absorbs—a switch that could trigger a potent feedback loop, making warming ever worse. And researchers studying glacial cores from the Greenland Ice Sheet recently concluded that local climate shifts have occurred with incredible rapidity in the past—18° in one three-year stretch. Other scientists worry that such a shift might be enough to flood the oceans with fresh water and reroute or shut off currents like the Gulf Stream and the North Atlantic, which keep Europe far warmer than it would otherwise be. (See "The Great Climate Flip-flop," by William H. Calvin, January *Atlantic*.) In the words of Wallace Broecker, of Columbia University, a pioneer in the field, "Climate is an angry beast, and we are poking it with sticks."

But we don't need worst-case scenarios: best-case scenarios make the point. The population of the earth is going to nearly double one more time. That will bring it to a level that even the reliable old earth we were born on would be hard-pressed to support. Just at the moment when we need everything to be working as smoothly as possible, we find ourselves inhabiting a new planet, whose carrying capacity we cannot conceivably estimate. We have no idea how much wheat this planet can grow. We don't know what its politics will be like: not if there are going to be heat waves like the one that killed more than 700 Chicagoans in 1995; not if rising sea levels and other effects of climate change create tens of millions of environmental refugees; not if a 1.5° jump in India's temperature could reduce the country's wheat crop by 10 percent or divert its monsoons....

We have gotten very large and very powerful, and for the foreseeable future we're stuck with the results. The glaciers won't grow back again anytime soon; the oceans won't drop. We've already done deep and systemic damage. To use a human analogy, we've already said the angry and unforgivable words that will haunt our marriage till its end. And yet we can't simply walk out the door. There's no place to go. We have to salvage what we can of our relationship with the earth, to keep things from getting any worse than they have to be.

If we can bring our various emissions quickly and sharply under control, we *can* limit the damage, reduce dramatically the chance of horrible surprises, preserve more of the biology we were born into. But do not under-

estimate the task. The UN's Intergovernmental Panel on Climate Change projects that an immediate 60 percent reduction in fossil-fuel use is necessary just to stabilize climate at the current level of disruption. Nature may still meet us halfway, but halfway is a long way from where we are now. What's more, we can't delay. If we wait a few decades to get started, we may as well not even begin. It's not like poverty, a concern that's always there for civilizations to address. This is a timed test, like the SAT: two or three decades, and we lay our pencils down. It's *the* test for our generations, and population is a part of the answer....

The numbers are so daunting that they're almost unimaginable. Say, just for argument's sake, that we decided to cut world fossil-fuel use by 60 percent—the amount that the UN panel says would stabilize world climate. And then say that we shared the remaining fossil fuel equally. Each human being would get to produce 1.69 metric tons of carbon dioxide annually—which would allow you to drive an average American car nine miles a day. By the time the population increased to 8.5 billion, in about 2025, you'd be down to six miles a day. If you carpooled, you'd have about three pounds of CO_2 left in your daily ration—enough to run a highly efficient refrigerator. Forget your computer, your TV, your stereo, your stove, your dishwasher, your water heater, your microwave, your water pump, your clock. Forget your light bulbs, compact fluorescent or not.

I'm not trying to say that conservation, efficiency, and new technology won't help. They will—but the help will be slow and expensive. The tremendous momentum of growth will work against it. Say that someone invented a new furnace tomorrow that used half as much oil as old furnaces. How many years would it be before a substantial number of American homes had the new device? And what if it cost more? And if oil stays cheaper per gallon than bottled water? Changing basic fuels—to hydrogen, say—would be even more expensive. It's not like running out of white wine and switching to red. Yes, we'll get new technologies. One day last fall *The New York Times* ran a special section on energy, featuring many up-and-coming improvements: solar shingles, basement fuel cells. But the same day, on the front page, William K. Stevens reported that international negotiators had all but given up on preventing a doubling of the atmospheric concentration of CO_2. The momentum of growth was so great, the negotiators said, that making the changes required to slow global warming significantly would be like "trying to turn a supertanker in a sea of syrup."

There are no silver bullets to take care of a problem like this. Electric cars won't by themselves save us, though they would help. We simply won't live efficiently enough soon enough to solve the problem. Vegetarianism won't cure our ills, though it would help. We simply won't live simply enough soon enough to solve the problem.

Reducing the birth rate won't end all our troubles either. That, too, is no silver bullet. But it would help.

There's no more practical decision than how many children to have. (And no more mystical decision, either.)

The bottom-line argument goes like this: The next fifty years are a special time. They will decide how strong and healthy the planet will be for centuries to come. Between now and 2050 we'll see the zenith, or very nearly, of human population. With luck we'll never see any greater production of carbon dioxide or toxic chemicals. We'll never see more species extinction or soil erosion. Greenpeace recently announced a campaign to phase out fossil fuels entirely by mid-century, which sounds utterly quixotic but could—if everything went just right—happen.

So it's the task of those of us alive right now to deal with this special phase, to squeeze us through these next fifty years. That's not fair—any more than it was fair that earlier generations had to deal with the Second World War or the Civil War or the Revolution or the Depression or slavery. It's just reality. We need in these fifty years to be working simultaneously on all parts of the equation—on our ways of life, on our technologies, and on our population.

As Gregg Easterbrook pointed out in his book *A Moment on the Earth* (1995), if the planet does manage to reduce its fertility, "the period in which human numbers threaten the biosphere on a general scale will turn out to have been much, much more brief" than periods of natural threats like the Ice Ages. True enough. But the period in question happens to be our time. That's what makes this moment special, and what makes this moment hard.

Bill McKibben is the author of several books about the environment, including *The End of Nature* (1989) and *Hope, Human and Wild* (1995). His article in this issue will appear in somewhat different form in his book *Maybe One: A Personal and Environmental Argument for Single-Child Families*, published in 1998 by Simon & Schuster.

America's STICKY Power

U.S. military force and cultural appeal have kept the United States at the top of the global order. But the hegemon cannot live on guns and Hollywood alone. U.S. economic policies and institutions act as "sticky power," attracting other countries to the U.S. system and then trapping them in it. Sticky power can help stabilize Iraq, bring rule of law to Russia, and prevent armed conflict between the United States and China.

By Walter Russell Mead

Since its earliest years, the United States has behaved as a global power. Not always capable of dispatching great fleets and mighty armies to every corner of the planet, the United States has nonetheless invariably kept one eye on the evolution of the global system, and the U.S. military has long served internationally. The United States has not always boasted the world's largest or most influential economy, but the country has always regarded trade in global terms, generally nudging the world toward economic integration. U.S. ideological impulses have also been global. The poet Ralph Waldo Emerson wrote of the first shot fired in the American Revolution as "the shot heard 'round the world," and Americans have always thought that their religious and political values should prevail around the globe.

Historically, security threats and trade interests compelled Americans to think globally. The British sailed across the Atlantic to burn Washington, D.C.; the Japanese flew from carriers in the Pacific to bomb Pearl Harbor. Trade with Asia and Europe, as well as within the Western Hemisphere, has always been vital to U.S. prosperity. U.S. President Thomas Jefferson sent the Navy to the Mediterranean to fight against the Barbary pirates to safeguard U.S. trade in 1801. Commodore Matthew Perry opened up Japan in

the 1850s partly to assure decent treatment for survivors of sunken U.S. whaling ships that washed up on Japanese shores. And the last shots in the U.S. Civil War were fired from a Confederate commerce raider attacking Union shipping in the remote waters of the Arctic Ocean.

The rise of the United States to superpower status followed from this global outlook. In the 20th century, as the British system of empire and commerce weakened and fell, U.S. foreign-policymakers faced three possible choices: prop up the British Empire, ignore the problem and let the rest of the world go about its business, or replace Britain and take on the dirty job of enforcing a world order. Between the onset of World War I and the beginning of the Cold War, the United States tried all three, ultimately taking Britain's place as the gyroscope of world order.

However, the Americans were replacing the British at a moment when the rules of the game were changing forever. The United States could not become just another empire or great power playing the old games of dominance with rivals and allies. Such competition led to war, and war between great powers was no longer an acceptable part of the international system. No, the United States was going to have to attempt something that no other nation had ever accomplished, something that many theorists of international relations would swear was impossible. The United States needed to build a system that could end thousands of years of great power conflicts, constructing a framework of power that would bring enduring peace to the whole world—repeating globally what ancient Egypt, China, and Rome had each accomplished on a regional basis.

To complicate the task a bit more, the new hegemon would not be able to use some of the methods available to the Romans and others. Reducing the world's countries and civilizations to tributary provinces was beyond any military power the United States could or would bring to bear. The United States would have to develop a new way for sovereign states to coexist in a world of weapons of mass destruction and of prickly rivalries among religions, races, cultures, and states.

In his 2002 book, *The Paradox of American Power: Why the World's Only Superpower Can't Go It Alone*, Harvard University political scientist Joseph S. Nye Jr. discusses the varieties of power that the United States can deploy as it builds its world order. Nye focuses on two types of power: hard and soft. In his analysis, hard power is military or economic force that coerces others to follow a particular course of action. By contrast, soft power—cultural power, the power of example, the power of ideas and ideals—works more subtly; it makes others want what you want. Soft power upholds the U.S. world order because it influences others to like the U.S. system and support it of their own free will [see sidebar: A Sticky History Lesson].

Nye's insights on soft power have attracted significant attention and will continue to have an important role in U.S. policy debates. But the distinction Nye suggests between two types of hard power—military and economic power—has received less consideration than it deserves. Traditional military power can usefully be called sharp power; those resisting it will feel bayonets pushing and prodding them in the direction they must go. This power is the foundation of the U.S. system. Economic power can be thought of as sticky power, which comprises a set of economic institutions and policies that attracts others toward U.S. influence and then traps them in it. Together with soft power (the values, ideas, habits, and politics inherent in the system), sharp and sticky power sustain U.S. hegemony and make something as artificial and historically arbitrary as the U.S.-led global system appear desirable, inevitable, and permanent.

SHARP POWER

Sharp power is a very practical and unsentimental thing. U.S. military policy follows rules that would have been understandable to the Hittites or the Roman Empire. Indeed, the U.S. military is the institution whose command structure is most like that of Old World monarchies—the president, after consultation with the Joint Chiefs, issues orders, which the military, in turn, obeys.

> # Like Samson in the temple of the Philistines, a collapsing U.S. economy would inflict enormous, unacceptable damage on the rest of the world.

Of course, security starts at home, and since the 1823 proclamation of the Monroe Doctrine, the cardinal principle of U.S. security policy has been to keep European and Asian powers out of the Western Hemisphere. There would be no intriguing great powers, no intercontinental alliances, and, as the United States became stronger, no European or Asian military bases from Point Barrow, Alaska, to the tip of Cape Horn, Chile.

The makers of U.S. security policy also have focused on the world's sea and air lanes. During peacetime, such lanes are vital to the prosperity of the United States and its allies; in wartime, the United States must control the sea and air lanes to support U.S. allies and supply military forces on other continents. Britain was almost defeated by Germany's U-boats in World War I and II; in today's world of integrated markets, any interruption of trade flows through such lanes would be catastrophic.

Finally (and fatefully), the United States considers the Middle East an area of vital concern. From a U.S. perspective, two potential dangers lurk in the Middle East. First, some outside power, such as the Soviet Union during the Cold War, can try to control Middle Eastern oil or at least interfere with secure supplies for the United States and its allies. Second, one country in the Middle East could take over the region and try to do the same thing. Egypt, Iran, and, most recently, Iraq have all tried and—thanks largely to U.S. policy—have all failed. For all its novel dangers, today's efforts by al Qaeda leader Osama bin Laden and his followers to create a theocratic power in the region that could control oil resources and extend dictatorial power throughout the Islamic world resembles other threats that the United States has faced in this region during the last 60 years.

As part of its sharp-power strategy to address these priorities, the United States maintains a system of alliances and bases intended to promote stability in Asia, Europe, and the Middle East. Overall, as of the end of September 2003, the United States had just over 250,000 uniformed military members stationed outside its frontiers (not counting those involved in Operation Iraqi Freedom); around 43 percent were stationed on NATO territory and approximately 32 percent in Japan and South Korea. Additionally, the United States has the ability to transport significant forces to these theaters and to the Middle East should tensions rise, and it preserves the ability to control the sea lanes and air corridors necessary to the security of its forward bases. Moreover, the United States maintains the world's largest intelligence and electronic surveillance organizations. Estimated to exceed $30 billion in 2003, the U.S. intelligence budget is larger than the individual military budgets of Saudi Arabia, Syria, and North Korea.

Over time, U.S. strategic thinking has shifted toward overwhelming military superiority as the surest foundation for national security. That is partly for the obvious reasons of greater security, but it is partly also because supremacy can be an important deterrent. Establishing an overwhelming military supremacy might not only deter potential enemies from military attack; it might also discourage other powers from trying to match the U.S. buildup. In the long run, advocates maintain, this strategy could be cheaper and safer than staying just a nose in front of the pack.

STICKY POWER

Economic, or sticky, power is different from both sharp and soft power—it is based neither on military compulsion nor on simple coincidence of wills. Consider the carnivorous sundew plant, which attracts its prey with a kind of soft power, a pleasing scent that lures insects toward its sap. But once the victim has touched the sap, it is stuck; it can't get away. That is sticky power; that is how economic power works.

Sticky power has a long history. Both Britain and the United States built global economic systems that attracted other countries. Britain's attracted the United States into participating in the British system of trade and investment during the 19th century. The London financial markets provided investment capital that enabled U.S. industries to grow, while Americans benefited from trading freely throughout the British Empire. Yet, U.S. global trade was in some sense hostage to the British Navy—the United States could trade with the world as long as it had Britain's friendship, but an interruption in that friendship would mean financial collapse. Therefore, a strong lobby against war with Britain always existed in the United States. Trade-dependent New England almost seceded from the United States during the War of 1812, and at every crisis in Anglo-American relations for the next century, England could count on a strong lobby of merchants and bankers who would be ruined by war between the two English-speaking powers.

The world economy that the United States set out to lead after World War II had fallen far from the peak of integration reached under British leadership. The two world wars and the Depression ripped the delicate webs that had sustained the earlier system. In the Cold War years, as it struggled to rebuild and improve upon the Old World system, the United States had to change both the monetary base and the legal and political framework of the world's economic system.

The United States built its sticky power on two foundations: an international monetary system and free trade. The Bretton Woods agreements of 1944 made the U.S. dollar

the world's central currency, and while the dollar was still linked to gold at least in theory for another generation, the U.S. Federal Reserve could increase the supply of dollars in response to economic needs. The result for almost 30 years was the magic combination of an expanding monetary base with price stability. These conditions helped produce the economic miracle that transformed living standards in the advanced West and in Japan. The collapse of the Bretton Woods system in 1973 ushered in a global economic crisis, but, by the 1980s, the system was functioning almost as well as ever with a new regime of floating exchange rates in which the U.S. dollar remained critical.

The progress toward free trade and economic integration represents one of the great unheralded triumphs of U.S. foreign policy in the 20th century. Legal and economic experts, largely from the United States or educated in U.S. universities, helped poor countries build the institutions that could reassure foreign investors, even as developing countries increasingly relied on state-directed planning and investment to jump-start their economies. Instead of gunboats, international financial institutions sent bankers and consultants around the world.

Behind all this activity was the United States' willingness to open its markets—even on a nonreciprocal basis—to exports from Europe, Japan, and poor nations. This policy, part of the overall strategy of containing communism, helped consolidate support around the world for the U.S. system. The role of the dollar as a global reserve currency, along with the expansionary bias of U.S. fiscal and monetary authorities, facilitated what became known as the "locomotive of the global economy" and the "consumer of last resort." U.S. trade deficits stimulated production and consumption in the rest of the world, increasing the prosperity of other countries and their willingness to participate in the U.S.-led global economy.

Opening domestic markets to foreign competitors remained (and remains) one of the most controversial elements in U.S. foreign policy during the Cold War. U.S. workers and industries facing foreign competition bitterly opposed such openings. Others worried about the long-term consequences of the trade deficits that transformed the United States into a net international debtor during the 1980s. Since the Eisenhower administration, predictions of imminent crises (in the value of the dollar, domestic interest rates, or both) have surfaced whenever U.S. reliance on foreign lending has grown, but those negative consequences have yet to materialize. The result has been more like a repetition on a global scale of the conversion of financial debt to political strength pioneered by the founders of the Bank of England in 1694 and repeated a century later when the United States assumed the debt of the 13 colonies.

In both of those cases, the stock of debt was purchased by the rich and the powerful, who then acquired an interest in the stability of the government that guaranteed the value of the debt. Wealthy Englishmen opposed the restoration of the Stuarts to the throne because they feared it would undermine the value of their holdings in the Bank of England.

A Sticky History Lesson

Germany's experience in World War I shows how "sticky power"—the power of one nation's economic institutions and policies—can act as a weapon. During the long years of peace before the war, Germany was drawn into the British-led world trading system, and its economy became more and more trade-dependent. Local industries depended on imported raw materials. German manufacturers depended on foreign markets. Germany imported wheat and beef from the Americas, where the vast and fertile plains of the United States and the pampas of South America produced food much more cheaply than German agriculture could do at home. By 1910, such economic interdependence was so great that many, including Norman Angell, author of *The Great Illusion*, thought that wars had become so ruinously expensive that the age of warfare was over.

Not quite. Sticky power failed to keep World War I from breaking out, but it was vital to Britain's victory. Once the war started, Britain cut off the world trade Germany had grown to depend upon, while, thanks to Britain's Royal Navy, the British and their allies continued to enjoy access to the rest of the world's goods. Shortages of basic materials and foods dogged Germany all during the war. By the winter of 1916-17, the Germans were seriously hungry. Meanwhile, hoping to even the odds, Germany tried to cut the Allies off from world markets with the U-boat campaigns in the North Atlantic. That move brought the United States into the war at a time when nothing else could have saved the Allied cause.

Finally, in the fall of 1918, morale in the German armed forces and among civilians collapsed, fueled in part by the shortages. These conditions, not military defeat, forced the German leadership to ask for an armistice. Sticky power was Britain's greatest weapon in World War I. It may very well be the United States' greatest weapon in the 21st century.

—W.R.M.

Likewise, the propertied elites of the 13 colonies came to support the stability and strength of the new U.S. Constitution because the value of their bonds rose and fell with the strength of the national government.

Similarly, in the last 60 years, as foreigners have acquired a greater value in the United States—government and private bonds, direct and portfolio private investments—more and more of them have acquired an interest in maintaining the strength of the U.S.-led system. A collapse of the U.S. economy and the ruin of the dollar would do more than dent the prosperity of the United States. Without their best customer, countries including China and Japan would fall into depressions. The financial strength of every country would be severely shaken should the United States collapse. Under those circumstances, debt becomes a strength, not a weakness, and other countries fear to break with the United States because they need its market and own its securities. Of course, pressed too far, a large national debt can turn from a source of strength to a crippling liability, and the United States must continue to justify other countries' faith by maintaining its long-term record of meeting its financial obligations. But, like Samson in the temple of the Philistines, a collapsing U.S. economy would

inflict enormous, unacceptable damage on the rest of the world. That is sticky power with a vengeance.

THE SUM OF ALL POWERS?

The United States' global economic might is therefore not simply, to use Nye's formulations, hard power that compels others or soft power that attracts the rest of the world. Certainly, the U.S. economic system provides the United States with the prosperity needed to underwrite its security strategy, but it also encourages other countries to accept U.S. leadership. U.S. economic might is sticky power.

How will sticky power help the United States address today's challenges? One pressing need is to ensure that Iraq's econome reconstruction integrates the nation more firmly in the global economy. Countries with open economies develop powerful trade-oriented businesses; the leaders of these businesses can promote economic policies that respect property rights, democracy, and the rule of law. Such leaders also lobby governments to avoid the isolation that characterized Iraq and Libya under economic sanctions. And looking beyond Iraq, the allure of access to Western capital and global markets is one of the few forces protecting the rule of law from even further erosion in Russia.

China's rise to global prominence will offer a key test case for sticky power. As China develops economically, it should gain wealth that could support a military rivaling that of the United States; China is also gaining political influence in the world. Some analysts in both China and the United States believe that the laws of history mean that Chinese power will someday clash with the reigning U.S. power.

Sticky power offers a way out. China benefits from participating in the U.S. economic system and integrating itself into the global economy. Between 1970 and 2003, China's gross domestic product grew from an estimated $106 billion to more than $1.3 trillion. By 2003, an estimated $450 billion of foreign money had flowed into the Chinese economy. Moreover, China is becoming increasingly dependent on both imports and exports to keep its economy (and its military machine) going. Hostilities between the United States and China would cripple China's industry, and cut off supplies of oil and other key commodities.

Sticky power works both ways, though. If China cannot afford war with the United States, the United States will have an increasingly hard time breaking off commercial relations with China. In an era of weapons of mass destruction, this mutual dependence is probably good for both sides. Sticky power did not prevent World War I, but economic interdependence runs deeper now; as a result, the "inevitable" U.S.-Chinese conflict is less likely to occur.

Sticky power, then, is important to U.S. hegemony for two reasons: It helps prevent war, and, if war comes, it helps the United States win. But to exercise power in the real world, the pieces must go back together. Sharp, sticky, and soft power work together to sustain U.S. hegemony. Today, even as the United States' sharp and sticky power

————[Want to Know More?]————

Joseph S. Nye Jr. introduced the concept of soft power in his seminal eassy "Soft Power" (FOREIGN POLICY, Fall 1990). His more recent **The Paradox of American Power: Why The World's Only Superpower Can't Go It Alone** (New York: Oxford University Press, 2002) is the best and most original book on how the United States can deploy different kinds of power.

The ancient Greek historian Thucydides is often credited with developing the basic doctrines of foreign policy realism in his classic **The History of the Peloponnesian War** (New York: Harper & brothers, 1836), but soft power played a major role in his account of the war between Athens and Sparta. Reading premodern histories, such as Livy's **History of Rome** (New York: Harper & brothers, 1958), with Nye's distinctions in mind will provide food for thought aobut the different forms that power can take.

For insight on the tangled relationship between Britian and the United States and U.S. global interests during the 18th and 19th century, see Walter Russell Mead's **Special Providence: American Foreign Policy and How It Changed the World** (New York: Knofp, 2001). Robert Jervis explains how the United States has always defined its interests globablly in **"The Comulsive Empire"** (FOREIGN POLICY, July/August 2003). Robert Skidelsky's **John Maynard Keynes: Fighting for Freedom, 1937-1946** (New York: Viking Press, 2001) describes the economic concerns of British and U.S. authorities as they looked to build a post-World War II global financial system. Charles P. Kindleberger's classic **A Financial History of Western Europe** (Boston: Allen & Unwin, 1984) illuninates Bretton Woods and a wide range of sticky-power issues. See also Carol C. Alderman's **"The Privatization of Foreign Aid: Reassessing National Largesse"** (*Foreign Affairs*, November/December 2003).

For links to relevant Web sites, access to the *FP* Archive, and a comprehensive index of related FOREIGN POLICY articles, go to `www.foreignpolicy.com`.

reach unprecedented levels, the rise of anti-Americanism reflects a crisis in U.S. soft power that challenges fundamental assumptions and relationships in the U.S. system. Resolving the tension so that the different forms of power reinforce one another is one of the principal challenges facing U.S. foreign policy in 2004 and beyond.

Walter Russell Mead is the Henry A. Kissinger senior fellow in U.S. foreign policy at the Council on Foreign Relations. This essay is adapted from his forthcoming book, Power, Terror, Peace, and War: America's Grand Strategy in a World at Risk *(New York: Knopf, 2004).*

Five Meta-Trends Changing the World

Global, overarching forces such as modernization and widespread interconnectivity are converging to reshape our lives. But human adaptability—itself a "meta-trend"—will help keep our future from spinning out of control, assures THE FUTURIST's lifestyles editor.

By David Pearce Snyder

Last year, I received an e-mail from a long-time Australian client requesting a brief list of the "meta-trends" having the greatest impact on global human psychology. What the client wanted to know was, which global trends would most powerfully affect human consciousness and behavior around the world?

The Greek root *meta* denotes a transformational or transcendent phenomenon, not simply a big, pervasive one. A meta-trend implies multidimensional or catalytic change, as opposed to a linear or sequential change.

What follows are five meta-trends I believe are profoundly changing the world. They are evolutionary, system-wide developments arising from the simultaneous occurrence of a number of individual demographic, economic, and technological trends. Instead of each being individual freestanding global trends, they are composites of trends.

Trend 1—Cultural Modernization

Around the world over the past generation, the basic tenets of modern cultures—including equality, personal freedom, and self-fulfillment—have been eroding the domains of traditional cultures that value authority, filial obedience, and self-discipline. The children of traditional societies are growing up wearing Western clothes, eating Western food, listening to Western music, and (most importantly of all) thinking Western thoughts. Most Westerners—certainly most Americans—have been unaware of the personal intensities of this culture war because they are so far away from the "battle lines." Moreover, people in the West regard the basic institutions of modernization, including universal education, meritocracy, and civil law, as benchmarks of social progress, while the defenders of traditional cultures see them as threats to social order.

Demographers have identified several leading social indicators as key measures of the extent to which a nation's culture is modern. They cite the average level of education for men and for women, the percentage of the salaried workforce that is female, and the percentage of population that lives in urban areas. Other indicators include the percentage of the workforce that is salaried (as opposed to self-employed) and the percentage of GDP spent on institutionalized socioeconomic support services, including insurance, pensions, social security, civil law courts, worker's compensation, unemployment benefits, and welfare.

As each of these indicators rises in a society, the birthrate in that society goes down. The principal measurable consequence of cultural modernization is declining fertility. As the world's developing nations have become better educated, more urbanized, and more institutionalized during the past 20 years, their birthrates have fallen dramatically. In 1988, the United Nations forecast that the world's population would double to 12 billion by 2100. In 1992, their estimate dropped to 10 billion, and they currently expect global population to peak at 9.1 billion in 2100. After that, demographers expect the world's population will begin to slowly decline, as has already begun to happen in Europe and Japan.

Three signs that a culture is modern: its citizens' average level of education, the number of working women, and the percentage of the population that is urban. As these numbers increase, the birthrate in a society goes down, writes author David Pearce Snyder.

The effects of cultural modernization on fertility are so powerful that they are reflected clearly in local vital statistics. In India, urban birthrates are similar to those in the United States, while rural birthrates remain unmanageably high. Cultural modernization is the linchpin of human sustainability on the planet.

The forces of cultural modernization, accelerated by economic globalization and the rapidly spreading wireless telecommunications infostructure, are likely to marginalize the world's traditional cultures well before the century is over. And because the wellsprings of modernization—secular industrial economies —are so unassailably powerful, terrorism is the only means by which the defenders of traditional culture can fight to preserve their values and way of life. In the near-term future, most observers believe that ongoing cultural conflict is likely to produce at least a few further extreme acts of terrorism, security measures not withstanding. But the eventual intensity and duration of the overt, vi-

olent phases of the ongoing global culture war are largely matters of conjecture. So, too, are the expert pronouncements of the probable long-term impacts of September 11, 2001, and terrorism on American priorities and behavior.

After the 2001 attacks, social commentators speculated extensively that those events would change America. Pundits posited that we would become more motivated by things of intrinsic value— children, family, friends, nature, personal self-fulfillment—and that we would see a sharp increase in people pursuing *pro bono* causes and public-service careers. A number of media critics predicted that popular entertainment such as television, movies, and games would feature much less gratuitous violence after September 11. None of that has happened. Nor have Americans become more attentive to international news coverage. Media surveys show that the average American reads less international news now than before September 11. Event-inspired changes in behavior are generally transitory. Even if current conflicts produce further extreme acts of terrorist violence, these seem unlikely to alter the way we live or make daily decisions. Studies in Israel reveal that its citizens have become habituated to terrorist attacks. The daily routine of life remains the norm, and random acts of terrorism remain just that: random events for which no precautions or mind-set can prepare us or significantly reduce our risk.

In summary, cultural modernization will continue to assault the world's traditional cultures, provoking widespread political unrest, psychological stress, and social tension. In developed nations, where the great majority embrace the tenets of modernization and where the threats from cultural conflict are manifested in occasional random acts of violence, the ongoing confrontation between tradition and modernization seems likely to produce security mea-

sures that are inconvenient, but will do little to alter our basic personal decision making, values, or day-to-day life. Developed nations are unlikely to make any serious attempts to restrain the spread of cultural modernization or its driving force, economic globalization.

Trend 2—Economic Globalization

On paper, globalization poses the long-term potential to raise living standards and reduce the costs of goods and services for people everywhere. But the short-term marketplace consequences of free trade threaten many people and enterprises in both developed and developing nations with potentially insurmountable competition. For most people around the world, the threat from foreign competitors is regarded as much greater than the threat from foreign terrorists. Of course, risk and uncertainty in daily life is characteristically high in developing countries. In developed economies, however, where formal institutions sustain order and predictability, trade liberalization poses unfamiliar risks and uncertainties for many enterprises. It also appears to be affecting the collective psychology of both blue-collar and white-collar workers—especially males— who are increasingly unwilling to commit themselves to careers in fields that are likely to be subject to low-cost foreign competition.

Strikingly, surveys of young Americans show little sign of xenophobia in response to the millions of new immigrant workers with whom they are competing in the domestic job market. However, they feel hostile and helpless at the prospect of competing with Chinese factory workers and Indian programmers overseas. And, of course, economic history tells us that they are justifiably concerned. In those job markets that supply untariffed international industries, a "comparable global wage" for comparable types of work can be expected to emerge worldwide. This will raise workers' wages for freely traded goods and services in developing nations, while depressing wages for comparable work in mature industrial economies. To earn more than the comparable global wage, labor in developed nations will have to perform *incomparable* work, either in terms of their productivity or the superior characteristics of the goods and services that they produce. The assimilation of mature information technology throughout all production and education levels should make this possible, but developed economies have not yet begun to mass-produce a new generation of high-value-adding, middle-income jobs.

Meanwhile, in spite of the undeniable short-term economic discomfort that it causes, the trend toward continuing globalization has immense force behind it. Since World War II, imports have risen from 6% of world GDP to more than 22%, growing steadily throughout the Cold War, and even faster since 1990. The global dispersion of goods production and the uneven distribution of oil, gas, and critical minerals worldwide have combined to make international interdependence a fundamental economic reality, and corporate enterprises are building upon that reality. Delays in globalization, like the September 2003 World Trade Organization contretemps in Cancun, Mexico, will arise as remaining politically sensitive issues are resolved, including trade in farm products, professional and financial services, and the need for corporate social responsibility. While there will be enormous long-term economic benefits from globalization in both developed and developing nations, the short-term disruptions in local domestic employment will make free trade an ongoing political issue that will be manageable only so long as domestic economies continue to grow.

Trend 3—Universal Connectivity

While information technology (IT) continues to inundate us with miraculous capabilities, it has given

us, so far, only one new power that appears to have had a significant impact on our collective behavior: our improved ability to communicate with each other, anywhere, anytime. Behavioral researchers have found that cell phones have blurred or changed the boundaries between work and social life and between personal and public life. Cell phones have also increased users' propensity to "micromanage their lives, to be more spontaneous, and, therefore, to be late for everything," according to Leysia Palen, computer science professor at the University of Colorado at Boulder.

Cell phones have blurred the lines between the public and the private. Nearly everyone is available anywhere, anytime—and in a decade cyberspace will be a town square, writes Snyder.

Most recently, instant messaging—via both cell phones and online computers—has begun to have an even more powerful social impact than cell phones themselves. Instant messaging initially tells you whether the person you wish to call is "present" in cyberspace—that is, whether he or she is actually online at the moment. Those who are present can be messaged immediately, in much the same way as you might look out the window and call to a friend you see in the neighbor's yard. Instant messaging gives a physical reality to cyberspace. It adds a new dimension to life: A person can now be "near," "distant," or "in cyberspace." With video instant messaging—available now, and widely available in three years—the illusion will be complete. We will have achieved what Frances Cairncross, senior editor of *The Economist*, has called "the death of distance."

Universal connectivity will be accelerated by the integration of the

telephone, cell phone, and other wireless telecom media with the Internet. By 2010, all long-distance phone calls, plus a third of all local calls, will be made via the Internet, while 80% to 90% of all Internet access will be made from Web-enabled phones, PDAs, and wireless laptops. Most important of all, in less than a decade, one-third of the world's population—2 billion people—will have access to the Internet, largely via Web-enabled telephones. In a very real sense, the Internet will be the "Information Highway"—the infrastructure, or infostructure, for the computer age. The infostructure is already speeding the adoption of flexplace employment and reducing the volume of business travel, while making possible increased "distant collaboration," outsourcing, and offshoring.

Corporate integrity and openness will grow steadily under pressure from watchdog groups and ordinary citizens demanding business transparency. The leader of tomorrow must adapt to this new openness or risk business disaster.

As the first marketing medium with a truly global reach, the Internet will also be the crucible from which a global consumer culture will be forged, led by the first global youth peer culture. By 2010, we will truly be living in a global village, and cyberspace will be the town square.

Trend 4—Transactional Transparency

Long before the massive corporate malfeasance at Enron, Tyco, and WorldCom, there was a rising global movement toward greater transparency in all private and public enterprises. Originally aimed at kleptocratic regimes in Africa and the former Soviet states, the movement has now become universal, with the establishment of more stringent international accounting standards and more comprehensive rules for corporate oversight and record keeping, plus a new UN treaty on curbing public-sector corruption. Because secrecy breeds corruption and incompetence, there is a growing worldwide consensus to expose the principal transactions and decisions of *all* enterprises to public scrutiny.

But in a world where most management schools have dropped all ethics courses and business professors routinely preach that government regulation thwarts the efficiency of the marketplace, corporate and government leaders around the world are lobbying hard against transparency mandates for the private sector. Their argument: Transparency would "tie their hands," "reveal secrets to their competition," and "keep them from making a fair return for their stockholders."

Most corporate management is resolutely committed to the notion that secrecy is a necessary concomitant of leadership. But pervasive, ubiquitous computing and comprehensive electronic documentation will ultimately make all things transparent, and this may leave many leaders and decision makers feeling uncomfortably exposed, especially if they were not provided a moral compass prior to adolescence. Hill and Knowlton, an international public-relations firm, recently surveyed 257 CEOs in the United States, Europe, and Asia regarding the impact of the Sarbanes-Oxley Act's reforms on corporate accountability and governance. While more than 80% of respondents felt that the reforms would significantly improve corporate integrity, 80% said they also believed the reforms would not increase ethical behavior by corporate leaders.

While most consumer and public-interest watchdog groups are demanding even more stringent regulation of big business, some corporate reformers argue that regu-

lations are often counterproductive and always circumventable. They believe that only 100% transparency can assure both the integrity and competency of institutional actions. In the world's law courts—and in the court of public opinion—the case for transparency will increasingly be promoted by nongovernmental organizations (NGOs) who will take advantage of the global infostructure to document and publicize environmentally and socially abusive behaviors by both private and public enterprises. The ongoing battle between institutional and socioecological imperatives will become a central theme of Web newscasts, Netpress publications, and Weblogs that have already begun to supplant traditional media networks and newspaper chains among young adults worldwide. Many of these young people will sign up with NGOs to wage undercover war on perceived corporate criminals.

In a global marketplace where corporate reputation and brand integrity will be worth billions of dollars, businesses' response to this guerrilla scrutiny will be understandably hostile. In their recently released *Study of Corporate Citizenship*, Cone/Roper, a corporate consultant on social issues, found that a majority of consumers "are willing to use their individual power to punish those companies that do not share their values." Above all, our improving comprehension of humankind's innumerable interactions with the environment will make it increasingly clear that total transparency will be crucial to the security and sustainability of a modern global economy. But there will be skullduggery, bloodshed, and heroics before total transparency finally becomes international law—15 to 20 years from now.

Trend 5—Social Adaptation

The forces of cultural modernization—education, urbanization, and institutional order—are producing social change in the developed world

as well as in developing nations. During the twentieth century, it became increasingly apparent to the citizens of a growing number of modern industrial societies that neither the church nor the state was omnipotent and that their leaders were more or less ordinary people. This realization has led citizens of modern societies to assign less weight to the guidance of their institutions and their leaders and to become more self-regulating. U.S. voters increasingly describe themselves as independents, and the fastest-growing Christian congregations in America are nondenominational.

Since the dawn of recorded history, societies have adapted to their changing circumstances. Moreover, cultural modernization has freed the societies of mature industrial nations from many strictures of church and state, giving people much more freedom to be individually adaptive. And we can be reasonably certain that modern societies will be confronted with a variety of fundamental changes in circumstance during the next five, 10, or 15 years that will, in turn, provoke continuous widespread adaptive behavior, especially in America.

Reaching retirement age no longer always means playing golf and spoiling the grandchildren. Seniors in good health who enjoy working probably won't retire, slowing the prophesied workforce drain, according to author David Pearce Snyder

---

During the decade ahead, *infomation*—the automated collection, storage, and application of electronic data—will dramatically reduce paperwork. As outsourcing and offshoring eliminate millions of U.S. middle-income jobs, couples are likely to work two lower-pay/lower-skill jobs to replace lost income. If our employers ask us to work from home to reduce the company's office rental costs, we will do so, especially if the arrangement permits us to avoid two hours of daily commuting or to care for our offspring or an aging parent. If a wife is able to earn more money than her spouse, U.S. males are increasingly likely to become house-husbands and take care of the kids. If we are in good health at age 65, and still enjoy our work, we probably won't retire, even if that's what we've been planning to do all our adult lives. If adult children must move back home after graduating from college in order to pay down their tuition debts, most families adapt accordingly.

Each such lifestyle change reflects a personal choice in response to an individual set of circumstances. And, of course, much adaptive behavior is initially undertaken as a temporary measure, to be abandoned when circumstances return to normal. During World War II, millions of women voluntarily entered the industrial workplace in the United States and the United Kingdom, for example, but returned to the domestic sector as soon as the war ended and a prosperous normalcy was restored. But the Information Revolution and the aging of mature industrial societies are scarcely temporary phenomena, suggesting that at least some recent widespread innovations in lifestyle —including delayed retirements and "sandwich households"—are precursors of long-term or even permanent changes in society.

The current propensity to delay retirement in the United States began in the mid-1980s and accelerated in the mid-1990s. Multiple surveys confirm that delayed retirement is much more a result of increased longevity and reduced morbidity than it is the result of financial necessity. A recent AARP survey, for example,

found that more than 75% of baby boomers plan to work into their 70s or 80s, regardless of their economic circumstances. If the baby boomers choose to age on the job, the widely prophesied mass exodus of retirees will not drain the workforce during the coming decade, and Social Security may be actuarially sound for the foreseeable future.

The Industrial Revolution in production technology certainly produced dramatic changes in society. Before the steam engine and electric power, 70% of us lived in rural areas; today 70% of us live in cities and suburbs. Before industrialization, most economic production was home- or family-based; today, economic production takes place in factories and offices. In preindustrial Europe and America, most households included two or three adult generations (plus children), while the great majority of households today are nuclear families with one adult generation and their children.

Current trends in the United States, however, suggest that the three great cultural consequences of industrialization—the urbanization of society, the institutionalization of work, and the atomization of the family—may all be reversing, as people adapt to their changing circumstances. The U.S. Census Bureau reports that, during the 1990s, Americans began to migrate out of cities and suburbs into exurban and rural areas for the first time in the twentieth century. Simultaneously, information work has begun to migrate out of offices and into households. Given the recent accelerated growth of telecommuting, self-employment, and contingent work, one-fourth to one-third of all gainful employment is likely to take place at home within 10 years. Meanwhile, growing numbers of baby boomers find themselves living with both their debt-burdened, underemployed adult

children and their own increasingly dependent aging parents. The recent emergence of the "sandwich household" in America resonates powerfully with the multigenerational, extended families that commonly served as society's safety nets in preindustrial times.

Leadership in Changing Times

The foregoing meta-trends are not the only watershed developments that will predictably reshape daily life in the decades ahead. An untold number of inertial realities inherent in the common human enterprise will inexorably change our collective circumstances—the options and imperatives that confront society and its institutions. Society's adaptation to these new realities will, in turn, create further changes in the institutional operating environment, among customers, competitors, and constituents. There is no reason to believe that the Information Revolution will change us any less than did the Industrial Revolution.

In times like these, the best advice comes from ancient truths that have withstood the test of time. The Greek philosopher-historian Heraclitus observed 2,500 years ago that "nothing about the future is inevitable except change." Two hundred years later, the mythic Chinese general Sun Tzu advised that "the wise leader exploits the inevitable." Their combined message is clear: "The wise leader exploits change."

David Pearce Snyder is the lifestyles editor of THE FUTURIST and principal of The Snyder Family Enterprise, a futures consultancy located at 8628 Garfield Street, Bethesda, Maryland 20817. Telephone 301-530-5807; e-mail davidpearcesnyder@earthlink.net; Web site www.the-futurist.com.

Holy Orders

Religious Opposition to Modern States

Mark Juergensmeyer

No one who watched in horror as the towers of the World Trade Center crumbled into dust on September 11, 2001, could doubt that the real target of the terrorist assault was US global power. Those involved in similar attacks and in similar groups have said as much. Mahmood Abouhalima, one of the Al Qaeda-linked activists convicted for his role in the 1993 attack on the World Trade Center, told me in a prison interview that buildings such as these were chosen to dramatically demonstrate that "the government is the enemy."

While the US government and its allies have been frequent targets of recent terrorist acts, religious leaders and groups are seldom targeted. An anomaly in this regard was the assault on the Shi'a shrine in the Iraqi city of Najaf on August 29, 2003, which killed more than 80 people including the venerable Ayatollah Mohammad Baqir al Hakim. The Al Qaeda activists who allegedly perpetrated this act were likely more incensed over the Ayatollah's implicit support for the US-backed Iraqi Governing Council than they were jealous of his popularity with Shi'a Muslims. Since the United Nations has also indirectly supported the US occupation of Iraq and Afghanistan, it too has been subject to Osama bin Laden's rage. This may well be the reason why the UN office in Baghdad was the target of the devastating assault on August 19, 2003, which killed the distinguished UN envoy Sergio Vieira de Mello. Despite the seeming diversity of the targets, the object of most recent acts of religious terror is an old foe of religion: the secular state.

Secular governments have been the objects of terrorism in virtually every religious tradition—not just Islam. A Christian terrorist, Timothy McVeigh, bombed the Oklahoma City Federal Building on April 19, 1995. A Jewish activist, Yigal Amir, assassinated Israel's Prime Minister Yitzhak Rabin. A Buddhist follower, Shoko Asahara, orchestrated the nerve gas attacks in the Tokyo subways near the Japanese parliament buildings. Hindu and Sikh militants have targeted government offices and

political leaders in India. In addition to government offices and leaders, symbols of decadent secular life have also been targets of religious terror. In August 2003, the Marriott Hotel in Jakarta, frequented by Westerners and Westernized Indonesians, was struck by a car bomb. The event resembled the December 2002 attacks on Bali nightclubs, whose main patrons were college-age Australians. In the United States, abortion clinics and gay bars have been targeted. The 2003 bombings in Morocco were aimed at clubs popular with tourists from Spain, Belgium, and Israel. Two questions arise regarding this spate of vicious religious assaults on secular government and secular life around the world. Why is religion the basis for opposition to the state? And why is this happening now?

Why Religion?

Religious activists are puzzling anomalies in the secular world. Most religious people and their organizations either firmly support the secular state or quiescently tolerate it. Bin Laden's Al Qaeda, like most of the new religious activist groups, is a small group at the extreme end of a hostile subculture that is itself a small minority within the larger Muslim world. Bin Laden is no more representative of Islam than McVeigh is of Christianity or Asahara of Buddhism.

Still, it is undeniable that the ideals of activists like bin Laden are authentically and thoroughly religious. Moreover, even though their network consists of only a few thousand members, they have enjoyed an increase in popularity in the Muslim world after September 11, 2001, especially after the US-led occupations of Afghanistan and Iraq. The authority of religion has given bin Laden's cadres the moral legitimacy to employ violence in assaulting symbols of global economic and political power. Religion has also provided them the metaphor of cosmic war, an image of spiritual struggle that every religion contains within its repository of symbols, seen as the fight between good and bad, truth and evil. In this sense, attacks such as those on the World Trade Center and UN headquarters in Baghdad were very religious. They were meant to be catastrophic acts of biblical proportions.

From Worldly Struggles to Sacred Battles

Although recent acts of religious terrorism such as the attacks on the World Trade Center and United Nations had no obvious military goal, they were intended to make an impact on the public consciousness. They are a kind of perverse performance of power meant to ennoble the perpetrators' views of the world while drawing viewers into their notions of cosmic war. In my 2003 study of the global rise of religious violence, *Terror in the Mind of God*, I found a strikingly familiar pattern. In almost every recent case of religious violence, concepts of cosmic war have been accompanied by claims of moral justification. It is not so much that religion has become politicized but that politics has become religionized. Through enduring absolutism, worldly struggles have been lifted into the high proscenium of sacred battle.

This is what makes religious warfare so difficult to address. Enemies become satanized, and thus compromise and negotiation become difficult. The rewards for those who fight for the cause are trans-temporal, and the timelines of their struggles are vast. Most social and political struggles look for conclusions within the lifetimes of their participants, but religious struggles can take generations to succeed.

I once had the opportunity to point out the futility—in secular military terms—of the radical Islamic struggle in Palestine to Dr. Abdul Aziz Rantisi, the head of the political wing of the Hamas movement. It seemed to me that Israel's military force was strong enough that a Palestinian military effort could never succeed. Dr. Rantisi assured me that "Palestine was occupied before, for two hundred years." He explained that he and his Palestinian comrades "can wait again—at least that long." In his calculation, the struggles of God can endure for eons before their ultimate victory.

Insofar as the US public and its leaders embraced the image of war following the September 11 attacks, the US view of the war was also prone to religionization. "God Bless America" became the country's unofficial national anthem. US President George Bush spoke of defending America's "righteous cause" and of the "absolute evil" of its enemies. However, the US military engagement in the months following September 11 was primarily a secular commitment to a definable goal and largely restricted to objectives in which civil liberties and moral rules of engagement still applied.

In purely religious battles waged in divine time and with heavenly rewards, there is no need to compromise goals. There is also no need to contend with society's laws and limitations when one is obeying a higher authority. In spiritualizing violence, religion gives the act of violence remarkable power.

Ironically, the reverse is also true: terrorism can empower religion. Although sporadic acts of terrorism do not lead to the establishment of new religious states, they make the political potency of religious ideology impossible to ignore. The first wave of religious activism, from the Islamic revolution in Iran in 1978 to the emergence of Hamas during the Palestinian *intifada* in the early 1990s, focused on religious nationalism and the vision of individual religious states. Now religious activism has an increasingly global vision. The Christian militia, the Japanese Aum Shinrikyo, and the Al Qaeda network all target what they regard as a repressive and secular form of global culture and control.

Part of the attraction of religious ideologies is that they are so personal. They impart a sense of redemption and dignity to those who uphold them, often men who feel marginalized from public life. One can view their efforts to demonize their enemies and embrace ideas of cosmic war as attempts at ennoblement and empowerment. Such efforts would be poignant if they were not so horribly destructive.

Yet they are not just personal acts. These violent efforts of symbolic empowerment have an effect beyond whatever personal satisfaction and feelings of potency they impart to those who support and conduct them. The very act of killing on behalf of a moral code is a political statement. Such acts break the state's monopoly on morally sanctioned killing. By putting the right to take life in their own hands, the perpetrators of religious violence make a daring claim of power on behalf of the powerless—a basis of legitimacy for public order other than that on which the secular state relies.

Coincidence of Globalization and Modernization

These recent acts of religious violence are occurring in a way different from the various forms of holy warfare that have occurred throughout history. They are responses to a contemporary theme in the world's political and social life: globalization. The World Trade Center symbolized bin Laden's hatred of two aspects of secular government—a certain kind of modernization and a certain kind of globalization— even though the Al Qaeda network was itself both modern and transnational. Its members were often highly sophisticated and technically skilled professionals, and its organization was composed of followers of various nationalities who moved effortlessly from place to place with no obvious nationalist agenda or allegiance. In a sense, they were not opposed to modernity and globalization, so long as it fit their own design. But they loathed the Western-style modernity that they perceived secular globalization was forcing upon them.

Some 23 years earlier, during the Islamic revolution in Iran, Ayatollah Khomeini rallied the masses with the similar notion that the United States was forcing its economic exploitation, political institutions, and secular culture on an unknowing Islamic society. The Ayatollah accused urban Iranians of having succumbed to "Westoxification"—an inebriation with Western culture and ideas. The many strident movements of religious nationalism that have erupted around the world in the more than two decades following the Iranian revolution have echoed this cry. This anti-Westernism

RE-EVALUATING RELIGION

In an age of globalization, pre-modernists, modernists, and post-modernists offer contrasting perspectives on the role of religion.

Pre-Modernist Perspective

- Views religious organizations as essential to effective opposition of communist and authoritarian regimes
- Relies on past historical experience
- Thinks that the spread of secularization by means of globalization will cause only negative effects by destroying the power of religious organizations to check the power of government

Modernist Perspective

- Believes globalization is a drive force in the secularization of society and slow disappearance of religious groups throughout the world
- Argues that religions that abide in the modern age exist as marginal communities that sometimes initiate conflict against secularization
- Holds that religious organizations can play the positive role of correcting accidental distortions or perversions of the generally beneficial course of modernization

Post-Modernist Perspective

- Rejects traditional, pre-modern religions
- Allows for "spiritual experiences" to occur without religious constraints
- Considers expressive individualism to be a core value. Globalization brings about the success of expressive individualism breaking up all traditional, local, and governmental structures.

Foreign Policy Research Institute

has at heart an opposition to a certain kind of modernism that is secular, individualistic, and skeptical. Yet, in a curious way, by accepting the modern notion of the nation-state and adopting the technological and financial instruments of modern society, many of these movements of religious nationalism have claimed a kind of modernity on their own behalf.

Religious politics could be regarded as an opportunistic infection that has set in at the present weakened stage of the secular nation-state. Globalization has crippled secular nationalism and the nation-state in several ways. It has weakened them economically, not only through the global reach of transnational businesses, but also by the transnational nature of their labor supply, currency, and financial instruments. Globalization has eroded their sense of national identity and unity through the expansion of media and communications, technology, and popular culture, and through the unchallenged military power of the United States. Some of the most intense movements for ethnic

and religious nationalism have arisen in states where local leaders have felt exploited by the global economy, unable to gain military leverage against what they regard as corrupt leaders promoted by the United States, and invaded by images of US popular culture on television, the Internet, and motion pictures.

Other aspects of globalization—the emergence of multicultural societies through global diasporas of peoples and cultures and the suggestion that global military and political control might fashion a "new world order"—has also elicited fear. Bin Laden and other Islamic activists have exploited this specter, and it has caused many concerned citizens in the Islamic world to see the US military response to the September 11 attacks as an imperialistic venture and a bully's crusade, rather than the righteous wrath of an injured victim. When US leaders included the invasion and occupation of Iraq as part of its "war against terror," the operation was commonly portrayed in the Muslim world as a ploy for the United States to expand its global reach.

"BY ADOPTING THE ... INSTRUMENTS OF MODERN SOCIETY, MANY OF THESE MOVEMENTS OF RELIGIOUS NATIONALISM HAVE CLAIMED A KIND OF MODERNITY ON THEIR OWN BEHALF.

This image of a sinister US role in creating a new world order of globalization is also feared in some quarters of the West. Within the United States, for example, the Christian Identity movement and Christian militia organizations have been alarmed over what they imagine to be a massive global conspiracy of liberal US politicians and the United Nations to control the world. Timothy McVeigh's favorite book, *The Turner Diaries*, is based on the premise that the United States has already unwittingly succumbed to a conspiracy of global control from which it needs to be liberated through terrorist actions and guerilla bands. In Japan, a similar conspiracy theory motivated leaders of the Aum Shinrikyo religious movement to predict a catastrophic World War III, and attempted to simulate Armageddon with their 1995 nerve gas attack in a Tokyo subway train.

Identity and Control

As far-fetched as the idea of a "new world order" of global control may be, there is some truth to the notion that the integration of societies and the globalization of culture have brought the world closer together. Although it is unlikely that a cartel of malicious schemers designed this global trend, the effect of globalization on local societies and national identities has nonetheless been profound. It has undermined the modern idea of the state by providing non-national and transnational forms of economic, social, and cultural interaction. The global economic and social ties of the inhabitants of contemporary global cities are intertwined in a way that supercedes the idea of a national social contract—the Enlightenment notion that peoples in particular regions are naturally linked together in a specific country. In a global world, it is hard to say where particular regions begin and end. For that matter, in multicultural societies, it is hard to say how the "people" of a particular nation should be defined.

This is where religion and ethnicity step in to redefine public communities. The decay of the nation-state and disillusionment with old forms of secular nationalism have produced both the opportunity and the need for nationalisms. The opportunity has arisen because the old orders seem so weak, yet the need for national identity persists because no single alternative form of social cohesion and affiliation has yet appeared to dominate public life the way the nation-state did in the 20th century. In a curious way, traditional forms of social identity have helped to rescue one of Western modernity's central themes: the idea of nationhood. In the increasing absence of any other demarcation of national loyalty and commitment, these old staples—religion, ethnicity, and traditional culture—have become resources for national identification.

Consequently, religious and ethnic nationalism has provided a solution in the contemporary political climate to the perceived insufficiencies of Western-style secular politics. As secular ties have begun to unravel in the post- Soviet and post-colonial era, local leaders have searched for new anchors with which to ground their social identities and political loyalties. What is significant about these ethno-religious movements is their creativity—not just their use of technology and mass media, but also their appropriation of national and global networks. Although many of the framers of the new nationalisms have reached back into history for ancient images and concepts that will give them credibility, theirs are not simply efforts to resuscitate old ideas from the past. These are contemporary ideologies that meet present-day social and political needs.

In the context of Western modernism, the notion that indigenous culture can provide the basis for new political institutions, including resuscitated forms of the nation-state, is revolutionary. Movements that support ethno-religious nationalism are therefore often confrontational and sometimes violent. They reject the intervention of outsiders and their ideologies and, at the risk of being intolerant, pander to their indigenous cultural bases and enforce traditional social boundaries. It is thus no surprise that they clash with each other and with defenders of the secular state. Yet even such conflicts serve a purpose for the movements: they help define who they are as a people and who they are not. They are not, for instance, secular modernists.

Understandably, then, these movements of anti-Western modernism are ambivalent about modernity, unsure whether it is necessarily Western and always evil. They are also ambivalent about globalization, the most recent stage of modernity. On one hand, these political movements of anti-modernity are reactions to the globalization of Western culture. They are responses to the insufficiencies of what is often touted as the world's global standard: the elements of secular, Westernized urban society that are found not only in the West but in many parts of the former Third World, seen by their detractors as vestiges of colonialism. On the other hand, these new ethno-religious identities are alternative modernities with international and supernational aspects of their own. This means that in the future, some forms of anti-modernism will be global, some will be virulently antiglobal, and yet others will be content with creating their own alternative modernities in ethno-religious nation-states.

Each of these forms of religious anti-modernism contains a paradoxical relationship between forms of globalization and emerging religious and ethnic nationalisms. One of history's ironies is that the globalism of culture and the emergence of transnational political and economic institutions enhance the need for local identities. They also promote a more localized form of authority and social accountability.

The crucial problems in an era of globalization are identity and control. The two are linked in that a loss of a sense of belonging leads to a feeling of powerlessness. At the same time, what has been perceived as a loss of faith in secular nationalism is experienced as a loss of agency as well as selfhood. For these reasons, the assertion of traditional forms of religious identities are linked to attempts to reclaim personal and cultural power. The vicious outbreaks of antimodernist religious terrorism in the first few years of the 21st century can be seen as tragic attempts to regain social control through acts of violence. Until there is a surer sense of citizenship in a global order, religious visions of moral order will continue to appear as attractive, though often disruptive, solutions to the problems of authority, identity, and belonging in a globalized world.

MARK JUERGENSMEYER is Professor of Sociology and Director of Global and International Studies at the University of California, Santa Barbara.

UNIT 2
Population and Food Production

Unit Selections

5. **The Big Crunch**, Jeffrey Kluger
6. **Scary Strains**, Anne Underwood and Jerry Adler
7. **Bittersweet Harvest: The Debate Over Genetically Modified Crops**, Honor Hsin

Key Points to Consider

- What are the basic characteristics and trends of the world's population? How many people are there? How long do people typically live?

- How fast is the world's population growing? What are the reasons for this growth? How do population dynamics vary from one region to the next?

- How does rapid population growth affect the quality of the environment, social structures, and the ways in which humanity views itself?

- What are some of the threats to the human population that might cause it to decline or collapse?

- There is a growing debate about genetically modified food. What are the differing perspectives on this debate and what issues are likely to be contested in the near future?

- How can economic and social policies be changed in order to reduce the impact of population growth on environmental quality?

- In an era of global interdependence, how much impact can individual governments have on demographic changes?

DUSHKIN ONLINE **Links: www.dushkin.com/online/**
These sites are annotated in the World Wide Web pages.

The Hunger Project
http://www.thp.org
Penn Library: Resources by Subject
http://www.library.upenn.edu/cgi-bin/res/sr.cgi
World Health Organization
http://www.who.int
WWW Virtual Library: Demography & Population Studies
http://demography.anu.edu.au/VirtualLibrary/

After World War II, the world's population reached an estimated 2 billion people. It had taken 250 years to triple to that level. In the 60 years since the end of World War II, the population has tripled again to 6 billion. When the typical reader of this book reaches the age of 50, experts estimate that the global population will have reached 8 1/2 billion! By 2050, or about 100 years after World War II, some experts forecast that 10 to 12 billion people may populate the world. A person born in 1946 (a so-called baby boomer) who lives to be 100 could see a six-fold increase in population.

Nothing like this has ever occurred before. To state this in a different way: In the next 50 years there will have to be twice as much food grown, twice as many schools and hospitals available, and twice as much of everything else— just to maintain the current and rather uneven standard of living. We live in an unprecedented time in human history.

One of the most interesting aspects of this population growth is that there is little agreement about whether this situation is good or bad. The government of China, for example, has a policy that encourages couples to have only one child. In contrast, there are a few governments that use various financial incentives to promote large families.

In the first decade of the new millennium, there are many population issues that transcend numerical or economic considerations. The disappearance of indigenous cultures is a good example of the pressures of population growth on people who live on the margins of modern society. Finally, while demographers develop various scenarios forecasting population growth,

it is important to remember that there are circumstances that could lead, not to growth, but to a significant decline in global population. The spread of AIDS and other infectious diseases reveals that confidence in modern medicine's ability to control these scourges may be premature. Nature has its own checks and balances to the population dynamic that are not policy instruments of some international organization. This factor is often overlooked in an age of technological optimism.

The lead article in this section provides an overview of the general demographic trends in the contemporary world. The unit continues with an overview of the evolution of new strains of infectious diseases and the threats they pose to populations.

There is, of course, no greater check on population growth than the ability to produce an adequate food supply. Some experts question whether current technologies are sustainable over the long run. How much food are we going to need in the decades to come, and how are farmers and fishermen going to produce it? The debate about genetically modified crops is the lightening rod for this critical topic.

Making predictions about the future of the world's population is a complicated task, for there are a variety of forces at work and considerable variation from region to region. The danger of over-simplification must be overcome if governments and international organizations are going to respond with meaningful policies. Perhaps one could say that there is not a global population problem but rather many population challenges that vary from country to country and region to region.

THE BIG CRUNCH

Birthrates are falling, but it may be a half-century before the number of people—and their impact—reaches a peak

By Jeffrey Kluger

ODDS ARE YOU'LL NEVER MEET ANY OF THE ESTIMATED 247 HUMAN BEINGS WHO WERE BORN IN THE PAST MINUTE. IN A POPULATION OF 6 BILLION, 247 IS A DEMOGRAPHIC HICCUP. IN THE MINUTE BEFORE LAST, HOWEVER, THERE WERE ANOTHER 247. IN THE MINUTES TO COME THERE WILL be another, then another, then another. By next year at this time, all those minutes will have produced nearly 130 million newcomers to the great human mosh pit. That kind of crowd is awfully hard to miss.

For folks inclined to fret that the earth is heading for the environmental abyss, the population problem has always been one of the biggest causes for worry—and with good reason. The last time humanity celebrated a new century there were 1.6 billion people here for the party—or a quarter as many as this time. In 1900 the average life expectancy was, in some places, as low as 23 years; now it's 65, meaning the extra billions are staying around longer and demanding more from the planet. The 130 million or so births registered annually—

even after subtracting the 52 million deaths—is still the equivalent of adding nearly one new Germany to the world's population each year.

But things may not be as bleak as they seem. Lately demographers have come to the conclusion that the population locomotive—while still cannonballing ahead—may be chugging toward a stop. In country after country, birthrates are easing, and the population growth rate is falling.

To be sure, this kind of success is uneven. For every region in the world that has brought its population under control, there's another where things are still exploding. For every country that has figured out the art of sustainable agriculture, there are others that have worked their land to exhaustion. The population bomb may yet go off before governments can snuff the fuse, but for now, the news is better than it's been in a long time. "We could have an end in sight to population growth in the next century," says Carl Haub, a demographer with the nonprofit Population Research Bureau. "That's a major change."

Cheering as the population reports are becoming today, for much of the past 50 years, demographers were bearers of mostly bad tidings. In census after census, they reported that humanity was not just settling the planet but smothering it. It was not until the century was nearly two-thirds over that scientists and governments finally bestirred themselves to do something about it. The first great brake on population growth came in the early 1960s, with the development of the birth-control pill, a magic pharmacological bullet that made contraception easier—not to mention tidier—than it had ever been before. In 1969 the United Nations got in on the population game, creating the U.N. Population Fund, a global organization dedicated to bringing family-planning techniques to women who would not otherwise have them. In the decades that followed, the U.N. increased its commitment, sponsoring numerous global symposiums to address the population problem further. The most significant was the 1994 Cairo conference, where attendees pledged $5.7 billion to reduce birthrates in the developing world and acknowledged that giving women more education and reproductive freedom was the key to accomplishing that goal. Even a global calamity like AIDS has yielded unexpected dividends, with international campaigns to promote condom use and abstinence helping to prevent not only disease transmission but also conception.

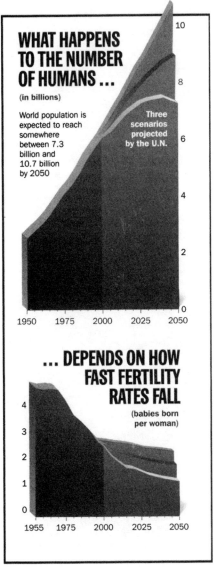

WHAT HAPPENS TO THE NUMBER OF HUMANS ...

(in billions)

World population is expected to reach somewhere between 7.3 billion and 10.7 billion by 2050

Three scenarios projected by the U.N.

10
8
6
4
2
0

1950   1975   2000   2025   2050

... DEPENDS ON HOW FAST FERTILITY RATES FALL

(babies born per woman)

4
3
2
1
0

1955   1975   2000   2025   2050

Source: United Nations

Such efforts have paid off in a big way. According to U.N. head counters, the average number of children produced per couple in the developing world—a figure that reached a whopping 4.9 earlier this century—has plunged to just 2.7. In many countries, including Spain, Slovenia, Greece and Germany, the fertility rate is well below 1.5, meaning parents are pro-

ducing 25% fewer offspring than would be needed to replace themselves—in effect, throwing the census into reverse. A little more than 30 years ago, global population growth was 2.04% a year, the highest in human history. Today it's just 1.3%. "It was a remarkable century," says Joseph Chamie of the U.N. Population Division. "We quadrupled the population in 100 years, but that's not going to happen again."

Sunny as the global averages look, however, things get a lot darker when you break them down by region. Even the best family-planning programs do no good if there is neither the money nor governmental expertise to carry them out, and in less-developed countries—which currently account for a staggering 96% of the annual population increase—both are sorely lacking. In parts of the Middle East and Africa, the fertility rate exceeds seven babies per woman. In India, nearly 16 million births are registered each year, for a growth rate of 1.8%. While Europe's population was three times that of Africa in 1950, today the two continents have about the same count. At the current rate, Africa will triple Europe in another 50 years.

Many of the countries in the deepest demographic trouble have imposed aggressive family-planning programs, only to see them go badly—even criminally—awry. In the 1970s, Indian Prime Minister Indira Gandhi tried to reduce the national birthrate by offering men cash and transistor radios if they would undergo vasectomies. In the communities in which those sweeteners failed, the government resorted to coercion, putting millions of males—from teenage boys to elderly men—on the operating table. Amid the popular backlash that followed, Gandhi's government was turned out of office, and the public rejected family planning.

China's similarly notorious one-child policy has done a better job of slowing population growth but not without problems. In a country that values boys over girls, one-child rules have led to abandonments, abortions and infanticides, as couples limited to a single offspring keep spinning the reproductive wheel until it comes up male. "We've learned that there is no such thing as 'population control,'" says Alex Marshall of the U.N. Population Fund. "You don't control it. You allow people to make up their own mind."

That strategy has worked in many countries that once had runaway population growth. Mexico, one of Latin America's population success stories, has

made government-subsidized contraception widely available and at the same time launched public-information campaigns to teach people the value of using it. A recent series of ads aimed at men makes the powerful point that there is more machismo in clothing and feeding offspring than in conceiving and leaving them. In the past 30 years, the average number of children born to a Mexican woman has plunged from seven to just 2.5. Many developing nations are starting to recognize the importance of educating women and letting them—not just their husbands—have a say in how many children they will have.

But bringing down birthrates loses some of its effectiveness as mortality rates also fall. At the same time Mexico reduced its children-per-mother figure, for example, it also boosted its average life expectancy from 50 years to 72—a wonderful accomplishment, but one that offsets part of the gain achieved by reducing the number of births.

When people live longer, populations grow not just bigger but also older and frailer. In the U.S. there has been no end of hand wringing over what will happen when baby boomers—who owe their very existence to the procreative free-for-all that followed World War II—retire, leaving themselves to be supported by the much smaller generation they produced. In Germany there are currently four workers for every retired person. Before long that ratio will be down to just 2 to 1.

STATE OF THE PLANET

Humans already use 54% of Earth's rainfall, says the U.N. report, and 70% of that goes to agriculture

For now, the only answer may be to tough things out for a while, waiting for the billions of people born during the great population booms to live out their long life, while at the same time continuing to reduce birthrates further so that things don't get thrown so far out of kilter again. But there's no telling if the earth—already worked to exhaustion feeding the 6 billion people currently here—can take much more. People in the richest countries consume a disproportionate share of the world's resources, and as poorer nations push to catch up, pressure on the planet will keep growing. "An ecologist looks at the population size relative to the carrying capacity of Earth," says Lester Brown, president of the Worldwatch Institute. "Looking at it that way, things are much worse than we expected them to be 20 years ago."

How much better they'll get will be decided in the next half-century (*see chart*). According to three scenarios published by the U.N., the global population in the year 2050 will be somewhere between 7.3 billion and 10.7 billion, depending on how fast the fertility rate falls. The difference between the high scenario and the low scenario? Just one child per couple. With the species poised on that kind of demographic knife edge, it pays for those couples to make their choices carefully.

—Reported by William Dowell/New York, Meenakshi Ganguly/New Delhi and Dick Thompson/Washington

SCARY STRAINS

What keeps infectious-disease experts up at night? A novel and virulent flu getting loose in a crowded world, where Tokyo, Nairobi and Moscow are a plane ride apart.

By Anne Underwood and Jerry Adler

THE FIRST NEWSPAPER STORIES showed up in the summer of 1997, buried on the inside pages: an influenza virus designated H5N1, known to be fatal to chickens but never before seen in humans, had killed a 3-year-old child in Hong Kong. BOY'S DEATH STUMPS EXPERTS, read the small headline in The Atlanta Journal-Constitution. Then, in December, a 2-year-old in a different part of Hong Kong fell ill from the same virus, but recovered. "To me," remarked Robert G. Webster, a leading virologist, "the startling thing about the second case is that there *is* a second case." More cases appeared—18 in all, of which six were fatal—and Hong Kong prepared to slaughter every chicken within its borders. The danger passed, then flared up in 2001 and 2002, when chickens in the Hong Kong markets began dying again, although this time there were no known human victims. Millions of fowl were killed in an effort to eradicate the virus, but it cropped up again in 2003, this time in seven countries, including Thailand and Vietnam. The toll this year stands at 31 human deaths out of 43 cases, a fatality rate of 75 percent. Only a few diseases, such as untreated AIDS, have been as lethal. (SARS, an unrelated respiratory disease that caused so much panic last year, killed 774 out of about 8,100 victims.) So far, virtually all the victims of H5N1 appear to have caught it directly from contact with fowl. Should the virus somehow acquire the ability to spread from person to person, the result

could be a global catastrophe. Says Dr. John Treanor, an infectious-disease expert at the University of Rochester, "I don't think we completely understand why it hasn't happened already."

This year's vaccine shortage may cost lives (and the jobs of those responsible), but it falls under the heading of administrative nightmares, not medical ones. Infectious-disease experts have a different scenario to keep them awake: a novel influenza virus to which people have never been exposed getting loose in a crowded world crisscrossed by jet planes that bring Tokyo, Nairobi and Moscow within, figuratively, spitting distance of one another. Three times in the last century—in 1918, 1957 and 1968—such a virus emerged, and the result each time was a "pandemic," an epidemic that's unusually widespread. The worst by far was in 1918, when at least 20 million people died. It was called the "Spanish flu," but a new book by John M. Barry, "The Great Influenza," argues that it actually emerged in rural Kansas, a place where, as in parts of present-day Asia, people lived in close proximity to domestic fowl and pigs.

The 1918 flu—variant H1N1— spread with terrifying speed; in six days at a single Army base, Barry writes, the hospital census went from 610 to more than 4,000. It killed with devastating swiftness: pedestrians literally collapsed in the street; people woke up healthy and were dead by nightfall. It attacked multi-

ple organs in the body, but always the respiratory system first, laying waste to the defenses by which the body keeps pathogens out of the lungs. Most victims succumbed to a secondary infection of bacterial pneumonia, for which there was no treatment in 1918. But in other cases, the virus was fatal in itself. Multiplying explosively throughout the respiratory tract, it provoked an immune response so furious that it devastated the lung's delicate tissues. And it was those deaths that explained H1N1's unique terror. Influenza typically kills the very

A Fearsome Flu

Each year, 36,000 Americans die from influenza; 200,000 are hospitalized. The vaccine shortage won't help matters, but it's hardly a worst-case scenario, as the avian flu could be. Here's why:

SOURCE: Flu normally attacks animals, but it can also infect us. New strains arise when (and where) infected birds and humans cross paths.
MUTATION: When common flu and bird flu invade a human cell, they shuffle genes and can produce a bug with mutated proteins that fit human receptors.
VICTIM: Human immune cells can't produce antibodies because they haven't been primed to fight this new strain.

A Brief History of Influenza

More than 85 years ago, a bird flu caused one of the deadliest pandemics in human history. A look at the successes and scares since then:

1918: In less than a year, the Spanish flu kills 500,000 in the United States and at least 20 million worldwide.

1933: Three English physicians isolate the influenza A virus and successfully infect a ferret.

1941: To immunize WWII troops, the U.S. Army develops the first flu vaccine.

1957: Spreading from China, the Asian flu claims nearly 70,000 U.S. lives before researchers develop a vaccine.

1976: A rare swine flu infects soldiers at Fort Dix, N.J.; 40 million Americans get vaccinated.

1997: Hong Kong slaughters all its chickens after losing six of 18 residents infected with bird flu.

2004: Forty-three people in Thailand and Vietnam catch bird flu; 31 die.

young and the old, whose immune systems are too weak to fight it off, but Spanish flu killed young men and women in the prime of life.

The exceptional virulence of Spanish flu is not fully explained, although Jeffery Taubenberger, a military pathologist, has been reconstructing the viral genome from Army autopsy samples and a corpse preserved in the Alaskan permafrost. (His preliminary finding is that it represented only minor changes from a variety known to infect birds.) But researchers understand this much: the H1 subtype was new to most humans then alive, who as a result were almost defenseless against it. The same was true of the H2 subtype in 1957, and of H3 in 1968. "Whenever a virus

shows up in humans with a new [H factor], you get a pandemic," says Treanor. That is one reason why doctors are so wary of an H5 virus: hardly anyone alive today has been exposed to one. "We thought we knew the rules," says Stephen Morse, director of the Center for Public Health Preparedness at Columbia University, "and one of those rules was that H1, H2 and H3 cause flu in humans, not H5. This is like the clock striking 13."

There is, fortunately, one big barrier that H5N1 apparently hasn't crossed yet: efficient human-to-human transmission. No one knows what molecular rearrangement might permit it to crash this barrier, but it could happen in a couple of ways: by random mutation, or by exchanging genetic information with a mammalian virus, either in a human host or another animal, such as a pig. Many flu variants are believed to have emerged in that way, and there are plenty of opportunities. "The ominous aspect of avian flu is how widespread the virus is among poultry in Eastern Asia," says Dr. Julie Gerberding, director of the Centers for Disease Control and Prevention. "From a statistical perspective, we are looking at a very worrisome situation." A 1999 study by the CDC forecast that a new flu pandemic of "average" lethality could cause as many as 207, 000 deaths in this country alone, with up to 100 million people infected. Direct medical costs could reach $166 billion.

Of course, an H5N1 pandemic could be much worse, or it could be mild. Some researchers, like Paul Ewald, a biologist at the University of Louisville, doubt that another 1918 is at hand, even if the virus mutates. There were unique circumstances that year; the war forced millions of young men into close quarters in barracks, troopships and trenches, where there was no escape from a sick

comrade. In normal circumstances, Ewald speculates, the spread of Spanish flu would have been limited by its very virulence. People too sick to get up and move around—not to mention those who die—don't infect others, unless they're sleeping inches apart. In the absence of war, he presumes the flu would have either burned itself out quickly or evolved to a less virulent form that could spread more efficiently.

A new pandemic could kill 2 million to 7.4 million people worldwide

Still, that's just a hypothesis, and not one to which you'd choose to entrust your grandmother's life. So researchers are trying to decode the H5N1 genome, looking for clues to its virulence and its possible vulnerabilities. And while they do, trillions of blindly replicating viruses are mutating and combining in unpredictable ways. In fact, as Richard Webby of St. Jude Children's Research Hospital in Memphis, Tenn., points out, of the eight gene segments in the virus, only one—the H5 gene itself—remains from the version that first began killing people in 1997. "It's a completely different beast," he says, although whether it's becoming more deadly, he cannot say. But just last week, 30 captive Bengal tigers on a farm in Thailand died after being fed raw chickens contaminated with avian flu. Thoroughly cooking poultry kills the virus, but the keepers weren't aware the chickens were infected. Or perhaps they just weren't concerned, because flu viruses were almost never deadly to cats. Until this one.

With Debra Rosenberg, Washington and Alexandra Seno, Hong Kong

Bittersweet Harvest

The Debate Over Genetically Modified Crops

Honor Hsin

In 1982 scientists on the 4th floor of the Monsanto Company U Building successfully introduced a foreign gene into a plant cell for the first time in history. These plants were genetically modified: they continued to express the new gene while exhibiting normal plant physiology and producing normal offspring. This breakthrough spawned the field of genetically modified (GM) crop production. Since the discovery, however, the international response to GM crops has been mixed. Along with the tremendous potential that lies vested in this technology, there are many risks and uncertainties involved as well. Arguments have centered on the health implications and environmental impact of cultivating GM crops and have raised disputes over national interests, global policy, and corporate agendas. Although there are many sides to this debate, discussions on GM crop regulation should be held within the context of scientific evidence, coupled with a careful weighing of present and future agricultural prospects.

Benefits and Costs

The possibility of environmental benefits first spurred the development of GM crops. The environmental issues at stake can be illustrated by one example of a potent genetic modification, the introduction of an endotoxin gene from *Bacillus thuringiensis* (Bt), a soil microorganism used for decades by organic growers as an insecticide, into soybeans, corn, and cotton. These GM crops promise to reduce the need to spray large amounts of chemicals into a field's ecosystem since the toxins are produced by

the plants themselves. The Bt crops pose environmental risks, however, and could possibly harm other organisms. Bt corn was shown to harm monarch butterfly caterpillars in the laboratory, although later studies performed with more realistic farming conditions found this result conclusively only with Syngenta Company's Bt maize, which expressed up to 40 percent more toxin than other brands. Another pertinent environmental issue is the possible evolution of Bt resistance in pests. Since the Bt toxin expressed by the crops is ubiquitous in the field, there is positive selection for resistance against it, which would quickly make Bt's effect obsolete. Experimentation has begun, however, that involves regulating the percentage of Bt crops in a field so that a balance can be achieved between high yields and survival of Bt-sensitive pests. Although there are still multiple layers of ecosystem complexity that need to be considered, careful scientific research can begin to address these questions.

Another potential area of risk that needs to be analyzed is the effect of GM crops on human health. A possible consequence of Bt expression in crops is the development of allergic reactions in farmers since the toxin is more highly concentrated in the crops than in the field. Furthermore, the method used to insert foreign genes into GM crops always risks manipulation of unknown genes in the plant, resulting in unforeseen consequences. The effects of GM crops on humans therefore must be tested rigorously. Fortunately, no solid evidence yet exists for adverse physiological reactions to GM crops in humans,

and some scientists argue that these same genetic-modification techniques are also currently being used in the development of pharmaceutical and industrial products.

A prevailing theme in the GM debate is that when discrepancies between scientific consensus and government policy result in unwanted consequences, the blame is often placed directly on GM crop technology itself. In 2000 about 300,000 acres of StarLink corn, a Bt crop produced by Aventis CropScience, were being cultivated in the United States. Since the US Environmental Protection Agency had declared its uncertainty over the allergenic potential of StarLink, the crops were grown with the understanding that they would be used solely as animal feed. Later that year news broke that StarLink corn had found its way into numerous taco food products around the world. This incident received wide press coverage and brought instant attention to the debate over GM crop safety. More at issue, though, were the United States' lax policies of GM crop approval and regulation. For nearly a decade, the US government made no distinction between GM crops and organically grown crops, and allergenicity safety tests were not mandatory. Only recently has the US Food and Drug Administration begun to reconsider its policies.

Canada is another leading producer of GM crops, with regulatory policies similar to those of the United States. Recent controversy surrounding Canada's cultivation of GM rapeseed, or canola, brought attention to another major environmental risk of GM crops. Unlike wheat and soybeans, which can self-pollinate to reproduce, the pollen of rapeseed plants spreads up to 800 meters beyond the field. There have been concerns in Ottawa over the government's refusal to reveal the location of ongoing GM wheat testing by Monsanto, resulting in fear of unwanted pollen spreading. This issue demonstrates one of the most potent risks of GM crops: uncontrolled breeding and the introduction of foreign genes into the natural ecosystem. An example of such an incident is Mexico's discovery of transgenic genes in non-GM strains of maize, although this result is still under scrutiny. More measures must be tested to restrain these possibilities. Current research on introducing the foreign genes into chloroplasts, which are only carried in the maternal line and not in pollen, offers a promising example.

Unfortunately, activist organizations rarely cite credible scientific evidence in their positions and have won much public sympathy by exploiting popular fears and misconceptions about genetic-engineering technology.

Europe's policy toward GM crops lies on the opposite end of the spectrum. In 1996 Europe approved the import of Monsanto's Roundup Ready soybeans and in 1997 authorized the cultivation of GM corn from Novartis. At around this time, however, there were rising concerns in Britain over BSE (bovine spongiform encephalopathy), or mad cow disease, which was thought to have killed more than two dozen people and cost the country the equivalent of billions of US dollars. The public was enraged over what it believed was a failure of government regulation, and in 1998 the European Commission voted to ban the import and cultivation of new GM crops. Besides the disappointment of private GM corporations like Monsanto, the United States claims to have lost US$600 million in corn exports to the

European Union. Recently, several European countries have considered lifting the ban contingent on the establishment of adequate labeling practices. The United States has complained to EU officials that labeling requirements discriminate against its agricultural exports, bringing the GM debate into the midst of a world trade dispute. In late January 2000, a tentative agreement was reached on the Montreal Biosafety Protocol in which the United States, Canada, Australia, Argentina, Uruguay, and Chile agreed to preliminary labeling of international exports and a precautionary principle allowing EU countries to reject imports if a scientific risk assessment of the imported crop is provided. This agreement, however, does not override decisions made by the World Trade Organization.

Corporate Control

The European public's anti-GM crop stance stems primarily from the success of environmental advocacy groups such as Greenpeace and Friends of the Earth. Numerous demonstrations have occurred throughout Britain, France, and other EU countries where GM crops have been uprooted and destroyed. Unfortunately, activist organizations rarely cite credible scientific evidence in their positions and have won much public sympathy by exploiting popular fears and misconceptions about genetic-engineering technology.

One issue they highlight that might prove significant, however, is the role of corporate interests in the GM-crop debate. A few years ago, Monsanto's attempt to acquire the "terminator" technology sparked tremendous controversy. This patent consisted of an elaborate genetically engineered control system designed to inhibit the generation of fertile seeds from crops. In essence, it was developed so that farmers would need to purchase new GM seeds each year, although arguments were raised that this technology could help prevent uncontrolled GM crop breeding. After much pressure from the nonprofit advocacy group Rural Advancement Foundation International, however, Monsanto announced in late 1999 that it would not market the "terminator" technology.

The "terminator" ordeal attracted so much attention because it placed Monsanto's corporate interest directly against the strongest argument in favor of genetic-engineering technology: potential cost savings and nutritional value of GM crops to developing countries. The UN Development Programme recently affirmed that GM crops could be the key to alleviating global hunger. Although the United Nations has expressed concern over precautionary testing of crops (through agencies like the World Health Organization), some contend that Western opposition to this technology ignores concerns of sub-Saharan and South Asian countries where malnutrition and poverty are widespread.

India is among those nations that could benefit from GM-crop technology. India's population has been growing by 1.8 percent annually; by 2025 India will need to produce 30 percent more grain per year to feed the twenty million new mouths added to its population. The need for higher food productivity is highlighted by incidents of poor farmers in Warangal and Punjab who have committed suicide when faced with devastated crops and huge debts on pesticides. The Indian government has approved several GM crops for commercial production, and testing has also commenced on transgenic cotton, rice, maize, tomato, and cauliflower, crops that would reduce the need for pesticides. A recent furor erupted over the discovery of around 11,000 hectares of illegal Bt cotton in Gujarat. The Gujarat administration responded immediately by ordering the fields stripped, the crops burned, and the seeds destroyed. There is still uncertainty over who will repay the farmers, who claim that Mahyco, a Monsanto subsidiary, is attempting to monopolize the distribution of Bt crops in India, and that the Indian government is also yielding to pressure from pesticide manufacturers. Corporate battles still abound in a nation where many farmers appear to be in need of agricultural change.

Feed the World

Many opponents of GM crops argue that the technology is not needed to help solve the problem of world hunger, with 800 million people who do not have enough to eat. They often argue that the world produces enough food to feed nine billion people while there are only six billion people today, implying that global hunger is simply a matter of distribution and not food productivity. Unfortunately, fixing the distribution problem is a complex issue. Purchasing power would need to increase in developing countries, coupled with increased food production in both developing and developed countries so that crops can be marketed at a price the underprivileged can afford. Since land for farming is limited, the remaining option for increasing crop productivity is to increase yield. While GM-crop technology is not the only method that can be used to achieve this end, it can contribute greatly toward it.

On Dr. Shiva's argument for supporting local knowledge in agricultural practices, Dr. Prakash argues that, from experience, "[local knowledge] is losing one third of your children before they hit the age of three. Is that the local knowledge that you want to keep reinforcing and keep perpetuating?"

Some consider GM crops part of a series of corporate attempts to control markets in developing countries and thus they brand GM technology another globalization "evil." Dr. Vandana Shiva of the Research Foundation for Science, Technology, and Ecology argues that globalization has pressured farmers in developing countries to grow monocultures—single-crop farming—instead of fostering sustainable agricultural diversity. Genetic engineering, in this view, is the next industrialization effort after chemical pesticides, and would also bear

no greater benefit than indigenous polycultural farming. The Food and Agricultural Organization of the United Nations also notes the leaning of research investment toward monocultures, spurred on by the profit potential of GM crops.

On the other hand, GM-crop technology serves to increase crop yield on land already in use for agricultural purposes, thereby preserving biodiversity in unused land. In the words of Dr. C. S. Prakash "using genetics helped [to] save so much valuable land from being under the plow." On Shiva's argument for supporting local knowledge in agricultural practices, Dr. Prakash argues that, from experience, "[local knowledge] is losing one third of your children before they hit the age of three. Is that the local knowledge that you want to keep reinforcing and keep perpetuating?"

Continuing along these lines and bringing GM technology in developing countries into the broader context of morality, leaders including Per Pinstrup-Andersen, director of the International Food Policy Research Institute, and Hassan Adamu, Nigeria's minister of agriculture, emphasize the importance of providing freedom of access, education, and choice in GM technology to the individual farmer himself. In Africa, for example, many local farmers have benefited from hybrid seeds obtained from multinational corporations. On a larger scale, however, Africa's agricultural production per unit area is among the lowest in the world, and great potential lies in utilizing GM crops to help combat pestilence and drought problems. On the issue of local knowledge, Dr. Florence Wambugu of the International Service for the Acquisition of Agribiotech Applications in Kenya (ISAAA) asserts that GM crops consist of "packaged technology in the seed" that can yield benefits without a change in local agricultural customs.

On another front of the world hunger debate, a promising benefit that GM-crop technology brings to developing countries is the introduction or enhancement of nutrients in crops. The first prod-

uct to address this was "golden rice," an engineered form of rice that expresses high levels of beta-carotene, a precursor of Vitamin A, which could be used to combat Vitamin A deficiency found in over 120 million children worldwide. Although many advocacy groups claim that the increased levels of Vitamin A from a golden rice diet are not high enough to fully meet recommended doses of Vitamin A, studies suggest that a less-than-full dose can still make a difference in an individual whose Vitamin A intake is already deficiently low. Currently the International Rice Research Institute is evaluating environmental and health concerns. After such tests are completed, however, there remains one final hurdle in the marketing process that advocates on both sides of the GM debate do agree on: multilateral access and sharing between public and private sectors. The International Undertaking on Plant Genetic Resources was established to foster such relationships for the world's key crops, but more discussions will have to take place on the intellectual-property rights of GM-crop patents.

Science First

Monsanto recently drafted a pledge of Five Commitments: Respect, Transparency, Dialogue, Sharing, and Benefits. These are qualities that all multinational organizations should bring to the debate over GM crops. In the meantime, the technology of genetic engineering has already emerged and bears promising potential. On the question of world hunger, GM crops are not the full solution, but they can play a part in one. There are possible risks which must be examined and compared to the risks associated with current agricultural conditions, and progress must not be sought too hastily. It is important to base considerations of the benefits and risks of GM crops on careful scientific research, rather than corporate interest or public fears.

HONOR HSIN, Staff Writer, *Harvard International Review*

UNIT 3

The Global Environment and Natural Resources Utilization

Unit Selections

Key Points to Consider

- What are the basic environmental challenges that confront both governments and individual consumers?

- What are the dimensions to the debate about global warming and the use of fossil fuel?

- What transformations will societies, that are heavy users of fossil fuels, undergo in order to meet future energy needs?

- Has the international community adequately responded to problems of pollution and threats to our common natural heritage? Why or why not?

- What is the natural resource picture going to look like 30 years from now?

- How is society, in general, likely to respond to the conflicts between lifestyle and resource conservation?

- Can a sustainable economy be organized and what changes in behavior and values are necessary to accomplish this?

 **Links: www.dushkin.com/online/**
These sites are annotated in the World Wide Web pages.

National Geographic Society
http://www.nationalgeographic.com

National Oceanic and Atmospheric Administration (NOAA)
http://www.noaa.gov

SocioSite: Sociological Subject Areas
http://www.pscw.uva.nl/sociosite/TOPICS/

United Nations Environment Programme (UNEP)
http://www.unep.ch

Beginning in the eighteenth century, the modern nation-state was conceived, and over many generations it has evolved to the point where it is now difficult to imagine a world without national governments. These legal entities have been viewed as separate, self-contained units that independently pursue their "national interests." Scholars often described the world as a political community of independent units that interact with each other (a concept that has been described as a billiard ball model).

This perspective of the international community as comprised of self-contained and self-directed units has undergone major rethinking in the past 35 years. One reason for this is the international consequences of the growing demands being placed on natural resources. The Middle East, for example, contains a large majority of the world's known oil reserves. The United States, Western Europe, China and Japan are very dependent on this vital source of energy. This unbalanced supply and demand equation has created an unprecedented lack of self-sufficiency for the world's major economic powers.

The increased interdependence of countries is further illustrated by the fact that air and water pollution do not respect political boundaries. One country's smoke is often another country's acid rain. The concept that independent political units control their own destiny, in short, makes less sense than it may have 100 years ago. In order to more fully understand why this is so, one must first look at how Earth's natural resources are being utilized and how this may be affecting the global environment.

The initial article in the unit examines the broad dimensions of the uses and abuses of natural resources. The central theme in this article is whether or not human activity is in fact bringing about fundamental changes in the functioning of Earth's self-regulating ecological systems. In many cases an unsustainable rate of usage is under way, and, as a consequence, an alarming decline in the quality of the natural resource base is taking place.

An important conclusion resulting from this analysis is that contemporary methods of resource utilization often create problems that transcend national boundaries. Global climate changes, for example, will affect everyone, and if these changes are to be successfully addressed, international collaboration will be required. The consequences of basic human activities such as growing and cooking food are profound when multiplied billions of times every day. A single country or even a few countries working together cannot have a significant impact on redressing these problems. Solutions will have to be conceived that are truly global in scope. Just as there are shortages of natural resources, there are also shortages of new ideas for solving many of these problems.

Unit 3 continues by examining specific case studies that explore in greater detail the issues raised in the first article. Implicit in this discussion is the challenge of moving from the perspective of the environment as primarily an economic resource to be consumed to a perspective that has been defined as "sustainable development." This change is easily called for, but in fact it goes to the core of social values and basic economic activities. Developing sustainable practices, therefore, is a challenge of unprecedented magnitude.

Nature is not some object "out there" to be visited at a national park. It is the food we eat and the energy we consume. Human beings are joined in the most intimate of relationships with the natural world in order to survive from one day to the next. It is ironic how little time is spent thinking about this relationship. This lack of attention, however, is not likely to continue, for rapidly growing numbers of people and the increased use of energy-consuming technologies are placing unprecedented pressures on Earth's carrying capacity.

DEFLATING THE WORLD'S BUBBLE ECONOMY

"Unless the damaging trends that have been set in motion are reversed quickly, we could see vast numbers of environmental refugees abandoning areas scarred by depleted aquifers and exhausted soils...."

By Lester R. Brown

THROUGHOUT HISTORY, humans have lived on the Earth's sustainable yield—the interest from its natural endowment. Now, however, we are consuming the endowment itself. In ecology, as in economics, we can consume principal along with interest in the short run, but, for the long term, that practice leads to bankruptcy. By satisfying our excessive demands through overconsumption of the Earth's natural assets, we are in effect creating a global bubble economy. Bubble economies are not new. American investors got an up-close view of this when the bubble in high-tech stocks burst in 2000, and the Nasdaq, an indicator of the value of these stocks, declined by some 75%. Japan had a similar experience in 1989 when its real estate bubble collapsed, depreciating assets by 60%. The Japanese economy has been reeling ever since.

These two events primarily affected those living in the U.S. and Japan, but the global bubble economy that is based on the overconsumption of the Earth's natural capital will affect the entire planet. The trouble is, since Sept. 11, 2001, political leaders, diplomats, and the media have been preoccupied with terrorism

and, more recently, the conflict in Iraq. These certainly are matters of concern, but if they divert us from addressing the environmental trends that are undermining our future, Osama bin Laden and his followers will have achieved their goal of disrupting our way of life in a way they could not have imagined.

Of all the sectors affected by the bubble economy, food may be the most vulnerable. Today's farmers are dealing with major new challenges: their crops must endure the highest temperatures in 11,000 years as well as widespread aquifer depletion and the resulting loss of irrigation water unknown to previous generations. The average global temperature has risen in each of the last three decades. The 16 warmest years since record-keeping began in 1880 have occurred since 1980. With the three warmest years on record—1998, 2001, and 2002—coming in the last five years, crops are facing unprecedented heat stress. Higher temperatures reduce yields through their effect on photosynthesis, moisture balance, and fertilization. As the temperature rises above 34° Celsius (94° Fahrenheit), evaporation increases and photosynthesis and fer-

tilization are impeded. Scientists at the International Rice Research Institute in the Philippines and at the U.S. Department of Agriculture together have developed a rule of thumb that each 1° Celsius rise in temperature above the optimum during the growing season reduces grain yields by 10%.

Findings indicate that if the temperature reaches the lower end of the range projected by the Intergovernmental Panel on Climate Change, grain harvests in tropical regions could be reduced by an average of five percent by 2020 and 11% by 2050. At the upper end of the range, yields could drop 11% by 2020 and 46% by 2050. Avoiding these declines will be difficult unless scientists can develop crop strains that are not vulnerable to thermal stress.

The second challenge facing farmers—falling water tables—also is a recent phenomenon. Using traditional animal- or human-powered waterlifting devices, it was virtually impossible to exhaust aquifers. With the spread of powerful diesel and electric pumps during the last half-century, however, overuse has become commonplace. As the world demand for water has climbed, water tables have fallen in scores of countries, including China, India, and the U.S., which together produce nearly half of the world's grain. Many other nations are straining their water reserves, too, setting the stage for dramatic cutbacks in water resources. The more populous among these are Pakistan, Iran, and Mexico. Overpumping creates an illusion of food security, enabling farmers to support a growing population with a practice that virtually ensures an eventual decline in food production and skyrocketing prices.

Food is fast becoming a national security issue as growth in the world harvest slows and falling water tables and rising temperatures hint at upcoming shortages. More than 100 countries import wheat. Some 40 import rice. While a handful of nations are only marginally dependent on imports, many could not survive without them. Iran and Egypt, for example, rely on imports for 40% of their grain supply. Algeria, Japan, South Korea, and Taiwan each import 70% or more. For Israel and Yemen, over 90%. Six countries—the U.S., Canada, France, Australia, Argentina, and Thailand—supply 90% of grain exports. The U.S. alone controls close to half the planet's grain exports, a larger share than Saudi Arabia does of oil.

Thus far, the countries that import heavily are small and midsized. China, however, the most populous nation, soon is likely to turn to international markets in a major way. When the former Soviet Union unexpectedly moved in that direction in 1972 for roughly one-tenth of its grain supply following a weather-reduced harvest, wheat prices climbed from $1.90 to $4.89 a bushel. Bread prices soon rose, too. A politics of food scarcity emerged. Pressure from within grain-exporting countries to restrict exports in order to check the rise in domestic food prices was common.

If China depletes its wheat reserves and looks elsewhere to cover the shortfall, now 40,000,000 tons per year,

the situation could destabilize overnight, because it would mean petitioning the U.S., thus presenting a potentially delicate geopolitical situation in which 1,300,000,000 Chinese consumers boasting a $100,000,000,000 trade surplus with the U.S. will be competing with U.S. consumers for American grain. If that leads to rising food prices in this country, how will the government respond? In times past, it could have restricted exports, even imposing a trade embargo, as it did with soybeans to Japan in 1974. Today, though, the U.S. has a huge stake in a politically stable China. Growing at seven to eight percent a year, China is the engine that is powering not only the Asian economy but, to some degree, the global economy as well.

For the world's poor—the millions living in cities on one dollar per day or less and already spending 70% of their income on food—escalating grain prices would be life-threatening. A doubling of prices could impoverish vast numbers in a shorter period of time than any event in history. With desperate individuals holding their governments responsible, such a price spike also could destabilize governments of low-income, grain-importing nations.

Historically, there were two food reserves: the global carryover stocks of grain and the cropland idled under the U.S. farm policy to limit production. The latter could be cultivated within a year. Since the U.S. land set-aside initiative ended in 1996, however, there have been only carryover stocks as a reserve.

Food security has changed in other ways. Traditionally, it was largely an agricultural matter. Now, though, it is something that our entire society is responsible for. National population and energy policies may have a greater impact on food security than agricultural policies do. With most of the 3,000,000,000 additional individuals forecasted by 2050 being born in countries already facing water shortages, population control may have a larger influence on food security

than crop planting proposals. Achieving an acceptable balance between food and consumers depends on family planners and farmers working together.

Climate change is the wild card in the food security deck. It is perhaps a measure of the complexity of our time that decisions reached in the Ministry of Energy may have more to do with future food security than those in the Ministry of Agriculture. The effects of population and energy policies on food security differ in one important respect: population stability can be achieved by a country acting unilaterally; climate stability cannot.

While the food sector may be the first to reveal the true size of the bubble economy, other wake-up calls, including more destructive storms, deadly heat waves, and collapsing fisheries, also could signal the extent to which we have overshot our ecological limitations. Unless the damaging trends that have been set in motion are reversed quickly, we could see vast numbers of environmental refugees abandoning areas scarred by depleted aquifers and exhausted soils, as well as fleeing advancing deserts and rising seas. In a world where civilization is being squeezed between expanding deserts from the interior continents and rising seas on the periphery, refugees are likely to number not in the millions, but in the tens of millions.

Preventing the bubble from bursting will require an unprecedented degree of international cooperation. Indeed, in both scale and urgency, the effort required is comparable to the U.S. mobilization during World War II. Rapid systemic change—alteration based on market signals that tell the ecological truth—is needed. This means lowering income taxes while raising tariffs on environmentally destructive activities, such as fossil fuel burning, to incorporate the ecological costs. Unless the market can be made to send signals that reflect reality, we will continue making faulty decisions as consumers, corporate planners, and government policymakers.

Stabilizing world population at around 7,500,000,000 is central to avoiding economic breakdowns in countries with large projected population increases that are already overconsuming their natural capital assets. No less than 36 nations, all in Europe (except Japan), essentially have done so. The challenge is to create the economic and social conditions—and to adopt the priorities—that will lead to population stability in the remaining lands. The keys here are offering primary education to every child, providing vaccinations along with basic and reproductive health care, and offering family planning services.

Shifting from a carbon- to a hydrogen-based economy to stabilize climate is quite feasible. Advances in wind turbine design and solar cell manufacturing, the availability of hydrogen generators, and the evolution of fuel cells provide the technologies necessary to build a climate-benign hydrogen economy. Moving quickly to renewable energy sources and improving efficiency depend on incorporating the indirect costs of burning fossil fuels into the market price.

On the energy front, Iceland is the first nation to adopt a national plan to convert its carbon-based energy economy to one of hydrogen. It is starting with the conversion of the Reykjavik bus fleet to fuel cell engines, then will proceed with converting automobiles, and, eventually, the fishing fleet. Iceland's first hydrogen service station opened in April.

Denmark and Germany, meanwhile, are leading proponents of wind power. Denmark, the pioneer, gets 18% of its electricity from turbines and plans to upgrade to 40% by 2030. Germany has developed some 12,000 megawatts of wind-generating capacity. Its northernmost state of Schleswig-Holstein receives 28% of its electricity in that fashion. Spain also is on the fast track in this area.

Japan has emerged as the number-one manufacturer and consumer of solar cells. With its commercialization of a solar roofing material, it now leads the world in electricity generated from solar cells and is well positioned to assist in the electrification of villages in developing areas.

"Preventing the bubble from bursting will require an unprecedented degree of international cooperation. Indeed, in both scale and urgency, the effort required is comparable to the U.S. mobilization during World War II.

The Netherlands leads the industrial world in utilizing the bicycle as an alternative to the automobile. In Amsterdam's pedal-friendly environment, up to 40% of all trips are taken by that mode of transportation. This reflects the priority given to bikes in the design and operation of the country's urban transport systems. At many traffic signals, for example, cyclists are allowed to go first when the light changes.

The Canadian province of Ontario is one of the leaders in phasing out coal. It plans to replace its five coal-fired power plants with gas-fired plants, wind farms, and efficiency gains. This initiative calls for the first plant to close in 2005 and the last one by 2015. The resulting reduction in carbon emissions will be the equivalent of taking 4,000,000 cars off the road. This approach is a model for local and national governments everywhere.

Meanwhile, in pioneering drip irrigation technology, Israel has become the world leader in the efficient use of agricultural water. This unusually labor-intensive irrigation practice is ideally suited where water is scarce and labor is abundant. Water pricing also can be effective in encouraging efficiency. In South Africa, for example, households receive a fixed amount of water for basic needs at a low price, but when water use exceeds this level, the price escalates. This helps ensure that basic needs are met while discouraging waste. Doesn't it make sense to reduce urban and industrial water demand by managing waste without discharging it into the local environment, thereby allowing water to be recycled indefinitely?

In stabilizing soils, South Korea stands out, as its once denuded mountainsides and hills are now covered with trees. The nation's level of flood control, water storage, and hydrological stability is an example for other countries. Although the two Koreas are separated by just a narrow, demilitarized zone, the contrast between them is stark. In North Korea, where little permanent vegetation remains, droughts and floods alternate and hunger is chronic.

The U.S. record in soil conservation is an impressive one. Beginning in the late 1980s, American farmers systematically retired roughly 10% of the most erodible cropland, planting grass on the bulk of it. In addition, they've adopted various soil-conserving initiatives, including minimum- and no-till practices. Consequently, the U.S. has reduced soil erosion by almost 40% in less than two decades.

There is a growing sense among the more thoughtful political and opinion leaders worldwide that business as usual no longer is a viable option. Unless we respond to the social and environmental issues that are undermining our future, we may not be able to avoid economic decline and social disintegration. The prospect of weakened states is growing as the HIV epidemic, water shortages, and land hunger threaten to overwhelm countries on the lower rungs of the global economic ladder. Failed states are a matter of concern not only because of the social costs to their people, but because they serve as ideal bases for international terrorist organizations.

It is easy to spend hundreds of billions in response to terrorist threats, but the reality is that the resources needed to disrupt a modern economy are small, and a Department of Homeland Security, however heavily funded, provides only minimum protection from suicidal extremists. The challenge is not just to provide a high-tech military response to terrorism, but to build a global society that is environmentally sustainable, socially equitable, and democratically based—where there is hope for everyone. Such an effort would more effectively undermine the spread of terrorism than a doubling of military expenditures.

We can construct an economy that does not destroy its natural support systems, a community where the basic needs of all the Earth's people are satisfied, and a world that will allow us to think of ourselves as civilized. The choice is ours—yours and mine. We can stay with the status quo and preside over a global bubble economy that will keep expanding until it finally bursts, or we can be the generation that stabilizes population, eradicates poverty, and alleviates climate change. Historians will record the choice, but it is ours to make.

Lester R. Brown is president of the Earth Policy Institute, Washington, D.C., and author of Plan B: Rescuing a Planet Under Stress and a Civilization in Trouble.

Shifting the pain

World's resources feed California's growing appetite

Tom Knudson

Half a hemisphere separates the headwaters of the Amazon River and the frostbitten northern latitudes of Canada.

But the two landscapes have one thing in common.

You can see it along a muddy rain-forest road in Ecuador, in the silver glint of a pipeline snaking through the grass. North of Edmonton, Alberta, a different sight catches your eye: an old-growth forest of spruce, pine and aspen shredded by a dusty maze of logging roads.

That oil pipeline and those logging roads are linked, via quiet rivers of commerce, to the largest concentration of consumers in North America, to a culture that proudly protects its own coastline and forests from exploitation while using more gasoline, wood and paper than any other state in America: California.

With 34 million people and the world's fifth-largest economy, California has long consumed more than it produces. But today, its passion for protecting natural resources at home while importing them in record quantities from afar is backfiring on the world's environment.

It is exporting the pain of producing natural resources—polluted water, pipeline accidents, piecemeal forests and human conflicts—to the far corners of the planet, to places out of sight and out of mind. California is the state of denial.

"There is a disconnect going on," said William Libby, a professor emeritus of forestry at the University of California, Berkeley, who lectures and consults on forest issues around the globe. "We consume like mad. And we preserve like mad."

Since the days of John Muir—the California naturalist whose writings and ramblings helped kindle the conservation movement just over a century ago—concern for the environment has been a cornerstone of California life.

And seldom has conservation touched California so deeply as during the past 10 years. Since 1992, environmental rules have eliminated or sharply reduced logging on 10 million acres of national forest land in the state—an area 13 times larger than Yosemite National Park. In the Mojave and Great Basin deserts, 3.5 million acres were declared wilderness in 1994—an expanse half again the size of Yellowstone National Park.

And while that conservation legacy will enrich Californians—and California ecosystems—for generations to come, its reach also extends far beyond the Golden State.

Libby was one of the first to notice, while on sabbatical in New Zealand in 1992. As the volume of wood cut from California forests dropped due to regulations to protect spotted owls, the demand for logs in New Zealand soared—making loggers there happy.

"Prices were insane," Libby said. "The New Zealanders wanted me to get them a dead spotted owl so they could stuff it, put it in the lobby and genuflect to it."

He soon discovered logging was on the rise in other places, too, and has since published several articles that link preservation of California forests with species extinctions elsewhere.

"We Californians are really not very good conservationists—we're very good preservationists," he said. "Conservation means you use resources well and responsibly. Preservation means you are rich enough to set aside things you want and buy them from someone else."

A half-century ago, California was self-sufficient in wood. Today, the state imports 80 percent of what it uses. Follow some of that wood back to its source and you find yourself in the northern boreal forest, where Canada allows trees to be cut in ways not permitted in California.

On average, nine of every 10 acres logged in Canada are clear-cut—the contentious practice of leveling large patches of the forest. And more than two-thirds of Canadian logging takes place in stands that have never been nicked by a chain saw—virgin forests that in California would be regarded as sanctuaries.

"Many Americans believe Canada is this incredible wilderness, but it's not true," said Richard Thomas, an Edmonton consultant and author of a 1998 provincial study critical of logging practices in Alberta. "We are very much like a Third World country when it comes to our resources. We just let other countries have at it."

Six thousand miles south, a wave of development for another resource crucial to California—crude oil—is inflicting similarly serious wounds across Ecuador's Amazon. Rain forests that were home to kaleidoscopic displays of plant and animal life in the 1970s and '80s now are showcases of pollution and poverty.

Every day, an average of 235,000 barrels of oil is pumped from the region for export to world markets. The largest portion—65,000 to 85,000 barrels a day—is shipped to refineries in Los Angeles and San Francisco.

The discovery of more reserves in the Amazon is setting off a new wave of controversy and threatening the cultural survival of semi-nomadic rain-forest tribes. Still, the country's new president, Lucio Gutiérrez, assured financiers in New York earlier this year that he supports more drilling because Ecuador is deeply in debt and needs foreign investment.

"The historical challenges for my government are very clear," Gutiérrez said at the time.

California's hunger for the planet's natural resources need not stir up trouble, if a system were in place to prevent it. You can find such a safeguard in the storm-tossed North Pacific, where Canadian fishermen, working under a federal plan that gives them an ownership stake in fish, are harvesting millions of pounds of rockfish every year for California without hurting the environment.

"Everybody is quite conscientious," said Jim Harris, a Canadian trawler. "We've got a fishery that is going to be here for the duration."

The clash of conservation and consumption in California may be large, but it is not unique in this country.

"We're the largest consuming nation basically of everything," said James Bowyer, a professor at the University of Minnesota who specializes in conservation policy and natural resource consumption.

"Yet we find every reason in the world why we shouldn't mine steel, why we shouldn't drill for oil," Bowyer said. "It's ironic because we are transferring the impacts to someplace else. And then we are telling ourselves what we are doing is good for the environment.

"And not only are we transferring those impacts, we are magnifying them by turning to nations that don't have the stringent environmental controls that we do."

No government agency maps the global impact of California consumers. But a small Oakland think tank, Redefining Progress, has assembled estimates of the mountain of resources, from wood to fossil fuel, fresh water to seafood, consumed by 146 nations—and some California counties—a yardstick it calls an "ecological footprint."

The United States, a world leader in the conservation of natural resources, has a larger footprint (24 acres per person) than all nations except the United Arab Emirates (with 25 acres). Do the math and you find America's 291 million people draw upon a 7 billion-acre chunk of the planet—an area roughly three times the size of the United States.

An assessment for Marin County—the pricey, conservation-minded San Francisco suburb—found citizens there eat, drink, spend and drive their way through even more of the planet's natural wealth: 27 acres per person a year—the largest ecological footprint ever calculated.

The group's footprints have attracted attention from scientists and policy-makers around the world. And although some people criticize the methods as imprecise, none denies the basic premise.

"The idea is right," said Libby, the forestry professor.

Last year, Libby found some Californians are not eager to hear about the global consequences of conservation and consumption.

At a conference on Sierra Nevada forest management, held in Nevada City, Libby asked the 250 people attending how many of them lived in houses made of wood.

Almost everyone did. Then he asked how many had houses built with alternatives such as used tires and straw bales. Only two or three people responded.

A few moments later, Libby recalled, he asked, "How many people are comfortable with species going extinct somewhere else because we're not going to cut any wood on the Tahoe National Forest?'"

At that point, Libby said, "Somebody in the audience shouted: 'We don't like your question.'"

Water scarcity could overwhelm the next generation

Janet L. Sawin

Up to 7 billion people in 60 countries—more than the whole present population of the world—will face water scarcity within the next half-century, according to the United Nations' *World Water Development Report* released in March. The UN report is the most complete appraisal of water resources to date. Other reports released the same month offer dire projections of the impacts of water scarcity on human health, the environment, and global political stability. [1]

Demand for fresh water has tripled in the past 50 years, and is continuing to rise as a result of population growth and economic development. About 70 percent of the demand is for agriculture. But increasing water use is not just a function of the greater number of people needing to eat and drink; it also results from pollution and misuse of available water supplies, both directly via dumping or runoff of effluents into water, and indirectly via pollution of air and soil. The damage is accelerated by wetlands destruction and other abuses—including human-caused global warming. According to the UN report, "recent estimates suggest that climate change will account for about 20 percent of the increase in global water scarcity" in coming decades.

Water mismanagement has become a crisis of governance that will impact heavily on public health and the environment, while heightening tensions and conflicts over declining resources. Worldwide, the greatest impacts will be on the poor, who are most vulnerable to water-borne illness—which further perpetuates their poverty. At present, 1.1 billion people lack access to clean water and 2.4 billion lack adequate sanitation. In 2000, an estimated 2.2 million people, most of them infants or children under five, died from water-related diseases.

It is not only the poor who are at risk, however. Around the same time the UN report was issued, two reports published in New Jersey revealed that the state's drinking water supply contains hundreds of chemicals, including steroids, detergents, pesticides, preservatives, and prescription medications. While all were found only in low concentrations, researchers cannot rule out possible health risks of ingesting low levels of multiple chemicals over long time periods. The water bodies studied contained antibiotics that could breed drug-resistant bacteria, and endocrine disruptors believed to cause neurological and reproductive birth defects. Other studies in the U.S. and Europe have yielded compa-

rable results, with hundreds of lakes, streams, and rivers testing positive for traces of similar chemicals.

The UN report concludes that a more positive outcome is possible, *if* the global community improves infrastructure, takes full advantage of conservation and efficiency technologies, and enacts appropriate water pricing plans and water treaties. In the global North, upgrading water treatment facilities to deal with contaminants will be costly, but necessary to safeguard human health. One implication of the report is that attempting to rely only on water treatment is a losing battle and that it is becoming increasingly urgent, in all countries, to reduce or eliminate toxins at their source.

Notes

1. The report estimates that humans need 20 to 50 liters of water free from harmful contaminants daily to meet basic needs. For monitoring purposes, the World Health Organixation/UN Children's Fund defines reasonable access to water as at least 20 liters per day, from an improved source within one kilometer of the user's dwelling.

From *World Watch*, Vol. 16 No. 4. © 2003 by Worldwatch Institute, www.worldwatch.org.

VANISHING ALASKA

Global warming is flooding villages along the coast.
Should they stand and fight—or surrender and move?

By Margot Roosevelt

Shishmaref is melting into the ocean. Over the past 30 years, the Inupiaq Eskimo village, perched on a slender barrier island 625 miles north of Anchorage, has lost 100 ft. to 300 ft. of coastline—half of it since 1997. As Alaska's climate warms, the permafrost beneath the beaches is thawing and the sea ice is thinning, leaving its 600 residents increasingly vulnerable to violent storms. One house has collapsed, and 18 others had to be moved to higher ground, along with the town's bulk-fuel tanks.

Giant waves have washed away the school playground and destroyed $100,000 worth of boats, hunting gear and fish-drying racks. The remnants of multimillion-dollar seawalls, broken up by the tides, litter the beach. "It's scary," says village official Luci Eningowuk. "Every year we agonize that the next storm will wipe us out."

The erosion of the island, now only a quarter-mile wide, is not the only ominous sign that large changes are afoot. The ice-fishing season that used to start in October has moved to December because the ocean freezes later each year. Berry picking begins in July instead of August. Most distressing for the Inupiaq is that thin ice makes it harder to hunt *oogruk*—the bearded seal that is a staple of their diet and culture.

At the Nayokpuk Trading Co., where infant formula sells for $21 a package and the only eggs for sale, sent by bush plane, sit broken in their shells, the talk is of the disruption of nature's rhythms. "When was the last time we went hunting on snow machines?" owner Percy Nayokpuk asks a customer. "About 15 years," answers Reuben Weyiouanna. Because a loaded snowmobile would break through the ice, the elders these days have to drag their boats seven miles across the ice to go hunting—and the season begins in May instead of June. "If the weather keeps changing," says Nayokpuk, "it will mean the end of Shishmaref."

The fate of one stubborn little village normally wouldn't make much of a splash. But Shishmaref and other Alaskan settlements are attracting national attention because scientists see them as gloomy harbingers. "Shishmaref is the canary in the coal mine—an indicator of what's to come elsewhere," says Gunter Weller, director of the University of Alaska's Cooperative Institute for Arctic Research.

Global warming, caused in part by the burning of oil and gas in factories and cars, is traumatizing polar regions, where the complex meteorological processes associated with snow, permafrost and ice magnify its effects. A study published in Science last week reported that glaciers in West Antarctica are thinning twice as fast as they did in the 1990s. In Alaska the annual mean air temperature has risen 4°F to 5°F in the past three decades—compared with an average of just under 1°F worldwide. As a result, the state's glaciers are melting; insects are destroying vast swaths of forest; and thawing permafrost is sinking roads, pipelines and homes. Arctic Ocean ice has shrunk 5% to 10%, at an accelerating rate. Says Weller: "There is natural variability, but the evidence is overwhelming that humanity has altered the climate."

It must be said that if Shishmaref sinks beneath the waves, it won't be much of a loss to global tourism. The village is so remote that no road connects it to the outside world. The occasional barge unloads fuel after the ice breaks up, and when the weather is good, battered bush planes ferry in DVDs and cartons of Cheetos from the Sam's Club in Fairbanks. Visually, this village is nothing like the romantic images of Eskimos in igloos from old *National Geographic* magazines. Weathered clapboard houses, surrounded by rusty engine parts, sit helter-skelter along muddy paths. Indoor plumbing is rare, and drinking water collects in plastic buckets under rain gutters. Empty

Coke cans and cigarette packets litter the streets. In the ramshackle town hall, a sign reads, CITY OF SHISHMAREF BINGO WILL NOT BE ACCEPTING ANY MORE PERSONAL CHECKS. Another warns against siphoning gasoline from the village fire truck.

THE ICE-FISHING SEASON HERE USED TO START IN OCTOBER. NOW IT STARTS IN DECEMBER

Still, like many of Alaska's native villages, Shishmaref clings to its subsistence culture. The town supports 10 dog teams, and a local musher, Herbie Nayokpuk, is known statewide as the Shishmaref Cannonball for his top-place finishes in the Iditarod race. Walrustusk carving is taught in school, along with the Inupiaq language. And if the town itself is ugly, it is balanced by the desolate beauty of the slate-colored sea, the ducks flying in formation over the lagoon and the musk ox roaming in emerald meadows dotted with wild cotton. Some two-thirds of the local diet still derives from hunting and fishing. In the diamond light of late summer, whole families forage for salmonberries, which the elders eat mixed with grated caribou fat. ("Eskimo ice cream," they call it.) The kids prefer it with Cool Whip.

"This is our grocery store," says Tony Weyiouanna, pulling shimmering white fish from his gill net.

But up and down Alaska's coast, alarm is spreading that the natural bounty on which the culture is built is at risk. At Point Hope, a bowhead-whaling village that dates from 600 B.C., flooding seawater threatens the airport runway and a seven-mile evacuation road. "During storms, people begin to panic," says town official Rex Rock. In the Pribilof Islands, villagers blame global warming along with industrial contaminants for the decline of 20 species, ranging from kelp to sea lion. In Barrow, capital of the oil-rich North Slope Borough, sandbags and dredging haven't protected $500 million in infrastructure. "We are at a crossroads," says Mayor Edith Vorderstrasse. "Is it practical to stand and fight our Mother Ocean? Or do we surrender and move?"

The prospect of relocating whole Eskimo villages—global warming's first American refugees—is gathering political support. Last January, Shishmaref citizens voted to move to a site called Tin Creek, 12 miles away, across a lagoon. And last June, Alaska's powerful Senator, Appropriations Committee Chairman Ted Stevens, convened federal, state and local officials for a two-day hearing in Anchorage to hear impassioned pleas from village leaders who want help repairing their infrastructure or relocating. Among the most eloquent was Eningowuk, 54, a mother of six who heads the Shishmaref Erosion and Relocation Coalition. "Shishmaref is where it is because of what the ocean, rivers, streams and the land provide to us," she testified. "We are hunters, and we are gatherers. We have been here for countless generations. We value our way of life. It provides for our very existence."

But moving Shishmaref to a more protected location could be prohibitively expensive, especially given the high cost of building in the Arctic. When the U.S. Army Corps of Engineers looked at relocating Kivalina, a nearby village of 380 people, the price tag was $100 million to $400 million—roughly $1 million for each resident.

And it wouldn't necessarily stop there. A recent Government Accountability Office (GAO) study found four villages, including Shish-maref, to be in "imminent danger" and 20 others to have serious problems. Overall, 184 out of 213 Alaska native villages face some flooding and erosion, the GAO report noted, although how much is due to global warming and how much to the natural movement of rivers and coasts is uncertain.

Although most Shishmaref residents want to relocate, they also are worried about moving inland. Nayokpuk fears that the cost of living will double if fuel has to be transported over land. And Stanley Tocktoo, the vice mayor, says that it will be harder to dig the ice cellars the villagers use for fermenting their meat in the mud beneath the Tin Creek site than it was in Shishmaref's sand. As his son Harvey, 11, watches a Jackie Chan movie and picks fermented-walrus morsels off his father's plate, Tocktoo reflects that the farther away the village has to move from the ocean, the more trouble it will be "to get access to all this good food."

An expensive precedent may be set here. If global warming ever begins washing away coastal towns in the rest of the U.S., the cost of mass relocations would be unimaginable. But Shishmaref's villagers are adamant about their need to stay together, and they greet with horror any suggestion that they be dispersed to Nome or Kotzebue. The village—where everyone knows everyone else's name, and everyone is more or less related to everyone else—must relocate as a whole, or it would be the "annihilation of our community by dissemination," says Eningowuk. Whatever the solution, the Inupiaq are looking for it to be paid for by the folks who sent them global warming in the first place. And who would that be? "The Nalauqmiu—white people," says Eningowuk with a rueful smile.

UNIT 4
Political Economy

Unit Selections

Key Points to Consider

- Are those who argue that there is in fact a process of globalization overly optimistic? Why or why not?

- What are some of the impediments to a truly global political economy?

- How has globalization accelerated the growth of criminal behavior?

- How are the political economies of traditional societies different from those of the consumer-oriented societies?

- What are some of the barriers that make it difficult for non-industrial countries to develop?

- How are China and other emerging countries trying to alter their ways of doing business in order to meet the challenges of globalization? Are they likely to succeed?

- What economic challenges do countries like Japan and the U.S. face in the years to come?

DUSHKIN ONLINE **Links: www.dushkin.com/online/**
These sites are annotated in the World Wide Web pages.

Belfer Center for Science and International Affairs (BCSIA)
http://ksgwww.harvard.edu/csia/
U.S. Agency for International Development
http://www.info.usaid.gov
The World Bank Group
http://www.worldbank.org

A defining characteristic of the twentieth century was the intense struggle between proponents of two economic ideologies. At the heart of the conflict was the question of what role government should play in the management of a country's economy. For some the dominant capitalist economic system appeared to be organized primarily for the benefit of a few wealthy people. From their perspective, the masses were trapped in poverty, supplying cheap labor to further enrich the privileged elite. These critics argued that the capitalist system could be changed only by gaining control of the political system and radically changing the ownership of the means of production. In striking contrast to this perspective, others argued that the best way to create wealth and eliminate poverty was through the profit motive, which encouraged entrepreneurs to create new products and businesses. An open and competitive marketplace, from this point of view, minimized government interference and was the best system for making decisions about production, wages, and the distribution of goods and services.

Both academic discourse and often violent conflict characterized the contest of capitalism versus socialism/communism. The Russian and Chinese revolutions overthrew the old social order and created radical changes in the political and economic systems in these two important countries. The political structures that were created to support new systems of agricultural and industrial production (along with the centralized planning of virtually all aspects of economic activity) eliminated most private ownership of property. These two revolutions were, in short, unparalleled experiments in social engineering.

The collapse of the Soviet Union and the dramatic reforms that have taken place in China have recast the debate about how to best structure contemporary economic systems. Some believe, that with the end of communism and the resulting participation of hundreds of millions of new consumers in the global market, an unprecedented new era has been entered. Many have noted that this process of "globalization" is being accelerated by a revolution in communication and computer technologies. Proponents of this view argue that a new global economy is emerging that will ultimately eliminate national economic systems.

Others are less optimistic about the prospects of globalization. They argue that the creation of a single economic system where there are no boundaries to impede the flow of both capital and goods and services does not mean a closing of the gap between the world's rich and poor. Rather, they argue that giant corporations will have fewer legal constraints on their behavior, and this will lead to greater exploitation of workers and the accelerated destruction of the environment. Further, these critics point out that the unintended globalization of drug trafficking and other criminal behaviors is developing more rapidly than appropriate remedies can be developed.

The use of the term "political economy" for the title of this unit recognizes that economic and political systems are not separate. All economic systems have some type of marketplace where goods and services are bought and sold. Government (either national or international) regulates these transactions to some degree; that is, government sets the rules that regulate the marketplace.

One of the most important concepts in assessing the contemporary political economy is "development." For the purposes of this unit, the term "development" is defined as an improvement in the basic aspects of life: lower infant mortality rates, longer life expectancy, lower disease rates, higher rates of literacy, healthier diets, and improved sanitation. Judged by these standards, some countries are more "developed" than others. A fundamental question that a thoughtful reader must consider is whether globalization is resulting in increased development not only for a few people but also for all of those participating in the global political economy.

The unit is organized into two sections. The first is a general discussion of the concept of globalization. How is the concept defined, and what are the differing perspectives on it? For example, is the idea of a global economy merely wishful thinking by those who sit on top of the power hierarchy, self-deluded into believing that globalization is an inexorable force that will evolve in its own way, following its own rules? Or will there continue to be the traditional tensions of power politics, that is, between the powerful and those who are either ascending or descending in power?

The second subsection is a selection of case studies that focus on specific countries or economic sectors. These case studies have been selected to challenge the reader to develop his or her own conclusions about the positive and negative consequences of the globalization process. Does the contemporary global political economy result in an age-old system of winners and losers, or can everyone positively benefit from its system of wealth creation and distribution?

The Complexities and Contradictions of Globalization

Globalization, we are told, is what every business should be pursuing, and what every nation should welcome. But what, exactly, is it? James Rosenau offers a nuanced understanding of a process that is much more real, and transforming, than the language of the marketplace expresses.

JAMES N. ROSENAU

The mall at Singapore's airport has a food court with 15 food outlets, all but one of which offering menus that cater to local tastes; the lone standout, McDonald's, is also the only one crowded with customers. In New York City, experts in *feng shui*, an ancient Chinese craft aimed at harmonizing the placement of man-made structures in nature, are sought after by real estate developers in order to attract a growing influx of Asian buyers who would not be interested in purchasing buildings unless their structures were properly harmonized.

Most people confronted with these examples would probably not be surprised by them. They might even view them as commonplace features of day-to-day life late in the twentieth century, instances in which local practices have spread to new and distant sites. In the first case the spread is from West to East and in the second it is from East to West, but both share a process in which practices spread and become established in profoundly different cultures. And what immediately comes to mind when contemplating this process? The answer can be summed up in one word: globalization, a label that is presently in vogue to account for peoples, activities, norms, ideas, goods, services, and currencies that are decreasingly confined to a particular geographic space and its local and established practices.

Indeed, some might contend that "globalization" is the latest buzzword to which observers resort when things seem different and they cannot otherwise readily account for them. That is why, it is reasoned, a great variety of activities are labeled as globalization, with the result that no widely accepted formulation of the concept has evolved. Different observers use it to describe different phenomena, and often there is little overlap among the various usages. Even worse, the elusiveness of the concept of globalization is seen as underlying the use of a variety of other, similar terms—world society, interdependence, centralizing tendencies, world system, globalism, universalism, internationalization, globality—that come into play when efforts are made to grasp why public affairs today seem significantly different from those of the past.

Such reasoning is misleading. The proliferation of diverse and loose definitions of globalization as well as the readiness to use a variety of seemingly comparable labels are not so much a reflection of evasive confusion as they are an early stage in a profound ontological shift, a restless search for new ways of understanding unfamiliar phenomena. The lack of precise formulations may suggest the presence of buzzwords for the inexplicable, but a more convincing interpretation is that such words are voiced in so many different contexts because of a shared sense that the human condition is presently undergoing profound transformations in all of its aspects.

WHAT IS GLOBALIZATION?

Let us first make clear where globalization fits among the many buzzwords that indicate something new in world affairs that is moving important activities and concerns beyond the national seats of power that have long served as the foundations of economic, political, and social life. While all the buzzwords seem to cluster around the same dimension of the present human condition, useful distinctions can be drawn among them.

Most notably, if it is presumed that the prime characteristic of this dimension is change—a transformation of practices and norms—then the term "globalization" seems appropriate to denote the "something" that is changing humankind's preoccupation with territoriality and the traditional arrangements of the state system. It is a term that directly implies change, and thus differentiates the phenomenon as a process rather than as a prevailing condition or a desirable end state.

Conceived as an underlying process, in other words, globalization is not the same as globalism, which points to aspirations for a state of affairs where values are shared by or pertinent to all the world's more than 5 billion people, their environment, and their role as citizens, consumers, or producers with an interest in collective action to solve common problems. And it can also be distinguished from universalism, which refers to those values that embrace all of humanity (such as the values that science or religion draws on), at any time or place. Nor is it coterminous with complex interdependence, which signifies structures that link people and communities in various parts of the world.

Although related to these other concepts, the idea of globalization developed here is narrower in scope. It refers neither to values nor to structures, but to sequences that unfold either in the mind or in behavior, to processes that evolve as people and organizations go about their daily tasks and seek to realize their particular goals. What distinguishes globalizing processes is that they are not hindered or prevented by territorial or jurisdictional barriers. As indicated by the two examples presented at the outset, such processes can readily spread in many directions across national boundaries, and are capable of reaching into any community anywhere in the world. They consist of all those forces that impel individuals, groups, and institutions to engage in similar forms of behavior or to participate in more encompassing and coherent processes, organizations, or systems.

Contrariwise, localization derives from all those pressures that lead individuals, groups, and institutions to narrow their horizons, participate in dissimilar forms of behavior, and withdraw to less encompassing processes, organizations, or systems. In other words, any technological, psychological, social, economic, or political developments that foster the expansion of interests and practices beyond established boundaries are both sources and expressions of the processes of globalization, just as any developments in these realms that limit or reduce interests are both sources and expressions of localizing processes.

Note that the processes of globalization are conceived as only capable of being worldwide in scale. In fact, the activities of no group, government, society, or company have never been planetary in magnitude, and few cascading sequences actually encircle and encompass the entire globe. Televised events such as civil wars and famines in Africa or protests against governments in Eastern Europe may sustain a spread that is worldwide in scope, but such a scope is not viewed as a prerequisite of globalizing dynamics. As long as it has the potential of an unlimited spread that can readily transgress national jurisdictions, any interaction sequence is considered to reflect the operation of globalization.

Obviously, the differences between globalizing and localizing forces give rise to contrary conceptions of territoriality. Globalization is rendering boundaries and identity with the land less salient while localization, being driven by pressures to narrow and withdraw, is highlighting borders and intensifying the deep attachments to land that can dominate emotion and reasoning.

In short, globalization is boundary-broadening and localization is boundary-heightening. The former allows people, goods, information, norms, practices, and institutions to move about oblivious to despite boundaries. The boundary-heightening processes of localization are designed to inhibit or prevent the movement of people, goods, information, norms, practices, and institutions. Efforts along this line, however, can be only partially successful. Community and state boundaries can be heightened to a considerable extent, but they cannot be rendered impervious. Authoritarian governments try to make them so, but their policies are bound to be undermined in a shrinking world with increasingly interdependent economies and communications technologies that are not easily monitored. Thus it is hardly surprising that some of the world's most durable tensions flow from the fact that no geographic borders can be made so airtight to prevent the infiltration of ideas and goods. Stated more emphatically, some globalizing dynamics are bound, at least in the long run, to prevail.

The boundary-expanding dynamics of globalization have become highly salient precisely because recent decades have witnessed a mushrooming of the facilities, interests, and markets through which a potential for worldwide spread can be realized. Likewise, the boundary-contracting dynamics of localization have also become increasingly significant, not least because some people and cultures feel threatened by the incursions of globalization. Their jobs, their icons, their belief systems, and their communities seem at risk as the boundaries that have sealed them off from the outside world in the past no longer assure protection. And there is, of course, a basis of truth in these fears. Globalization does intrude; its processes do shift jobs elsewhere; its norms do undermine traditional mores. Responses to these threats can vary considerably. At one extreme are adaptations that accept the boundary-broadening processes and make the best of them by integrating them into local customs and practices. At the other extreme are responses intended to ward off the globalizing processes by resort to ideological purities, closed borders, and economic isolation.

THE DYNAMICS OF FRAGMEGRATION

The core of world affairs today thus consists of tensions between the dynamics of globalization and localization. Moreover, the two sets of dynamics are causally linked, almost as if every increment of globalization gives rise to an increment of localization, and vice versa. To account for these tensions I have long used the term "fragmegration," an awkward and perhaps even grating label that has the virtue of capturing the pervasive interactions between the fragmenting forces of localization and the integrative forces of globalization.1 One can readily observe the unfolding of fragmegrative dynamics in the struggle of the European Union to cope with proposals for monetary unification or in the electoral campaigns and successes of Jean-Marie Le Pen in France, Patrick Buchanan in the United States, and Pauline Hanson in Australia—to mention only three examples.

It is important to keep in mind that fragmegration is not a single dynamic. Both globalization and localization are clusters of forces that, as they interact in different ways and through different channels, contribute to more encompassing processes in the case of globalization and to less encompassing processes in the case of localization. These various dynamics, moreover, operate in all realms of human activity, from the cultural and social to the economic and political.

In the political realm, globalizing dynamics underlie any developments that facilitate the expansion of authority, policies, and interests beyond existing socially constructed territorial boundaries, whereas the politics of localization involves any trends in which the scope of authority and policies undergoes contraction and reverts to concerns, issues, groups, and institutions that are less extensive than the prevailing socially constructed territorial boundaries. In the economic realm, globalization encompasses the expansion of production, trade, and investments beyond their prior locales, while localizing dynamics are at work when the activities of producers and consumers are constricted to narrower boundaries. In the social and cultural realms, globalization operates to extend ideas, norms, and practices beyond the settings in which they originated, while localization highlights or compresses the original settings and thereby inhibits the inroad of new ideas, norms, and practices.

It must be stressed that the dynamics unfolding in all these realms are long-term processes. They involve fundamental human needs and thus span all of human history. Globalizing dynamics derive from peoples' need to enlarge the scope of their self-created orders so as to increase the goods, services, and ideas available for their well-being. The agricultural revolution, followed by the industrial and postindustrial transformations, are among the major sources that have sustained globalization. Yet even as these forces have been operating, so have contrary tendencies toward contraction been continuously at work. Localizing dynamics derive from people's need for the psychic comforts of close-at-hand, reliable support—for the family and neighborhood, for local cultural practices, for a sense of "us" that is distinguished from "them." Put differently, globalizing dynamics have long fostered large-scale order, whereas localizing dynamics have long created pressure for small-scale order. Fragmegration, in short, has always been an integral part of the human condition.

GLOBALIZATION'S EVENTUAL PREDOMINANCE

Notwithstanding the complexities inherent in the emergent structures of world affairs, observers have not hesitated to anticipate what lies beyond fragmegration as global history unfolds. All agree that while the contest between globalizing and localizing dynamics is bound to be marked by fluctuating surges in both directions, the underlying tendency is for the former to prevail over the latter. Eventually, that is, the dynamics of globalization are expected to serve as the bases around which the course of events is organized.

Consensus along these lines breaks down, however, over whether the predominance of globalization is likely to have desirable or noxious consequences. Those who welcome globalizing processes stress the power of economic variables. In this view the globalization of national economies through the diffusion of technology and consumer products, the rapid transfer of financial resources, and the efforts of transnational companies to extend their market shares is seen as so forceful and durable as to withstand and eventually surmount any and all pressures toward fragmentation. This line acknowledges that the diffusion that sustains the processes of globalization is a centuries-old dynamic, but the difference is that the present era has achieved a level of economic development in which it is possible for innovations occurring in any sector of any country's economy to be instantaneously transferred to and adapted in any other country or sector. As a consequence,

when this process of diffusion collides with cultural or political protectionism, it is culture and protectionism that wind up in the shop for repairs. Innovation accelerates. Productivity increases. Standards of living improve. There are setbacks, of course. The newspaper headlines are full of them. But we believe that the time required to override these setbacks has shortened dramatically in the developed world. Indeed, recent experience suggests that, in most cases, economic factors prevail in less than a generation....

Thus understood, globalization—the spread of economic innovations around the world and the political and cultural adjustments that accompany this diffusion—cannot be stopped.... As history teaches, the political organizations and ideologies that yield superior economic performance survive, flourish, and replace those that are less productive.[2]

While it is surely the case that robust economic incentives sustain and quicken the processes of globalization, this line of theorizing nevertheless suffers from not allowing for its own negation. The theory offers no alternative interpretations as to how the interaction of economic, political, and social dynamics will play out. One cannot demonstrate the falsity—if falsity it is—of the theory because any contrary evidence is seen merely as "setbacks," as expectable but temporary deviations from the predicted course. The day may come, of course, when event so perfectly conform to the predicted patterns of globalization that one is inclined to conclude that the theory has been affirmed. But in the absence of alternative scenarios, the theory offers little guidance as to how to interpret intervening events, especially those that highlight the tendencies toward fragmentation. Viewed in this way, it is less a theory and more an article of faith to which one can cling.

Other observers are much less sanguine about the future development of fragmegration. They highlight a litany of noxious consequences that they see as following from the eventual predominance of globalization: "its economism; its economic reductionism; its technological determinism; its political cynicism, defeatism, and immobilism; its de-socialization of the subject and resocialization of risk; its teleological subtext of inexorable global 'logic' driven exclusively by capital accumulation and the market; and its ritual exclusion of factors, causes, or goals other than capital accumulation and the market from the priority of values to be pursued by social action."[3]

Still another approach, allowing for either desirable or noxious outcomes, has been developed by Michael Zurn. He identifies a mismatch between the rapid extension of boundary-crossing activities and the scope of effective governance. Consequently, states are undergoing what is labeled "uneven denationalization," a primary process in which "the rise of international governance is still remarkable, but not accompanied by mechanisms for… democratic control; people, in addition, become alienated from the remote political process.… The democratic state in the Western world is confronted with a situation in which it is undermined by the process of globalization and overarched by the rise of international institutions."[4]

> *There is no inherent contradiction between localizing and globalizing tendencies.*

While readily acknowledging the difficulties of anticipating where the process of uneven denationalization is driving the world, Zurn is able to derive two scenarios that may unfold: "Whereas the pessimistic scenario points to instances of fragmentation and emphasizes the disruption caused by the transition, the optimistic scenario predicts, at least in the long run, the triumph of centralization." The latter scenario rests on the presumption that the increased interdependence of societies will propel them to develop ever more effective democratic controls over the very complex arrangements on which international institutions must be founded.

UNEVEN FRAGMEGRATION

My own approach to theorizing about the fragmegrative process builds on these other perspectives and a key presumption of my own—that there is no inherent contradiction between localizing and globalizing tendencies—to develop an overall hypothesis that anticipates fragmegrative outcomes and that allows for its own negation: the more pervasive globalizing tendencies become, the less resistant localizing reactions will be to further globalization. In other words, globalization and localization will coexist, but the former will continue to set the context for the latter. Since the degree of coexistence will vary from situation to situation (depending on the salience of the global economy and the extent to which ethnic and other noneconomic

factors actively contribute to localization), I refer, borrowing from Zurn, to the processes depicted by the hypothesis as uneven fragmegration. The hypothesis allows for continuing pockets of antagonism between globalizing and localizing tendencies even as increasingly (but unevenly) the two accommodate each other. It does not deny the pessimistic scenario wherein fragmentation disrupts globalizing tendencies; rather it treats fragmentation as more and more confined to particular situations that may eventually be led by the opportunities and requirements of greater interdependence to conform to globalization.

For globalizing and localizing tendencies to accommodate each other, individuals have to come to appreciate that they can achieve psychic comfort in collectivities through multiple memberships and multiple loyalties, that they can advance both local and global values without either detracting from the other. The hypothesis of uneven fragmegration anticipates a growing appreciation along these lines because the contrary premise, that psychic comfort can only be realized by having a highest loyalty, is becoming increasingly antiquated. To be sure, people have long been accustomed to presuming that, in order to derive the psychic comfort they need through collective identities, they had to have a hierarchy of loyalties and that, consequently, they had to have a highest loyalty that could only be attached to a single collectivity. Such reasoning, however, is a legacy of the state system, of centuries of crises that made people feel they had to place nation-state loyalties above all others. It is a logic that long served to reinforce the predominance of the state as the "natural" unit of political organization and that probably reached new heights during the intense years of the cold war.

But if it is the case, as the foregoing analysis stresses, that conceptions of territoriality are in flux and that the failure of states to solve pressing problems has led to a decline in their capabilities and a loss of legitimacy, it follows that the notion that people must have a "highest loyalty" will also decline and give way to the development of multiple loyalties and an understanding that local, national, and transnational affiliations need not be mutually exclusive. For the reality is that human affairs are organized at all these levels for good reasons; people have needs that can only be filled by close-at-hand organizations and other needs that are best served by distant entities at the national or transnational level.

In addition, not only is an appreciation of the reality that allows for multiple loyalties and memberships likely to grow as the effectiveness of states and the salience of national loyalties diminish, but it also seems likely to widen as the benefits of the global economy expand and people become increasingly aware of the extent to which their well-being is dependent on events and trends elsewhere in the world. At the same time, the distant economic processes serving their needs are impersonal and hardly capable of advancing the need to share with others in a collective affiliation. This need was long served by the nation-state, but with fragmegrative dynamics having undermined the national level as a source of psychic comfort and with transnational entities seeming too distant to provide the psychic benefits of affiliation, the satisfactions to be gained through more close-at-hand affiliations are likely to seem ever more attractive.

THE STAKES

It seems clear that fragmegration has become an enduring feature of global life; it is also evident that globalization is not merely a buzzword, that it encompasses pervasive complexities and contradictions that have the potential both to enlarge and to degrade our humanity. In order to ensure that the enlargement is more prevalent than the degradation, it is important that people and their institutions become accustomed to the multiple dimensions and nuances as our world undergoes profound and enduring transformations. To deny the complexities and contradictions in order to cling to a singular conception of what globalization involves is to risk the many dangers that accompany oversimplification.

NOTES

1. For an extensive discussion of the dynamics of fragmegration, see James N. Rosenau, *Along the Domestic-Foreign Frontier: Exploring Governance in a Turbulent World* (Cambridge: Cambridge University Press, 1997), ch. 6.
2. William W. Lewis and Marvin Harris, "Why Globalization Must Prevail," *The McKinsey Quarterly*, no. 2 (1992), p. 115.
3. Barry K. Gills, "Editorial: 'Globalization' and the 'Politics of Resistance,'" *New Political Economy*, vol. 2 (March 1997), p. 12.
4. Michael Zurn, "What Has Changed in Europe? The Challenge of Globalization and Individualization," paper presented at a meeting on What Has Changed? Competing Perspectives on World Order (Copenhagen, May 14–16, 1993), p. 40.

JAMES N. ROSENAU *is University Professor of International Affairs at George Washington University. His latest book is* Along the Domestic-Foreign Frontier: Exploring Governance in a Turbulent World *(Cambridge: Cambridge University Press, 1997). This article draws on the author's "New Dimensions of Security: The Interaction of Globalizing and Localizing Dynamics,"* Security Dialogue, *September 1994, and "The Dynamics of Globalization: Toward an Operational Formulation,"* Security Dialogue, *September 1996).*

Three Cheers for Global Capitalism

BY JOHAN NORBERG

Under what is rather barrenly termed "globalization"—the process by which people, information, trade, investments, democracy, and the market economy tend more and more to cross national borders—our options and opportunities have multiplied. We don't have to shop at the big local company; we can turn to a foreign competitor. We don't have to work for the village's one and only employer; we can seek alternative opportunities. We don't have to make do with local cultural amenities; the world's culture is at our disposal. Companies, politicians, and associations have to exert themselves to elicit interest from people who have a whole world of options. Our ability to control our own lives is growing, and prosperity is growing with it.

Free markets and free trade and free choices transfer power to individuals at the expense of political institutions. Because there is no central control booth, it seems unchecked, chaotic. Political theorist Benjamin Barber speaks for many critics when he bemoans the absence of "viable powers of opposing, subduing, and civilizing the anarchic forces of the global economy." "Globalization" conjures up the image of an anonymous, enigmatic, elusive force, but it is actually just the sum of billions of people in thousands of places making decentralized decisions about their own lives. No one is in the driver's seat precisely because all of us are steering.

No company would import goods from abroad if we didn't buy them. If we did not send e-mails, order books, and download music every day, the Internet would wither and die. We eat bananas from Ecuador, order magazines from Britain, work for export companies selling to Germany and Russia, vacation in Thailand, and save money for retirement by investing in South America and Asia. These things are carried out by businesses only because we as individuals want them to. Globalization takes place from the bottom up.

A recent book about the nineteenth-century Swedish historian Erik Geijer notes that he was able to keep himself up to date just by sitting in Uppsala reading the *Edinburgh Review* and the *Quarterly Review*. That is how simple and intelligible the world can be when only a tiny elite in the capitals of Europe makes any difference to the course of world events. How much more complex and confusing everything is now, with ordinary people having a say over their own lives. Elites may mourn that they have lost power, but everyday life has vastly improved now that inexpensive goods and outside information and different employment opportunities are no longer blocked by political barriers.

To those of us in rich countries, more economic liberty to pick and choose may sound like a trivial luxury, even an annoyance—but it isn't. Fresh options are invaluable for all of us. And the existence from which globalization delivers people in the Third World—poverty, filth, ignorance, and powerlessness—really is intolerable. When global capitalism knocks at the door of Bhagant, an elderly agricultural worker and "untouchable" in the Indian village of Saijani, it leads to his house being built of brick instead of mud, to shoes on his feet, and clean clothes—not rags—on his back. Outside Bhagant's house, the streets now have drains, and the fragrance of tilled earth has replaced the stench of refuse. Thirty years ago Bhagant didn't know he was living in India. Today he watches world news on television. The stand that we in the privileged world take on the burning issue of globalization can determine whether or not more people will experience the development that has taken place in Bhagant's village.

Critics of globalization often paint a picture of capitalist marauders secretly plotting for world mastery, but this notion is completely off the mark. It has mostly been pragmatic, previously socialist, politicians who fanned globalization in China, Latin America, and East Asia—after realizing that government control-freakery had ruined their societies. Any allegation of runaway capitalism has to be tempered by the observation that today we have the largest public sectors and highest taxes the world has ever known. The economic liberalization measures of the last quarter century may have abolished some of the recent past's centralist excesses, but they have hardly ushered in a system of laissez-faire.

What defenders of global capitalism believe in, first and foremost, is man's capacity for achieving great things by means of the combined force of market exchanges. It is not their intention to put a price tag on everything. The important things—love, family, friendship, one's own way of life—cannot be assigned a monetary value. Principled advocates of global economic liberty plead for a more open world because that setting unleashes individual creativity as none other can. At its core, the belief in capitalist freedom among nations is a belief in mankind.

People have a natural tendency to believe that everything is growing worse, and that things were better in the "old days." In 1014, Archbishop Wulfstan of York declared, "The world is in a rush and is getting close to its end."

Today, we hear that life is increasingly unfair amidst the market economy: "The rich are getting richer, and the poor are getting poorer." But if we look beyond the catchy slogans, we find that while many of the rich have indeed grown richer, so have most of the poor. Absolute poverty has diminished, and where it was greatest 20 years ago—in Asia—hundreds of millions of people have achieved a secure existence, even affluence, previously undreamed of. Global misery has diminished, and great injustices have started to unravel.

One of the most interesting books published in recent years is *On Asian Time: India, China, Japan 1966–1999*, a travelogue in which Swedish author Lasse Berg and photographer Stig Karlsson describe their visits to Asian countries during the 1960s. They saw poverty, abject misery, and imminent disaster. Like many, they could not bring themselves to hold out much hope for the future, and they thought that socialist revolution might be the only way out. Returning to India and China in the 1990s, they could not help seeing how wrong they were. More and more people have extricated themselves from poverty; the problem of hunger is steadily diminishing; the streets are cleaner. Squalid huts have given way to buildings wired for electricity with TV antennas on their roofs.

When Berg and Karlsson first visited Calcutta, fully one tenth of its inhabitants were homeless, and every morning trucks were sent by the public authorities or missionary societies to collect the bodies of those who had died in the night. Thirty years later, setting out to photograph people living on the streets, they had difficulty in *finding* such people. The rickshaw was disappearing from the urban scene, with people traveling by car, motorcycle, and subway instead.

When Berg and Karlsson showed young Indians photographs from the 1960s, they refused to believe it was even the same place. Could things really have been so dreadful? A striking illustration of the change is provided by a pair of photographs in their book. In the old picture, taken in 1976, a 12-year-old Indian girl named Satto holds up her hands. They are already furrowed and worn, prematurely aged by many years' hard work. The new picture shows Satto's 13-year-old daughter Seema, also holding up her hands. They are smooth and soft, the hands of a girl whose childhood has not been taken away from her.

The biggest change of all is in people's thoughts and dreams. Television and newspapers bring ideas and images from the other side of the globe, widening people's notions of the possible. Why make do with this kind of government when there are alternative political systems available?

Lasse Berg writes, self-critically:

> Reading what we observers, foreigners as well as Indians, wrote in the '60s and '70s, nowhere in these analyses do I see anything of present-day India. Often nightmare scenarios—overpopulation, tumult, upheaval or stagnation—but not this calm and steady forward-jogging, and least of all this modernization of thoughts and dreams. Who foresaw that consumerism would penetrate so deeply in and among the villages? Who foresaw that both the economy and general standard of living would do so well? Looking back, what the descriptions have in common is an overstatement of the extraordinary, frightening uncertainty, and an understatement of the force of normality.

Note that all of the dramatic development described by Berg has resulted from sharp moves over the past few decades toward international capitalism and trade.

This progress is all very well, many critics of globalization will argue, but even if the majority are better off, gaps have widened and wealthy people and countries have improved their lot more rapidly than others. The critics point out that 40 years ago the combined per capita GDP of the 20 richest countries was 15 times greater than that of the 20 poorest, and is now 30 times greater.

There are two reasons why this objection to globalization does not hold up. First, if everyone is better off, what does it matter that the improvement comes faster for some than for others? Only those who consider wealth a greater problem than poverty can find irritation in middle-class citizens becoming millionaires while the previously poverty-stricken become middle class.

Second, the allegation of increased inequality is simply wrong. The notion that global inequality has increased is largely based on figures from the U.N.'s 1999 *Human Development Report*. The problem with these figures is that they don't take into account what people can actually buy with their money. Without that "purchasing power" adjustment, the figures only show what a currency is worth on the international market, and nothing about local conditions. Poor people's actual living standards hinge on the cost of their food, their clothing, their housing—not what their money would get them while vacationing in Europe. That's why the U.N. uses purchasing-power-adjusted figures in other measures of living standards. It only resorts to the unadjusted figures, oddly, in order to present a theory of inequality.

A report from the Norwegian Institute for Foreign Affairs investigated global inequality by means of figures adjusted for purchasing power. Their data show that, contrary to conventional wisdom, inequality between countries has continuously *declined* ever since the end of the 1970s. This decline has been especially rapid since 1993, when globalization really gathered speed.

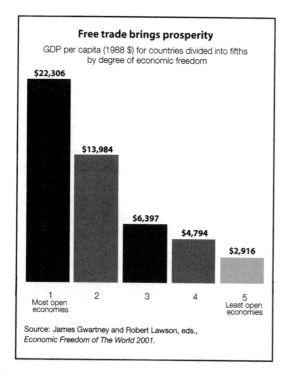

Free trade brings prosperity

GDP per capita (1988 $) for countries divided into fifths by degree of economic freedom

$22,306

$13,984

$6,397

$4,794

$2,916

1 Most open economies   2   3   4   5 Least open economies

Source: James Gwartney and Robert Lawson, eds., *Economic Freedom of The World 2001.*

More recently, similar research by Columbia University development economist Xavier Sala-i-Martin has confirmed those findings. He found that when U.N. figures are adjusted for purchasing power, they point to a sharp decline in world inequality. Sala-i-Martin and co-author Surjit Bhalla also found independently that if we focus on inequality between *persons*, rather than inequality between *countries*, global inequality at the end of 2000 was at its lowest point since the end of World War II.

Estimates that compare countries rather than individuals, both authors note, grossly overestimate real inequality because they allow gains for huge numbers of people to be outweighed by losses for far fewer. For instance, country aggregates treat China and Grenada as data points of equal weight, even though China's population is 12,000 times Grenada's. Once we shift our focus to people rather than nations, the evidence is overwhelming that the past 30 years have witnessed a strong shift toward global equalization.

One myth about trade is the notion that exports to other countries are a good thing, but that imports are somehow a bad thing. Many believe that a country grows powerful by selling much and buying little. The truth is that our standard of living will not rise until we use our money to buy more and cheaper things. One of the first trade theorists, James Mill, rightly noted in 1821 that "The benefit which derives from exchanging one commodity for another arises in all cases from the commodity *received*, not the commodity given." The only point of exports, in other words, is to enable us to get imports in return.

The absurdity of the idea that we must avoid cheap imports becomes clear if we imagine it applied to nonnational boundaries. If imports really were economically harmful, it would make sense for one city or state to prevent its inhabitants from buying from another. According to this logic, Californians would lose out if they bought goods from Texas, Brooklyn would gain by refusing to buy from Manhattan, and it would be better for a family to make everything itself instead of trading with its neighbors. It's obvious that such thinking would lead to a tremendous loss of welfare: The self-sufficient family would be hard pressed just to keep food on the table. When you go to the store, you "import" food—being able to do so cheaply is a benefit, not a loss. You "export" when you go to work and create goods and services. Most of us would prefer to "import" so cheaply that we could afford to "export" a little less.

Trade is not a zero-sum game in which one party loses what the other party gains. There would *be* no exchange if both parties did not feel that they benefited. The really interesting yardstick is not the "balance of trade" (where a "surplus" means that we are exporting more than we are importing) but the *quantity* of trade, since both exports and imports are gains. Imports are often feared as a potential cause of unemployment: If we import cheap toys and clothing from China, then toy and garment manufacturers here will have to scale down. But by obtaining cheaper goods from abroad, we save resources in the United States and can therefore invest in new industries and occupations.

Free trade brings freedom: freedom for people to buy and sell what they want. As an added benefit, this leads to the efficient use of resources. A company, or country, specializes where it can generate the greatest value.

Economic openness also leads to an enduring effort to improve production, because foreign competition forces firms to be as good and cheap as possible. As production in established industries becomes ever more efficient, resources are freed up for investment in new methods, inventions, and products. Foreign competition brings the same benefits that we recognize in economic competition generally; it simply extends competition to a broader field.

One of the most important but hard to measure benefits of free trade is that a country trading a great deal with the rest of the world imports new ideas and new techniques in the bargain. If the United States pursues free trade, our companies are exposed to the world's best ideas. They can then borrow those ideas, buy leading technology from elsewhere, and hire the best available manpower. This compels the companies to be more dynamic themselves. The world's output today is six times what it was 50 years ago, and world trade is 16 times greater. There is reason to believe that the trade growth drove much of the production growth. One comprehensive study of the effects of trade was conducted by Harvard economists Jeffrey Sachs and Andrew Warner. They examined the trade policies between 1970 and 1989 of 117 countries. The study reveals a statistically significant connection between free trade and economic growth. Growth was between three and six times *higher* in free-trade countries than in protectionist ones. Factors like improved education turned out to be vastly less important than trade in increasing economic progress.

Over those two decades, developing countries that practiced free trade had an average annual growth rate of 4.5 percent, while developing countries that practiced protectionism grew by only 0.7 percent. Among industrial countries, the free traders experienced annual growth of 2.3 percent, versus only 0.7 percent among the protectionists. It must be emphasized that this is not a matter of countries earning more because *others* opened to *their* exports. Rather, these countries earned more by keeping their own markets open.

If free trade is constantly making production more efficient, won't that result in the disappearance of job opportunities? When Asians manufacture our cars and South Americans produce our meat, auto workers and farmers in the United States lose their jobs and unemployment rises. Foreigners and developing countries will increasingly produce the things we need, until we don't have any jobs left. If increasing automation means everything we consume today will be able to be made by half the U.S. labor force in 20 years, doesn't that mean that the other half will be out of work? Such are the horror scenarios depicted in many anti-globalization writings.

The notion that a colossal unemployment crisis is looming began to grow popular in the mid 1970s. Since then, production has been streamlined and internationalized more than ever. Yet far more jobs have been created than have disappeared. We have more efficient production than ever before, but also more people at work. Between 1975 and 1998, employment in countries like the United States, Canada, and Australia rose by 50 percent.

And it is in the most internationalized economies, making the most use of modern technology, that employment has grown fastest. Between 1983 and 1995 in the United States, 24 million more job opportunities were created than disappeared. And those were not low-paid, unskilled jobs, as is often alleged. On the contrary, 70 percent of the new jobs carried a wage above the American median level. Nearly half the new jobs belonged to the most highly skilled, a figure which has risen even more rapidly since 1995.

So allegations of progressively fewer people being needed in production have no empirical foundation. And no wonder, for they are wrong in theory too. Imagine a pre-industrial economy where most everyone is laboring to feed himself. Then food production is improved by new technologies, new machines, foreign competition, and imports. That results in a lot of people being forced to leave the agricultural sector. Does that mean there is nothing for them to do, that consumption is constant? Of course not; the manpower which used to be required to feed the population shifts to clothing it, and providing better housing. Then improved transport, and entertainment. Then newspapers, telephones, and computers.

The notion that the quantity of employment is constant, that a job gained by one person is always a job taken from someone else, has provoked a variety of foolish responses. Some advocate that jobs must be shared. Others smash machinery. Many advocate raising tariffs and excluding immigrants. But the whole notion is wrong. The very process of a task being done

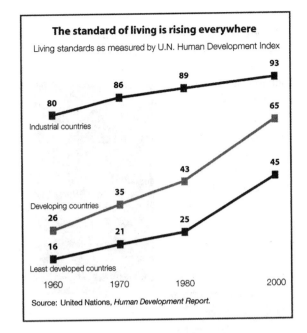

Source: United Nations, *Human Development Report.*

more efficiently, thus allowing jobs to be shed, enables new industries to grow, providing people with new and better jobs.

Efficiency does, of course, have a flip side. Economist Joseph Schumpeter famously described a dynamic market as a process of "creative destruction," because it destroys old solutions and industries, with a creative end in view. As the word "destruction" suggests, not everyone benefits from every market transformation in the short term. The process is painful for those who have invested in or are employed by less-efficient industries. Drivers of horse-drawn cabs lost out with the spread of automobiles, as did producers of paraffin lamps when electric light was introduced. In more modern times, manufacturers of typewriters were put out of business by the computer, and LP records were superseded by CDs. Painful changes of this kind happen all the time as a result of new inventions and methods of production. Unquestionably, such changes can cause trauma for those affected. But the most foolish way to counter such problems is to try to prevent them. It is generally fruitless; mere spitting into the wind. Besides, without "creative destruction," we would *all* be stuck with a lower standard of living.

The idea that we should halt change now is as misguided as the idea that we should have obstructed agricultural advances two centuries ago to protect the 80 percent of the population then employed on the land. A far better idea is to use the economic gains that flow from transformation to alleviate the consequences for those adversely affected. As a Chinese proverb has it, "When the wind of change begins to blow, some people build windbreaks while others build windmills."

But the troubles of change are seldom as widespread as a scan of the newspaper headlines might suggest. It is easy to report that 300 people lost their jobs in a car factory due to Japanese competition. It is less easy and less dramatic to report on the thousands of jobs that have been created because we were

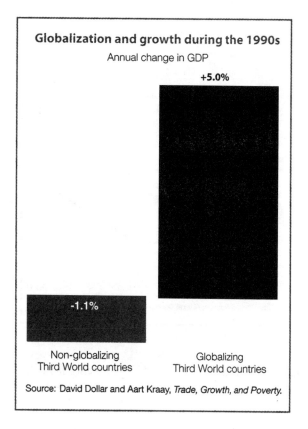

Globalization and growth during the 1990s

Annual change in GDP

+5.0%

-1.1%

Non-globalizing
Third World countries

Globalizing
Third World countries

Source: David Dollar and Aart Kraay, *Trade, Growth, and Poverty.*

have stagnated or fallen since the 1970s. A growing proportion of wages is now paid in non-money forms, such as health insurance, stocks, 401(k) contributions, day care, and so forth, to avoid taxation. When these benefits are included, American wages have risen right along with productivity. Among poor Americans, the proportion of consumption devoted to food, clothing, and housing has fallen since the 1970s from 52 to 37 percent, which clearly shows that they have money to spare for much more than the bare necessities of life.

The type of protectionism most fashionable among intellectuals today holds that we should not permit trade with countries that have poor working conditions or environmental protections. Since President Clinton proposed this kind of boycott at the World Trade Organization talks in Seattle in 1999, trade liberalization has largely deadlocked. Developing countries refuse to negotiate under such threats.

Whatever affluent demonstrators in economically powerful countries may believe, low wages and poor environmental conditions in developing countries are not the result of lack of enlightenment or stinginess. Generally the problem is that employers cannot afford to pay higher wages and provide better working conditions, because worker productivity is much lower in undeveloped countries. Wages can be raised as labor becomes more valuable, and that can be achieved only through better infrastructure, more education, new machinery, better organization, and increased investment. If we force these countries to raise wages before productivity has been improved, firms and consumers will be asked to pay more for their manpower than it is currently worth—denying those citizens any chance to work next to the more productive, better-paid workers of the Western world. Unemployment among the world's poor would swiftly rise.

Jesus Reyes-Heroles, Mexican ambassador to the United States, has explained, "In a poor country like ours the alternative to low-paying jobs isn't high-paying jobs—it's no jobs at all." In effect, labor and environmental provisions tell the developing countries: You are too poor to trade with us, and we are not going to trade with you until you have grown rich. The problem is that only through trade can they grow richer and thereby, step by step, improve their living standards and their social conditions. This is a catch-22.

Suppose this idea had been in vogue at the end of the nineteenth century. Britain and France would have noted that Swedish wages were only a fraction of theirs, that Sweden had a 12- or 13-hour working day and a six-day week, and that Swedes were chronically undernourished. Child labor was widespread in spinning mills, glassworks, and match and tobacco factories. One factory worker out of 20 was under 14 years old. Britain and France, accordingly, would have refused to trade with Sweden and closed their frontiers to Swedish cereals, timber, and iron ore.

Would Swedes have benefited? Hardly. Such a decision would have robbed them of earnings and blocked their industrial development. They would have been left with intolerable living conditions, children would have stayed in the factories, and perhaps to this day they would be eating tree-bark bread

able to use old resources more efficiently. Costs affecting a small group on an isolated occasion are simple to spotlight, while benefits that gradually accrue to nearly everyone creep up on us without our giving them a thought. Hardly any of the world's consumers have been informed by their news sources that wider selection, better quality, and lower prices spurred by competition following the Uruguay Round of trade liberalization have gained them between $100 billion and $200 billion annually. The difference is visible, though, in our refrigerators, our home electronics, and our wallets.

A review of more than 50 surveys of adjustments after trade liberalization in different countries shows clearly that adjustment problems are far milder than the conventional debate suggests. For every dollar of trade adjustment costs, roughly $30 is harvested in the form of welfare gains. A study of trade liberalization in 13 different countries showed that in all but one, industrial employment had already increased just one year after the liberalization. The process turns out to be far more creative than destructive.

If there are problems resulting from unshackled capitalism, they ought to be greatest in the United States, with its constant swirling economic transformations. But our job market is a bit like the Hydra in the legend of Hercules. Every time Hercules cuts off one of the beast's heads, two new ones appear. The danger of having to continue changing jobs all one's life is exaggerated: The average length of time an American stays in a particular job actually increased between 1983 and 1995, from 3.5 years to 3.8. Nor is it true, as many people believe, that more jobs are created in the United States only because real wages when the harvest failed. But that didn't happen. Trade with Sweden was allowed to grow uninterrupted, industrialization got under way, and the Swedish economy was revolutionized. This didn't just help Sweden, it also gave Britain and France a new peer for prosperous exchange.

If today, as a condition for trading with the developing countries, we require their mining industries to be as safe as the West's are now, we make demands that we ourselves did not meet when our own mining industries were developing. It was only after raising our incomes that we were able to develop the technology and afford the safety equipment we take for granted today. If we require developing countries to adopt those practices and equipment before they can afford them, their industries will be knocked out. That will not help the world's poor. And it will not help us gain new prosperous, stable, clean trading partners.

Advocates of protectionism often complain of "sweatshops" allegedly run by multinational corporations in the Third World. Let's look at the evidence: Economists have compared the conditions of people employed in American-owned facilities in developing countries with those of people employed elsewhere in the same country. In the poorest developing countries, the average employee of an American-affiliated company makes *eight times* the average national wage! In middle income countries, American employers pay *three times* the national average. Even compared with corresponding modern jobs in the same country, the multinationals pay about 30 percent higher wages. Marxists maintain that multinationals exploit poor workers. Are much higher wages "exploitation"?

The same marked difference can be seen in working conditions. The International Labor Organization has shown that multinationals, especially in the footwear and garment industries, are leading the trend toward better working conditions in the Third World. When multinational corporations accustom workers to better-lit, safer, and cleaner factories, they raise the general standard. Native firms then also have to offer better conditions, otherwise no one will work for them. Zhou Litai, one of China's foremost labor attorneys, has pointed out that Western consumers are the principal driving force behind the improvements of working conditions in China, and worries that "if Nike and Reebok go, this pressure evaporates."

One of the few Western participants in the globalization debate to have actually visited Nike's Asian subcontractors to find out about conditions there is Linda Lim of the University of Michigan. She found that in Vietnam, where the annual minimum wage was $134, Nike workers got $670. In Indonesia, where the minimum wage was $241, Nike's suppliers paid $720. If Nike were to withdraw on account of boycotts and tariffs imposed from the West, these employees would be put out of work and would move to more dangerous and less lucrative jobs in native industry or agriculture.

There are of course rogues among entrepreneurs, just as there are in politics, and all parts of life. But bad behavior by a few is no reason for banning large corporations from investing over-

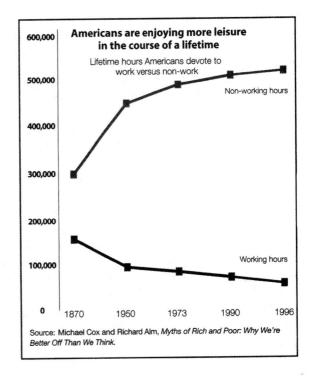

Americans are enjoying more leisure in the course of a lifetime

Lifetime hours Americans devote to work versus non-work

Source: Michael Cox and Richard Alm, *Myths of Rich and Poor: Why We're Better Off Than We Think.*

seas. That would make no more sense than disbanding the police because we find instances of police brutality.

It is commonly supposed that in order to cope with rising competition from the Third World and increasingly efficient machinery, we in the United States and Western Europe must work harder and put in longer and longer hours. Actually, the time we spend working has diminished, as rising prosperity resulting from growth has enabled us to earn the same pay by doing less work—if we want to. Compared with our parents' generation, most of today's workers go to work later, come home earlier, have longer lunch and coffee breaks, longer vacations, and more public holidays. In the U.S., working hours today are only about half of what they were a hundred years ago, and have diminished by about 10 percent just since 1973—a reduction equaling 23 days per year. On average, American workers have acquired five extra years of waking leisure time since 1973.

We start working progressively later in life, and retire earlier. Calculated over his lifetime, a Western worker in 1870 had only two hours off for each hour worked. By 1950 that figure had doubled to four hours off. Today it has doubled again to about eight hours off for each hour worked. Economic development, closely linked to an expansion of trade that has enabled us to specialize, makes it possible for us to reduce our working hours sharply even as we raise our material living standard.

It is natural for us in the affluent West to complain about stress. This is caused mostly by the fantastic growth of options. Pre-industrial citizens, spending all their lives in one place and perhaps meeting a hundred people in a lifetime, were less likely to feel that they did not have time for everything they wanted to do. People spent a lot of their non-working time sleeping.

Today, we can travel the world, read newspapers, see films from every corner of the globe, and meet a hundred people every day. We used to go to the mailbox and wait for the postman. Now e-mail sits in our inbox, waiting for us. We have a huge entertainment industry that offers an almost infinite number of ways to pass the time. No wonder the result is a certain frustration over not finding hours enough to do everything. But compared with the problems that earlier people had, and that most people in developing countries still have today, this kind of worry should be recognized for what it is—a luxury. Stress and burnout at work can be real. But they're not caused by globalization.

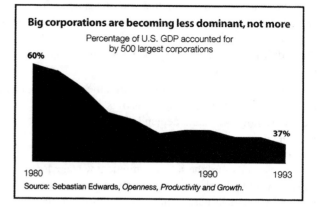

Big corporations are becoming less dominant, not more

Percentage of U.S. GDP accounted for by 500 largest corporations

60%

37%

1980   1990   1993

Source: Sebastian Edwards, *Openness, Productivity and Growth.*

Corporations have not acquired more power through free trade. Indeed, they used to be far more powerful—and still are in dictatorships and controlled economies. Large corporations have chances to corrupt or manipulate when power is distributed by public officials who can be hobnobbed over luncheons to give protection through monopolies, tariffs, or subsides. Free trade, on the other hand, exposes corporations to competition. Above all, it lets consumers ruthlessly pick and choose across national borders, rejecting companies that don't measure up.

People living in isolation are dependent on local enterprises and are forced to buy what they offer at the price they demand. Free trade and competition are the best guarantees of alternatives penetrating the market if the dominant firm does not satisfy. The only reason a lump of sugar costs two to three times as much in the E.U. as on the world's open markets is because European governments block trade with sugar tariffs. In this way, bureaucrats create far more monopolies than does capitalism.

Capitalists, however, aren't necessarily loyal adherents of capitalism: Often, they are happy to collaborate with governments to protect their privileges. A market economy and free trade are the best ways to force them to offer a better return for our resources. Free trade does not confer coercive power on anyone. The freedom of a business in a free market economy is like a waiter's freedom to offer the menu to a restaurant patron. And it entitles other waiters—foreign ones, even!—to come running up with rival menus. The loser in this process, if anyone, is the waiter who once had a monopoly.

Nothing forces people to accept new products. If they gain market share, it is because people want them. Even the biggest companies survive at the whim of customers. Mega-corporation Coca-Cola has to adapt the recipe for its drinks to different regions in deference to local variations in tastes. McDonald's sells mutton burgers in India, teriyaki burgers in Japan, and salmon burgers in Norway. TV mogul Rupert Murdoch failed to create a pan-Asian channel and instead has been forced to build different channels to suit local audiences.

Companies in free competition can grow large and increase their sales only by being better than others. Companies that fail to do so quickly go bust or get taken over by someone who can make better use of their capital, buildings, machinery, and employees. Capitalism is very tough—but mainly on firms offering outdated, poor-quality, or expensive goods and services.

Fear of established companies growing so large as to become unaccountable has absolutely no foundation in reality. In the U.S., the most capitalist large country in the world, the market share of the 25 biggest corporations has steadily dwindled over recent decades.

Freer markets make it easier for small firms with fresh ideas to compete with big corporations. Between 1980 and 1993, the 500 biggest American corporations saw their share of the country's total employment diminish from 16 to 11 percent. During the same period, the average personnel strength of American firms fell from 17 to 15 people, and the proportion of the population working in companies with more than 250 employees fell from 37 to 29 percent.

Of the 500 biggest enterprises in the United States in 1980, one third had disappeared by 1990. Another 40 percent had evaporated five years later. Whether they failed to grow enough to stay on the list, died, merged, or broke up, the key lesson is that big corporations have much less power over consumers than we sometimes imagine. Even the most potent corporation must constantly re-earn its stripes, or tumble fast.

Half the firms operating in the world today have fewer than 250 employees. The biggest brand logos are always flashed before our eyes, but we forget that they are constantly being joined by new ones. How many people recall that just a few years ago Nokia was a small Finnish firm making tires and boots?

Many people fear a "McDonaldization" or "Disneyfication" of the world, a creeping global homogeneity that leaves everyone wearing the same clothes, eating the same food, and seeing the same movies. But this portrayal does not accurately describe globalization. Anyone going out in the capitals of Europe today will have no trouble finding hamburgers and Coca-Cola, but he will just as easily find kebabs, sushi, Tex-Mex tacos, Peking duck, Thai lemongrass soup, and cappuccino. We know that Americans listen to Britney Spears and watch Adam Sandler films, but it's worth remembering that the United States is also a country with 1,700 symphony orchestras, 7.5 million annual trips to the opera, and 500 million museum visits a year. Globalization doesn't just give the world shlocky reality TV and overplayed music videos, but also classic films and documentaries on many new channels, news on the Internet, and masterpieces of music and literature in stores and on the Web.

The world is indeed moving toward a common objective, but that objective is not the predominance of a particular culture. Rather it is pluralism, the freedom to choose from a host of different paths and destinations. The market for experimental electronic music or film versions of novels by Dostoevsky may be small in any given place, so musicians and filmmakers producing such material could never produce anything without access to the much larger audience provided by globalization.

This internationalization is, ironically, what makes people believe that differences are vanishing. When you travel abroad, things look much the same as in your own country: The people there also have goods and chain stores from different parts of the globe. This phenomenon is not due to uniformity and the elimination of differences, but by the growth of pluralism everywhere.

Such opportunity can have negative effects in certain situations. When traveling to another country, we want to see something unique. Arriving in Rome and finding Hollywood films, Chinese food, Japanese Pokemon games, and Swedish Volvos, we miss the local color. And national specialties like pizza, pasta, and espresso are already familiar to us because we have them at home, too.

But this is another one of those luxury problems. I know a man from Prague who was visited by Czech friends who had settled abroad. The expatriates deplored the arrival of McDonald's in Prague because it threatened the city's distinctive charm. This response made the man indignant. How could they regard his home city as a museum, a place for them to visit now and then in order to avoid fast food restaurants? He wanted a real city, including the convenient and inexpensive food that these exile Czechs themselves had access to. A real, living city cannot be a "Prague summer paradise" for tourists. Other countries and their populations do not exist in order to give us picturesque holiday experiences. They, like us, are entitled to choose what suits their own tastes.

There's nothing new about cultures changing, colliding with each other, and cross-pollinating. Even the traditions we think of as most "authentic" have often resulted from cultural imports. One of the most sacred Christmas traditions in my home country of Sweden celebrates an Italian saint by adorning the hair of blond girls with lighted candles. Change and renewal are an inherent part of civilization. If we try to freeze certain cultural patterns in time they cease to be culture and instead become museum relics and folklore. There is nothing wrong with museums—but we can't live in one.

In the age of globalization, the ideas of freedom and individualism have attained tremendous force. There are few concepts as inspiring as that of self-determination. When people in other countries glimpse a chance to set their own course, it becomes almost irresistible. If there is any elimination of differences throughout the world, it has been the convergence of societies on the practice of allowing people to choose the sort of existence they please.

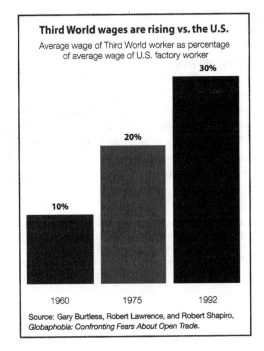

Third World wages are rising vs. the U.S.

Average wage of Third World worker as percentage of average wage of U.S. factory worker

30%

20%

10%

1960   1975   1992

Source: Gary Burtless, Robert Lawrence, and Robert Shapiro, *Globaphobia: Confronting Fears About Open Trade.*

Global commerce does undermine old economic interests, challenge cultures, and erode some traditional power centers. Advocates of globalization have to show that greater gains and opportunities counterbalance such problems. The antiglobalists are right that we could reject globalized trade if we adamantly insisted. Capital can be locked up, commercial flows blocked, and borders barricaded.

This happened across our planet at least once before. Decades of expanding economic liberty and globalization during the nineteenth century were replaced with nationalistic saber-rattling, centralization, and closed borders at the beginning of the twentieth century. The outbreak of World War I marked a new era.

Globalization resumed after World War II, but countries like Burma and North Korea show that it is still possible to cut oneself off from international trade, so long as one is prepared to pay heavily for doing so with political oppression and economic deprivation. It is not "necessary" that we follow the trend toward expanding global trade. It is highly desirable.

Globalization makes an excellent scapegoat. It contains all the anonymous forces that have served this purpose throughout history: foreign countries, other races and ethnic groups, the uncaring market. Globalization does not speak up for itself when politicians blame it for overturning economies, increasing poverty, enriching a tiny minority, polluting the environment, or cutting jobs. And globalization doesn't usually get any credit when good things happen—an economy running at high speed, poverty diminishing, clearing skies. So if the trend toward greater global interchange and liberty is to continue, an ideological defense is needed.

In 25 years there are likely to be 2 billion more of us on this planet, and 97 percent of that population increase will occur in the developing world. There are no automatic processes deciding what sort of world they will experience. Most of that will depend on what people like you and me believe, think, and fight for.

Lasse Berg and Stig Karlsson record Chinese villagers' descriptions of the changes they experienced since the 1960s: "The last time you were here, people's thoughts and minds were closed, bound up," stated farmer Yang Zhengming. But as residents acquired power over their own livelihoods they began to think for themselves. Yang explains that "a farmer could then own himself. He did not need to submit. He decided himself what he was going to do, how and when. The proceeds of his work were his own. It was freedom that came to us. We were allowed to own things for ourselves."

Coercion and poverty still cover large areas of our globe. But thanks to globalizing economic freedom, people know that living in a state of oppression is not natural or necessary. People who have acquired a taste of economic liberty and expanded horizons will not consent to be shut in again by walls or fences. They will work to create a better existence for themselves. The aim of our politics should be to give them that freedom.

Researcher Johan Norberg's books include In Defense of Global Capitalism, *which was published in English in September 2003.*

From *The American Enterprise*, June 2004. Copyright © 2004 by American Enterprise Institute. Reprinted by permission of *The American Enterprise*, a magazine of Politics, Business, and Culture. www.TAEmag.com

THE
Five Wars
OF GLOBALIZATION

The illegal trade in drugs, arms, intellectual property, people, and money is booming. Like the war on terrorism, the fight to control these illicit markets pits governments against agile, stateless, and resourceful networks empowered by globalization. Governments will continue to lose these wars until they adopt new strategies to deal with a larger, unprecedented struggle that now shapes the world as much as confrontations between nation–states once did.

By Moisés Naím

The persistence of al Qaeda underscores how hard it is for governments to stamp out stateless, decentralized networks that move freely, quickly, and stealthily across national borders to engage in terror. The intense media coverage devoted to the war on terrorism, however, obscures five other similar global wars that pit governments against agile, well-financed networks of highly dedicated individuals. These are the fights against the illegal international trade in drugs, arms, intellectual property, people, and money. Religious zeal or political goals drive terrorists, but the promise of enormous financial gain motivates those who battle governments in these five wars. Tragically, profit is no less a motivator for murder, mayhem, and global insecurity than religious fanaticism.

In one form or another, governments have been fighting these five wars for centuries. And losing them. Indeed, thanks to the changes spurred by globalization over the last decade, their losing streak has become even more pronounced. To be sure, nation-states have benefited from the information revolution, stronger political and economic linkages, and the shrinking importance of geographic distance. Unfortunately,

criminal networks have benefited even more. Never fettered by the niceties of sovereignty, they are now increasingly free of geographic constraints. Moreover, globalization has not only expanded illegal markets and boosted the size and the resources of criminal networks, it has also imposed more burdens on governments: Tighter public budgets, decentralization, privatization, deregulation, and a more open environment for international trade and investment all make the task of fighting global criminals more difficult. Governments are made up of cumbersome bureaucracies that generally cooperate with difficulty, but drug traffickers, arms dealers, alien smugglers, counterfeiters, and money launderers have refined networking to a high science, entering into complex and improbable strategic alliances that span cultures and continents.

Defeating these foes may prove impossible. But the first steps to reversing their recent dramatic gains must be to recognize the fundamental similarities among the five wars and to treat these conflicts not as law enforcement problems but as a new global trend that shapes the world as much as confrontations between nation-states did in the past. Customs officials, police officers, lawyers, and judges alone will never win these wars. Governments must recruit and deploy more spies, soldiers, diplomats, and economists who understand how to use incentives and regulations to steer markets away from bad social outcomes. But changing the skill set of government combatants alone will not end these wars. Their doctrines and institutions also need a major overhaul.

THE FIVE WARS

Pick up any newspaper anywhere in the world, any day, and you will find news about illegal migrants, drug busts, smuggled weapons, laundered money, or counterfeit goods. The global nature of these five wars was unimaginable just a decade ago. The resources—financial, human, institutional, technological—deployed by the combatants have reached unfathomable orders of magnitude. So have the numbers of victims. The tactics and tricks of both sides boggle the mind. Yet if you cut through the fog of daily headlines and orchestrated photo ops, one inescapable truth emerges: The world's governments are fighting a qualitatively new phenomenon with obsolete tools, inadequate laws, inefficient bureaucratic arrangements, and ineffective strategies. Not surprisingly, the evidence shows that governments are losing.

Drugs

The best known of the five wars is, of course, the war on drugs. In 1999, the United Nations' "Human Development Report" calculated the annual trade in illicit drugs at $400 billion, roughly the size of the Spanish economy and about 8 percent of world trade. Many countries are reporting an increase in drug use. Feeding this habit is a global supply chain that uses everything from passenger jets that can carry shipments of cocaine worth $500 million in a single trip to custom-built submarines that ply the waters between Colombia and Puerto Rico. To foil eavesdroppers, drug smugglers use "cloned" cell phones and broadband radio receivers while also relying on complex financial structures that blend legitimate and illegitimate enterprises with elaborate fronts and structures of cross-ownership.

The United States spends between $35 billion and $40 billion each year on the war on drugs; most of this money is spent on interdiction and intelligence. But the creativity and boldness of drug cartels has routinely outstripped steady increases in government resources. Responding to tighter security at the U.S.-Mexican border, drug smugglers built a tunnel to move tons of drugs and billions of dollars in cash until authorities discovered it in March 2002. Over the last decade, the success of the Bolivian and Peruvian governments in eradicating coca plantations has shifted production to Colombia. Now, the U.S.-supported Plan Colombia is displacing coca production and processing labs back to other Andean countries. Despite the heroic efforts of these Andean countries and the massive financial and technical support of the United States, the total acreage of coca plantations in Peru, Colombia, and Bolivia has increased in the last decade from 206,200 hectares in 1991 to 210,939 in 2001. Between 1990 and 2000, according to economist Jeff

DeSimone, the median price of a gram of cocaine in the United States fell from $152 to $112.

Even when top leaders of drug cartels are captured or killed, former rivals take their place. Authorities have acknowledged, for example, that the recent arrest of Benjamin Arellano Felix, accused of running Mexico's most ruthless drug cartel, has done little to stop the flow of drugs to the United States. As Arellano said in a recent interview from jail, "They talk about a war against the Arellano brothers. They haven't won. I'm here, and nothing has changed."

Arms Trafficking

Drugs and arms often go together. In 1999, the Peruvian military parachuted 10,000 AK-47s to the Revolutionary Armed Forces of Colombia, a guerrilla group closely allied to drug growers and traffickers. The group purchased the weapons in Jordan. Most of the roughly 80 million AK-47s in circulation today are in the wrong hands. According to the United Nations, only 18 million (or about 3 percent) of the 550 million small arms and light weapons in circulation today are used by government, military, or police forces. Illicit trade accounts for almost 20 percent of the total small arms trade and generates more than $1 billion a year. Small arms helped fuel 46 of the 49 largest conflicts of the last decade and in 2001 were estimated to be responsible for 1,000 deaths a day; more than 80 percent of those victims were women and children.

Small arms are just a small part of the problem. The illegal market for munitions encompasses top-of-the-line tanks, radar systems that detect Stealth aircraft, and the makings of the deadliest weapons of mass destruction. The International Atomic Energy Agency has confirmed more than a dozen cases of smuggled nuclear-weapons-usable material, and hundreds more cases have been reported and investigated over the last decade. The actual supply of stolen nuclear-, biological-, or chemical-weapons materials and technology may still be small. But the potential demand is strong and growing from both would-be nuclear powers and terrorists. Constrained supply and increasing demand cause prices to rise and create enormous incentives for illegal activities. More than one fifth of the 120,000 workers in Russia's former "nuclear cities"—where more than half of all employees earn less than $50 a month—say they would be willing to work in the military complex of another country.

Governments have been largely ineffective in curbing either supply or demand. In recent years, two countries, Pakistan and India, joined the declared nuclear power club. A U.N. arms embargo failed to prevent the reported sale to Iraq of jet fighter engine parts from Yugoslavia and the Kolchuga anti-Stealth radar system from Ukraine. Multilateral efforts to curb the manufacture and distribution of weapons are faltering, not least because some powers are unwilling to accept curbs on their own activities. In 2001, for example, the United States blocked a legally binding global treaty to control small arms in part because it worried about restrictions on its own citizens' rights to own guns. In the absence of effective international legislation and enforcement, the laws of economics dictate the sale of more weapons at cheaper prices: In 1986, an AK-47 in Kolowa, Kenya, cost 15 cows. Today, it costs just four.

Other Fronts

Drugs, arms, intellectual property, people, and money are not the only commodities traded illegally for huge profits by international networks. They also trade in human organs, endangered species, stolen art, and toxic waste. The illegal global trades in all these goods share several fundamental characteristics: Technological innovations and political changes open new markets, globalization is increasing both the geographical reach and the profit opportunities for criminal networks, and governments are on the losing end of the fight to stop them. Some examples:

Human organs: Corneas, kidneys, and livers are the most commonly traded human parts in a market that has boomed thanks to technology, which has improved preservation techniques and made transplants less risky. In the United States, 70,000 patients are on the waiting list for major organ transplants while only 20,000 of them succeed in getting the organ they need. Unscrupulous "organ brokers" partly meet this demand by providing, for a fee, organs and transplant services.

Some of the donors, especially of kidneys, are desperately poor. In India, an estimated 2,000 people a year sell their organs. Many organs, however, come from nonconsenting donors forced to undergo operations or from cadavers in police morgues. For example, medical centers in Germany and Austria were recently found to have used human heart valves taken without consent from the cadavers of poor South Africans.

Endangered species: From sturgeon for caviar in gourmet delicatessens to tigers or elephants for private zoos, the trade in endangered animals and plants is worth billions of dollars and includes hundreds of millions of plant and animal types. This trade ranges from live animals and plants to all kinds of wildlife products derived from them, including food products, exotic leather goods, wooden musical instruments, timber, tourist curiosities, and medicines.

Stolen art: Paintings and sculptures taken from museums, galleries, and private homes, from Holocaust victims, or from "cultural artifacts" poached from archeological digs and other ancient ruins are also illegally traded internationally in a market worth an

estimated $2 billion to $6 billion each year. The growing use of art-based transactions in money laundering has spurred demand over the last decade. The supply has boomed because the Soviet Union's collapse flooded the world's market with art that had been under state control. The Czech Republic, Poland, and Russia are three of the five countries most affected by art crime worldwide.

Toxic waste: Innovations in maritime transport, tighter environmental regulations in industrialized countries coupled with increased integration of poor countries to the global economy and better telecommunications have created a market where waste is traded internationally. Greenpeace estimates that during the 20 years prior to 1989, just 3.6 million tons of hazardous waste were exported; in the five years after 1989, the trade soared to about 6.7 billion tons. The environmental organization also reckons that 86 to 90 percent of all hazardous waste shipments destined for developing countries—purportedly for recycling, reuse, recovery, or humanitarian uses—are toxic waste.

—M.N.

Intellectual Property

In 2001, two days after recording the voice track of a movie in Hollywood, actor Dennis Hopper was in Shanghai where a street vendor sold him an excellent pirated copy of the movie with his voice already on it. "I don't know how they got my voice into the country before I got here," he wondered. Hopper's experience is one tiny slice of an illicit trade that cost the United States an estimated $9.4 billion in 2001. The piracy rate of business software in Japan and France is 40 percent, in Greece and South Korea it is about 60 percent, and in Germany and Britain it hovers around 30 percent. Forty percent of Procter & Gamble shampoos and 60 percent of Honda motorbikes sold in China in 2001 were pirated. Up to 50 percent of medical drugs in Nigeria and Thailand are bootleg copies. This problem is not limited to consumer products: Italian makers of industrial valves worry that their $2 billion a year export market is eroded by counterfeit Chinese valves sold in world markets at prices that are 40 percent cheaper.

The drivers of this bootlegging boom are complex. Technology is obviously boosting both the demand and the supply of illegally copied products. Users of Napster, the now defunct Internet company that allowed anyone, anywhere to download

and reproduce copyrighted music for free, grew from zero to 20 million in just one year. Some 500,000 film files are traded daily through file-sharing services such as Kazaa and Morpheus; and in late 2002, some 900 million music files could be downloaded for free on the Internet—that is, almost two and a half times more files than those available when Napster reached its peak in February 2001.

Global marketing and branding are also playing a part, as more people are attracted to products bearing a well-known brand like Prada or Cartier. And thanks to the rapid growth and integration into the global economy of countries, such as China, with weak central governments and ineffective laws, producing and exporting near perfect knockoffs are both less expensive and less risky. In the words of the CEO of one of the best known Swiss watchmakers: "We now compete with a product manufactured by Chinese prisoners. The business is run by the Chinese military, their families and friends, using roughly the same machines we have, which they purchased at the same industrial fairs we go to. The way we have rationalized this problem is by assuming that their customers and ours are different. The person that buys a pirated copy of one of our $5,000 watches for less than $100 is not a client we are losing. Perhaps it is a future client that some day will want to own the real thing instead of a fake. We may be wrong and we do spend money to fight the piracy of our products. But given that our efforts do not seem to protect us much, we close our eyes and hope for the better." This posture stands in contrast to that of companies that sell cheaper products such as garments, music, or videos, whose revenues are directly affected by piracy.

Governments have attempted to protect intellectual property rights through various means, most notably the World Trade Organization's Agreement on Trade-Related Aspects of Intellectual Property Rights (TRIPS). Several other organizations such as the World Intellectual Property Organization, the World Customs Union, and Interpol are also involved. Yet the large and growing volume of this trade, or a simple stroll in the streets of Manhattan or Madrid, show that governments are far from winning this fight.

Alien Smuggling

The man or woman who sells a bogus Hermes scarf or a Rolex watch in the streets of Milan is likely to be an illegal alien. Just as likely, he or she was transported across several continents by a trafficking network allied with another network that specializes in the illegal copying, manufacturing, and distributing of high-end, brand-name products.

Alien smuggling is a $7 billion a year enterprise and according to the United Nations is the fastest growing business of organized crime. Roughly 500,000 people enter the United States illegally each year—about the same number as illegally enter the European Union, and part of the approximately 150 million who live outside their countries of origin. Many of these backdoor travelers are voluntary migrants who pay smugglers up to $35,000, the top-dollar fee for passage from China to New York. Others, instead, are trafficked—that is, bought and sold internationally—as commodities. The U.S. Congressional Research Service reckons that each year between 1 million and 2 million people are trafficked across borders, the majority of whom are women and children. A woman can be "bought" in Timisoara, Romania, for between $50 and $200 and "resold" in Western Europe for 10 times that price. The United Nations Children's Fund estimates that cross-border smugglers in Central and Western Africa enslave 200,000 children a year. Traffickers initially tempt victims with job offers or, in the case of children, with offers of adoption in wealthier countries, and then keep the victims in subservience through physical violence, debt bondage, passport confiscation, and threats of arrest, deportation, or violence against their families back home.

Governments everywhere are enacting tougher immigration laws and devoting more time, money, and technology to fight the flow of illegal aliens. But the plight of the United Kingdom's government illustrates how tough that fight is. The British government throws money at the problem, plans to use the Royal Navy and Royal Air Force to intercept illegal immigrants, and imposes large fines on truck drivers who (generally unwittingly) transport stowaways. Still, 42,000 of the 50,000 refugees who have passed through the Sangatte camp (a main entry point for illegal immigration to the United Kingdom) over the last three years have made it to Britain. At current rates, it

will take 43 years for Britain to clear its asylum backlog. And that country is an island. Continental nations such as Spain, Italy, or the United States face an even greater challenge as immigration pressures overwhelm their ability to control the inflow of illegal aliens.

Money Laundering

The Cayman Islands has a population of 36,000. It also has more than 2,200 mutual funds, 500 insurance companies, 60,000 businesses, and 600 banks and trust companies with almost $800 billion in assets. Not surprisingly, it figures prominently in any discussion of money laundering. So does the United States, several of whose major banks have been caught up in investigations of money laundering, tax evasion, and fraud. Few, if any, countries can claim to be free of the practice of helping individuals and companies hide funds from governments, creditors, business partners, or even family members, including the proceeds of tax evasion, gambling, and other crimes. Estimates of the volume of global money laundering range between 2 and 5 percent of the world's annual gross national product, or between $800 billion and $2 trillion.

Smuggling money, gold coins, and other valuables is an ancient trade. Yet in the last two decades, new political and economic trends coincided with technological changes to make this ancient trade easier, cheaper, and less risky. Political changes led to the deregulation of financial markets that now facilitate cross-border money transfers, and technological changes made distance less of a factor and money less "physical." Suitcases full of banknotes are still a key tool for money launderers, but computers, the Internet, and complex financial schemes that combine legal and illegal practices and institutions are more common. The sophistication of technology, the complex web of financial institutions that crisscross the globe, and the ease with which "dirty" funds can be electronically morphed into legitimate assets make the regulation of international flows of money a daunting task. In Russia, for example, it is estimated that by the mid-1990s organized crime groups had set up 700 legal and financial institutions to launder their money.

Faced with this growing tide, governments have stepped up their efforts to clamp down on rogue international banking, tax havens, and money laundering. The imminent, large-scale introduction of e-money—cards with microchips that can store large amounts of money and thus can be easily transported outside regular channels or simply exchanged among individuals—will only magnify this challenge.

WHY GOVERNMENTS CAN'T WIN

The fundamental changes that have given the five wars new intensity over the last decade are likely to persist. Technology will continue to spread widely; criminal networks will be able to exploit these technologies more quickly than governments that must cope with tight budgets, bureaucracies, media scrutiny, and electorates. International trade will continue to grow, providing more cover for the expansion of illicit trade. International migration will likewise grow, with much the same effect, offering ethnically based gangs an ever growing supply of recruits and victims. The spread of democracy may also help criminal cartels, which can manipulate weak public institutions by corrupting police officers or tempting politicians with offers of cash for their increasingly expensive election campaigns. And ironically, even the spread of international law—with its growing web of embargoes, sanctions, and conventions—will offer criminals new opportunities for providing forbidden goods to those on the wrong side of the international community.

Even the spread of international law will offer criminals new opportunities for providing forbidden goods to those on the wrong side of the international community.

These changes may affect each of the five wars in different ways, but these conflicts will continue to share four common characteristics:

They are not bound by geography.

Some forms of crime have always had an international component: The Mafia was born in Sicily and exported to the United States, and smuggling has always been by definition international. But the five wars are truly global. Where is the theater or front line of the war on drugs? Is it Colombia or Miami? Myanmar (Burma) or Milan? Where are the battles against money launderers being fought? In Nauru or in London? Is China the main theater in the war against the infringement of intellectual property, or are the trenches of that war on the Internet?

They defy traditional notions of sovereignty.

Al Qaeda's members have passports and nationalities—and often more than one—but they are truly stateless. Their allegiance is to their cause, not to any nation. The same is also true of the criminal networks engaged in the five wars. The same, however, is patently *not* true of government employees—police officers, customs agents, and judges—who fight them. This asymmetry is a crippling disadvantage for governments waging these wars. Highly paid, hypermotivated, and resource-rich combatants on one side of the wars (the criminal gangs) can seek refuge in and take advantage of national borders, but combatants of the other side (the governments) have fewer resources and are hampered by traditional notions of sovereignty. A former senior CIA official reported that international criminal gangs are able to move people, money, and weapons globally faster than he can move resources inside his own agency, let alone worldwide. Coordination and information sharing among government agencies in different countries has certainly improved, especially after September 11. Yet these tactics fall short of what is needed to combat agile organizations that can exploit every nook and cranny of an evolving but imperfect body of international law and multilateral treaties.

They pit governments against market forces.

In each of the five wars, one or more government bureaucracies fight to contain the disparate, uncoordinated actions of thousands of independent, stateless organizations. These groups are motivated by large profits obtained by exploiting international price differentials, an unsatisfied demand, or the cost advantages produced by theft. Hourly wages for a Chinese cook are far higher in Manhattan than in Fujian. A gram of cocaine in Kansas City is 17,000 percent more expensive than in Bogotá. Fake Italian valves are 40 percent cheaper because counterfeiters don't have to cover the costs of developing the product. A well-funded guerrilla group will pay anything to get the weapons it needs. In each of these five wars, the incentives to successfully overcome government-imposed limits to trade are simply enormous.

They pit bureaucracies against networks.

The same network that smuggles East European women to Berlin may be involved in distributing opium there. The proceeds of the latter fund the purchase of counterfeit Bulgari watches made in China and often sold on the streets of Manhattan by illegal African immigrants. Colombian drug cartels make deals with Ukrainian arms traffickers, while Wall Street brokers controlled by the U.S.-based Mafia have been known to front for Russian money launderers. These highly decentralized groups and individuals are bound by strong ties of loyalty and common purpose and organized around semiautonomous clusters or "nodes" capable of operating swiftly and flexibly. John Arquilla and David Ronfeldt, two of the best known experts on these types of organizations, observe that networks often lack central leadership, command, or headquarters, thus "no precise heart or head that can be targeted. The network as a whole (but not necessarily each node) has little to no hierarchy; there may be multiple leaders…. Thus the [organization's] design may sometimes appear acephalous (headless), and at other times polycephalous (Hydra-headed)." Typically, governments respond to these challenges by forming interagency task forces or creating new bureaucracies. Consider the creation of the new Department of Homeland Security in the United States, which encompasses 22 former federal agencies and their 170,000 employees and is responsible for, among other things, fighting war on drugs.

RETHINKING THE PROBLEM

Governments may never be able to completely eradicate the kind of international trade involved in the five wars. But they can and should do better. There are at least four areas where efforts can yield better ideas on how to tackle the problems posed by these wars:

Develop more flexible notions of sovereignty.

Governments need to recognize that restricting the scope of multilateral action for the sake of protecting their sovereignty is often a moot point. Their sovereignty is compromised daily, not by nation-states but by stateless networks that break laws and cross borders in pursuit of trade. In May 1999, for example, the Venezuelan government denied U.S. planes authorization to fly over Venezuelan territory to monitor air routes commonly used by narcotraffickers. Venezuelan authorities placed more importance on the symbolic value of asserting sovereignty over air space than on the fact that drug traffickers' planes regularly violate Venezuelan territory. Without new forms of codifying and "managing" sovereignty, governments will continue to face a large disadvantage while fighting the five wars.

Strengthen existing multilateral institutions.

The global nature of these wars means no government, regardless of its economic, political, or military power, will make much progress acting alone. If this seems obvious, then why does Interpol, the multilateral agency in charge of fighting international crime, have a staff of 384, only 112 of whom are police officers, and an annual budget of $28 million, less than the price of some boats or planes used by drug traffickers? Similarly, Europol, Europe's Interpol equivalent, has a staff of 240 and a budget of $51 million.

One reason Interpol is poorly funded and staffed is because its 181 member governments don't trust each other. Many assume, and perhaps rightly so, that the criminal networks they are fighting have penetrated the police departments of other countries and that sharing information with such compromised officials would not be prudent. Others fear today's allies will become tomorrow's enemies. Still others face legal impediments to sharing intelligence with fellow nation-states or have intelligence services and law enforcement agencies with organizational cultures that make effective collaboration almost impossible. Progress will only be made if the world's governments unite behind stronger, more effective multilateral organizations.

Devise new mechanisms and institutions.

These five wars stretch and even render obsolete many of the existing institutions, legal frameworks, military doctrines, weapons systems, and law enforcement techniques on which governments have relied for years. Analysts need to rethink the concept of war "fronts" defined by geography and the definition of "combatants" according to the Geneva Convention. The functions of intelligence agents, soldiers, police officers, customs agents, or immigration officers need rethinking and adap-

tation to the new realities. Policymakers also need to reconsider the notion that ownership is essentially a physical reality and not a "virtual" one or that only sovereign nations can issue money when thinking about ways to fight the five wars.

Move from repression to regulation.

Beating market forces is next to impossible. In some cases, this reality may force governments to move from repressing the market to regulating it. In others, creating market incentives may be better than using bureaucracies to curb the excesses of these markets. Technology can often accomplish more than government policies can. For example, powerful encryption techniques can better protect software or CDs from being copied in Ukraine than would making the country enforce patents and copyrights and trademarks.

In all of the five wars, government agencies fight against networks motivated by the enormous profit opportunities created by other government agencies. In all cases, these profits can be traced to some form of government intervention that creates a major imbalance between demand and supply and makes prices and profit margins skyrocket. In some cases, these government interventions are often justified and it would be imprudent to eliminate them—governments can't simply walk away from the fight against trafficking in heroin, human beings, or weapons of mass destruction. But society can better deal with other segments of these kinds of illegal trade through regulation, not prohibition. Policymakers must focus on opportunities where market regulation can ameliorate problems that have defied approaches based on prohibition and armed interdiction of international trade.

Ultimately, governments, politicians, and voters need to realize that the way in which the world is conducting these five wars is doomed to fail—not for lack of effort, resources, or political will but because the collective thinking that guides government strategies in the five wars is rooted in wrong ideas, false assumptions, and obsolete institutions. Recognizing that governments have no chance of winning unless they change the ways they wage these wars is an indispensable first step in the search for solutions.

Moisés Naím is editor of FOREIGN POLICY _magazine._

From _Foreign Policy_, January/February 2003, pp. 28-37 © 2003 by the Carnegie Endowment for International Peace (www.foreignpolicy.com). Permission conveyed through Copyright Clearance Center, Inc.

Will
Globalization
Go Bankrupt?

Global integration is driven not by politics or the Internet or the World Trade Organization or even—believe it or not—McDonald's. No, throughout history, globalization has been driven primarily by monetary expansions. Credit booms spark periods of economic integration, while credit contractions quickly squelch them. Is today's world on the verge of another globalization bust?

By Michael Pettis

Only the young generation which has had a college education is capable of comprehending the exigencies of the times," wrote Alphonse, a third-generation Rothschild, in a letter to a family member in 1865. At the time the world was in the midst of a technological boom that seemed to be changing the globe beyond recognition, and certainly beyond the ability of his elders to understand. As part of that boom, capital flowed into remote corners of the earth, dragging isolated societies into modernity. Progress seemed unstoppable.

Eight years later, however markets around the world collapsed. Suddenly, investors turned away from foreign adventures and new technologies. In the depression that ensued, many of the changes eagerly embraced by the educated young—free markets, deregulated banks, immigration—seemed too painful to continue. The process of globalization, it seems, was neither inevitable nor irreversible.

What today we call economic globalization—a combination of rapid technological progress, large-scale capital flows, and burgeoning international trade—has happened many times before in the last 200 years. During each of these periods (including our own), engineers and entrepreneurs became folk heroes and made vast fortunes while transforming the world around them. They exploited scientific advances, applied a suc-

cession of innovations to older discoveries, and spread the commercial application of these technologies throughout the developed world [see box]. Communications and transportation were usually among the most affected areas, with each technological surge causing the globe to "shrink" further.

But in spite of the enthusiasm for science that accompanied each wave of globalization, as a historical rule it was primarily commerce and finance that drove globalization, not science or technology, and certainly not politics or culture. It is no accident that each of the major periods of technological progress coincided with an era of financial market expansion and vast growth in international commerce. Specifically, a sudden expansion of financial liquidity in the world's leading banking centers—whether an increase in British gold reserves in the 1820s or the massive transformation in the 1980s of illiquid mortgage loans into very liquid mortgage securities, or some other structural change in the financial markets—has been the catalyst behind every period of globalization.

If liquidity expansions historically have pushed global integration forward, subsequent liquidity contractions have brought globalization to an unexpected halt. Easy money had allowed investors to earn fortunes for their willingness to take risks, and the wealth generated by rising asset values and new investments made the liberal ideology behind the rapid market expansion seem unassailable. When conditions changed, however, the outflow of money from the financial centers was reversed. Investors rushed to pull their money out of risky ventures and into safer assets. Banks tightened up their lending requirements and refused to make new loans. Asset values collapsed. The costs of globalization, in the form of social disruption, rising income inequality, and domination by foreign elites, became unacceptable. The political and intellectual underpinnings of globalization, which had once seemed so secure, were exposed as fragile, and the popular counterattack against the logic of globalization grew irresistible.

THE BIG BANG

The process through which monetary expansions lead to economic globalization has remained consistent over the last two centuries. Typically, every few decades, a large shift in income, money supply, saving patterns, or the structure of financial markets results in a major liquidity expansion in the rich-country financial centers. The initial expansion can take a variety of forms. In England, for example, the development of joint-stock banking (limited liability corporations that issued currency) in the 1820s and 1830s—and later during the 1860s and 1870s—produced a rapid expansion of money, deposits, and bank credit, which quickly spilled over into speculative investing and international lending. Other monetary expansions were sparked by large increases in U.S. gold reserves in the early 1920s, or by major capital recyclings, such as the massive French indemnity payment after the Franco-Prussian War of 1870, the petrodollar recycling of the 1970s, or the recycling of Japan's huge trade surplus in the 1980s and 1990s. Monetary expansions also can result from the conversion of assets into more liquid instruments, such as with the explosion in U.S. speculative real-estate lending in the 1830s or the creation of the mortgage securities market in the 1980s.

The expansion initially causes local stock markets to boom and real interest rates to drop. Investors, hungry for high yields, pour money into new, nontraditional investments, including ventures aimed at exploiting emerging technologies. Financing becomes available for risky new projects such as railways, telegraph cables, textile looms, fiber optics, or personal computers, and the strong business climate that usually accompanies the liquidity expansion quickly makes these investments profitable. In turn, these new technologies enhance productivity and slash transportation costs, thus speeding up economic growth and boosting business profits. The cycle is self-reinforcing: Success breeds success, and soon the impact of rapidly expanding transportation and communication technology begins to cause a noticeable impact on social behavior, which adapts to these new technologies.

But it is not just new technology ventures that attract risk capital. Financing also begins flowing to the "peripheral" economies around the world, which, because of their small size, are quick to respond. These countries then begin to experience currency strength and real economic growth, which only reinforce

the initial investment decision. As more money flows in, local markets begin to grow. As a consequence of the sudden growth in both asset values and gross domestic product, political leaders in developing countries often move to reform government policies in these countries—whether reform consists of expelling a backward Spanish monarch in the 1820s, expanding railroad transportation across the Andes in the 1860s, transforming the professionalism of the Mexican bureaucracy in the 1890s, deregulating markets in the 1920s, or privatizing bloated state-owned firms in the 1990s. By providing the government with the resources needed to overcome the resistance of local elites, capital inflows enable economic-policy reforms.

This relationship between capital and reform is frequently misunderstood: Capital inflows do not simply respond to successful economic reforms, as is commonly thought; rather, they create the conditions for reforms to take place. They permit easy financing of fiscal deficits, provide industrialists who might oppose free trade with low-cost capital, build new infrastructure, and generate so much asset-based wealth as to mollify most members of the economic and political elite who might ordinarily oppose the reforms. Policymakers tend to design such reforms to appeal to foreign investors, since policies that encourage foreign investment seem to be quickly and richly rewarded during periods of liquidity. In reality, however, capital is just as likely to flow into countries that have failed to introduce reforms. It is not a coincidence that the most famous "money doctors"—Western-trained thinkers like French economist Jean-Gustave Courcelle-Seneuil in the 1860s, financial historian Charles Conant in the 1890s, and Princeton University economist Edwin Kemmerer in the 1920s, under whose influence many developing countries undertook major liberal reforms—all exerted their maximum influence during these periods. During the 1990s, their modern counterparts advised Argentina on its currency board, brought "shock therapy" to Russia, convinced China of the benefits of membership in the World Trade Organization, and everywhere spread the ideology of free trade.

Globalization takes place largely because sudden monetary expansions encourage investors to embrace new risks.

The pattern is clear: Globalization is primarily a monetary phenomenon in which expanding liquidity induces investors to take more risks. This greater risk appetite translates into the financing of new technologies and investment in less developed markets. The combination of the two causes a "shrinking" of the globe as communications and transportation technologies improve and investment capital flows to every part of the globe. Foreign trade, made easier by the technological advances, expands to accommodate these flows. Globalization takes place, in other words, largely because investors are suddenly eager to embrace risk.

THE BIG CRUNCH

As is often forgotten during credit and investment booms, however, monetary conditions contract as well as expand. In fact, the contraction is usually the inevitable outcome of the very conditions that prompted the expansion. In times of growth, financial institutions often overextend themselves, creating distortions in financial markets and leaving themselves vulnerable to external shocks that can force a sudden retrenchment in credit and investment. In a period of rising asset prices, for example, it is often easy for even weak borrowers to obtain collateral-based loans, which of course increases the risk to the banking system of a fall in the value of the collateral. For example, property loans in the 1980s dominated and ultimately brought down the Japanese banking system. As was evident in Japan, if the financial structure has become sufficiently fragile, a retrenchment can lead to a collapse that quickly spreads throughout the economy.

Since globalization is mainly a monetary phenomenon, and since monetary conditions eventually must contract, then the process of globalization can stop and even reverse itself. Historically, such reversals have proved extraordinarily disruptive. In each of the globalization periods before the 1990s, monetary contractions usually occurred when bankers and financial authorities began to pull back from market excesses. If liquidity contracts—in the context of a perilously overextended financial system—the likelihood of bank defaults and stock market instability is high. In 1837, for example, the U.S. and British banking systems, overdependent on real estate and commodity loans, collapsed in a series of crashes that left Europe's financial sector in tatters and the United States in the midst of bank failures and state government defaults.

The same process occurred a few decades later. Alphonse Rothschild's globalizing cycle of the 1860s ended with the stock market crashes that began in Vienna in May 1873 and spread around the world during the next four months, leading, among other things, to the closing of the New York Stock Exchange (NYSE) that September amid the near-collapse of American railway securities. Conditions were so bad that the rest of the decade after 1873 was popularly referred to in the United States as the Great Depression. Nearly 60 years later, that name was reassigned to a similar episode—the one that ended the Roaring Twenties and began with the near-breakdown of the U.S. banking system in 1930–31. The expansion of the 1960s was somewhat different in that it began to unravel during the early and mid-1970s when, thanks partly to the OPEC oil price hikes and subsequent petrodollar recycling, a second liquidity boom occurred, and lending to sovereign borrowers in the developing world continued through the end of the decade. However, the cycle finally broke down altogether when rising interest rates and contracting money engineered by then Federal Reserve Chairman Paul Volcker helped precipitate the Third World debt crisis of the 1980s. Indeed, with the exception of the globalization period of the early 1900s, which ended with the advent of World War I, each of these eras of international integration concluded with sharp monetary contractions that led to a banking system collapse or retrenchment, declining asset values, and a sharp reduction in both investor risk appetite and international lending.

Following most such market crashes, the public comes to see prevalent financial market practices as more sinister, and criticism of the excesses of bankers becomes a popular sport among politicians and the press in the advanced economies. Once capital stops flowing into the less developed, capital-hungry countries, the domestic consensus in favor of economic reform and international integration begins to disintegrate. When capital inflows no longer suffice to cover the short-term costs to the local elites and middle classes of increased international integration-including psychic costs such as feelings of wounded national pride-support for globalization quickly wanes. Populist movements, never completely dormant, become reinvigorated. Countries turn inward. Arguments in favor of protectionism suddenly start to sound appealing. Investment flows quickly become capital flight.

> **Following market crashes, the public comes to see financial markets as more sinister, and criticism of bankers becomes a popular sport among politicians and the press.**

This pattern emerged in the aftermath of the 1830s crash, when confidence in free markets nose-dived and the subsequent populist and nationalist backlash endured until the failure of the muchdreaded European liberal uprisings of 1848, which saw the earliest stirrings of communism and the publication of the *Communist Manifesto*. Later, in the 1870s, the economic depression that followed the mass bank closings in Europe, the United States, and Latin America was accompanied by an upsurge of political radicalism and populist outrage, along with bouts of protectionism throughout Europe and the United States by the end of the decade. Similarly, the Great Depression of the 1930s also fostered political instability and a popular revulsion toward the excesses of financial capitalism, culminating in burgeoning left-wing movements, the passage of anti-bank legislation, and even the jailing of the president of the NYSE.

PROFITS OF DOOM

Will these patterns manifest themselves again? Indeed, a new global monetary contraction already may be under way. In each of the previous contractions, stock markets fell, led by the collapse of the once-high-flying technology sector; lending to emerging markets dried up, bringing with it a series of sovereign defaults; and investors clamored for safety and security. Consider the crash of 1873, a typical case: Then, the equivalent of today's high-tech sector was the market for railway stocks and bonds, and the previous decade had seen a rush of new stock and bond offerings that reached near-manic proportions in the early 1870s. The period also saw rapid growth of lending to

Latin America, southern and eastern Europe, and the Middle East. Wall Street veterans had expressed nervousness about market excesses for years leading up to the crash, but the exuberance of investors who believed in the infinite promise of the railroads, at home and abroad, coupled with the rising prominence of bull-market speculators like Jay Gould and Diamond Jim Brady, swept them aside. When the market collapsed in 1873, railway securities were the worst hit, with many companies going bankrupt and closing their doors. Major borrowers from the developing world were unable to find new financing, and a series of defaults spread from the Middle East to Latin America in a matter of months. In the United States, the Congress and press became furious with the actions of stock market speculators and pursued financial scandals all the way to President Ulysses S. Grant's cabinet. Even Grant's brother-in-law was accused of being in cahoots with a notorious group that attempted a brutal gold squeeze.

Today, we see many of the same things. The technology sector is in shambles, and popular sentiment has turned strongly against many of the Wall Street heroes who profited most from the boom. Lending to emerging markets has all but dried up. As of this writing, the most sophisticated analysts predict that a debt default in Argentina is almost certain—and would unleash a series of other sovereign defaults in Latin America and around the world. The yield differences between risky assets and the safest and most liquid assets are at historical highs. In short, investors seem far more reluctant to take on risk than they were just a few years ago.

This lower risk tolerance does not bode well for poor nations. Historically, many developing countries only seem to experience economic growth during periods of heavy capital inflow, which in turn tend to last only as long as the liquidity-inspired asset booms in rich-country financial markets. Will the international consensus that supports globalization last when capital stops flowing? The outlook is not very positive. While there is still broad support in many circles for free trade, economic liberalization, technological advances, and free capital flows—even when the social and psychic costs are acknowledged—we already are witnessing a strong political reaction against globalization. This backlash is evident in the return of populist movements in Latin America; street clashes in Seattle, Prague, and Quebec; and the growing disenchantment in some quarters with the disruptions and uncertainties that follow in the wake of globalization.

The leaders now gathered in opposition to globalization—from President Hugo Chávez in Venezuela to Malaysian Prime Minister Mahathir bin Mohamad to anti-trade activist Lori Wallach in the United States—should not be dismissed too easily, no matter how dubious or fragile some of their arguments may seem. The logic of their arguments may not win the day, but rather a global monetary contraction may reverse the political consensus that was necessary to support the broad and sometimes disruptive social changes that accompany globalization. When that occurs, policy debates will be influenced by the less emotional and more thoughtful attacks on globalization by the likes of Robert Wade, a professor of political economy at the London School of Economics, who argues forcefully that glo-

balization has actually resulted in greater global income inequality and worse conditions for the poor.

Investing in the Future

In past periods of monetary expansion and globalization, Western societies have experienced the rapid development and commercial application of new technologies.

1822–37: expansion of canal building, first railway boom, application of steam power to the manufacturing process, advances in machine tool design, invention of McCormick's reaper, first gas-lighting enterprises, and development of the telegraph

1851–73: advances in mining, second railway boom, developments in shipping, and rapid growth in the number of corporations in continental Europe

1881–1914: explosive productivity growth in Europe and the United States, improvement in steel production and heavy chemical manufacturing, first power station, spread of electricity, development of the internal combustion engine, another railway boom, innovation in newspaper practice and technology, and developments in canning and refrigeration

1922–30: commercialization of automobiles and aircraft, new forms of mass media, rising popularity of cinema and radio broadcasting, spread of artificial fibers and plastics, widespread use of electricity in U.S. factories, the creation and sale of a variety of new electric appliances, and expanded telephone ownership

1960–73: development and application of transistor technology, advances in commercial flying and shipping, and the spread of telecommunications and software

1985–present: ubiquity of information processing, explosion in computer memory, advances in biotechnology and medical technologies, and commercial application of the Internet

If a global liquidity contraction is under way, antiglobalization arguments will resonate more strongly as many of the warnings about the greed of Wall Street and the dangers of liberal reform will seem to come true. Supposedly irreversible trends will suddenly reverse themselves. Further attempts to deepen economic reform, spread free trade, and increase capital and labor mobility may face political opposition that will be very difficult to overcome, particularly since bankers, the most committed supporters of globalization, may lose much of their prestige and become the target of populist attacks following a serious stock market decline. Because bankers are so identified with globalization, any criticism of Wall Street will also implicitly be a criticism of globalizing markets.

Financiers, after all, were not the popular heroes in the 1930s that they were during the 1920s, and current events seem to mirror past backlashes. Already the U.S. Securities and Exchange Commission, which was created during the Great Depression of the 1930s, is investigating the role of bankers and analysts in misleading the public on the market excesses of the 1990s. In June 2001, the industry's lobby group, the Securities Industry Association, proposed a voluntary code, euphemized as "a compilation of best practices… to ensure the ongoing integrity of securities research and analysis," largely to head off an expansion of external regulation. Increasingly, experts bewail the conflicts of interest inherent among the mega-banks that dominate U.S. and global finance.

Globalization itself always will wax and wane with global liquidity. For those committed to further international integration within a liberal economic framework, the successes of the recent past should not breed complacency since the conditions will change and the mandate for liberal expansion will wither. For those who seek to reverse the socioeconomic changes that globalization has wrought, the future may bring far more progress than they hoped. If global liquidity contracts and if markets around the world pull back, our imaginations will once again turn to the increasingly visible costs of globalization and away from the potential for all peoples to prosper. The reaction against globalization will suddenly seem unstoppable.

Michael Pettis is an investment banker and professor of finance at Columbia University. He is author of The Volatility Machine: Emerging Economies and the Threat of Financial Collapse *(New York: Oxford University Press, 2001).*

Want to Know More?

Charles P. Kindleberger's *A Financial History of Western Europe*, 2nd ed. (New York: Oxford University Press, 1993) is probably the best single volume for anyone interested in understanding the financial history of globalization. For a discussion of how structural changes in financial systems can lead to overextension and banking crises, consult Hyman P. Minsky's *Can "It" Happen Again? Essays on Instability and Finance* (Armonk: M.E. Sharpe, 1982). Also see Paul W. Drake's, ed., *Money Doctors, Foreign Debts, and Economic Reforms in Latin America from the 1890s to the Present* (Wilmington: SR Books, 1994) and Christian Suter's *Debt Cycles in the World Economy: Foreign Loans, Financial Crises, and Debt Settlements, 1820–1990* (Boulder: Westview Press, 1992). Michael Pettis's *The Volatility Machine: Emerging Economies and the Threat of Financial Collapse* (New York: Oxford University Press, 2001) identifies the specific events that set off liquidity booms in prior periods of global integration.

Frank Griffith Dawson's *The First Latin American Debt Crisis: The City of London and the 1822–25 Loan Bubble* (New Haven: Yale University Press, 1990) offers a wonderful account of an early era of globalization, although his description of events 180 years ago is too familiar for comfort. Matthew Josephson's famous *The Robber Barons: The Great American Capitalists, 1861–1901* (New York: Harcourt, Brace & World, 1962) is as good a place as any to start reading up on the railway booms of the late 19th century. Harold James's *The End of Globalization: Lessons from the Great Depression* (Cambridge: Harvard University Press, 2001) is one of the best recent books on economic history and discusses the globalizing period of the 1920s and the subsequent backlash.

Kevin H. O'Rourke and Jeffrey G. Williamson recently completed a major work on globalization at the end of the 19th century and the backlash against it in *Globalization and History: The Evolution of a Nineteenth-Century Atlantic Economy* (Cambridge: MIT Press, 1999). Finally, Niall Ferguson's *The House of Rothschild: Money's Prophets, 1798–1848* (New York: Viking Press, 1998) and *The World's Banker: The History of the House of Rothschild* (London: Weidenfeld & Nicolson, 1998) tell the story of a family intimately involved with every aspect of globalization.

• For links to relevant Web sites, as well as a comprehensive index of related FOREIGN POLICY articles, access **www.foreignpolicy.com**.

Soccer vs. McWorld

What could be more global than soccer? The world's leading professional players and owners pay no mind to national borders, with major teams banking revenues in every currency available on the foreign exchange and billions of fans cheering for their champions in too many languages to count. But in many ways, the beautiful game reveals much more about globalization's limits than its possibilities.

By Franklin Foer

Two omens of apocalypse, or perhaps global salvation: During the 2002 World Cup, the English midfielder David Beckham, famed bender of the ball, styled his hair in a mohawk. Almost instantly, Japanese adolescents appeared with tread marks on their shorn heads; professional women, according to the Japanese newsmagazine *Shukan Jitsuwa*, even trimmed their pubic hair in homage. A bit further west, in Bangkok, Thailand, the monks of the Pariwas Buddhist temple placed a Beckham statuette in a spot reserved for figures of minor deities.

It should surprise no one that this London cockney has replaced basketball icon Michael Jordan as the world's most transcendent celebrity athlete. After all, more than basketball or even the World Bank and the International Monetary Fund, soccer is the most globalized institution on the planet.

Soccer began to outgrow its national borders early in the post-World-War-II era. While statesman Robert Schuman was daydreaming about a common European market and government, European soccer clubs actually moved toward union. The most successful clubs started competing against one another in regular transnational tournaments, such as the events now known as Champions League and the Union of European Football Associations (UEFA) Cup. These tournaments were a fan's dream: the chance to see Juventus of Turin play Bayern Munich one week and FC Barcelona the next. But more important, they were an owner's dream: blockbuster fixtures that brought unprecedented gate receipts and an enormous infusion of television revenue. This transnational idea was such a good one that Latin America, Africa, and Asia quickly created their own knockoffs.

Once competition globalized, the hunt for labor resources quickly followed. Club owners scoured the planet for superstars that they could buy on the cheap. Spanish teams shopped for talent in former colonies such as Argentina and Uruguay.

Argentina plundered the leagues of poorer neighbors such as Paraguay. At first, this move toward an international market inspired a backlash. Politicians and sportswriters fretted that the influx from abroad would quash the development of young local talent. In Spain, for example, dictator Francisco Franco prohibited the importation of foreign players. Brazil's government declared Pelé a national treasure in 1961 and legally forbade his sale to a foreign team. But these stabs at nationalist economics could not ultimately stave off the seductive benefits of cheap, skilled labor from abroad. And, after a while, the foreign stars were needed to compete at the highest levels of European soccer. The game evolved to the point where an English club might field a team without any Englishmen.

By the 1990s, capital frictionlessly flowed across borders in the global soccer economy. European clubs not only posted scouts throughout the developing world, they also bought teams there. Ajax of Amsterdam acquired substantial shares of outfits in Cape Town and Ghana. Newcastle United began using China's Dalian Shide Football Club as a feeder. The biggest clubs started to think of themselves as multinational conglomerates. Organizations such as Manchester United and Real Madrid acquired a full portfolio of cable stations, restaurants, and megastores, catering to audiences as far away as Kuala Lumpur and Shanghai. Even with last year's dull markets, Manchester United's pretax profits for the 12 months ending on July 31, 2003, exceeded $65 million.

It is ironic, then, that soccer, for all its one-worldist features, doesn't evince the power of the new order as much as expose its limits. Manchester United and Real Madrid may embrace the ethos of globalization by accumulating wealth and diminishing national sovereignty. But a tangle of intensely local loyalties, identities, tensions, economies, and corruption endures—in some cases, not despite globalization, but because of it.

ENGLAND, HALF ENGLISH?

During Franco's rule, the clubs Athletic Bilbao and Real Sociedad were the only venues where Basque people could express their cultural pride without winding up in jail. In English industrial towns such as Coventry and Derby, soccer clubs ballasted together communities amid oppressive dinginess. It wasn't just that many clubs had deep cultural roots. Each nation evolved its own particular style of play—the Italian doctrine of *catenaccio* (or defensive lockdown), Brazilian samba soccer, and so on. In part, these were easy clichés. But to anyone who watched World Cups, they were also undeniably true clichés.

Three years ago, England, birthplace of the beautiful game, handed over its national team to a Swedish manager named Sven Goran Eriksson. It is difficult to convey just how shocked English fans felt. For much of the nation's soccer history, beloved, quintessentially English characters had run the team. These "lads," typically ex-players, often turned a blind eye when their squads drank lager on the eve

Fair Trade Soccer

Fans across cultures argue that soccer used to be a lot fairer. A middling team, fueled by gritty players and loyal fans, could emerge from nowhere to hoist the championship trophy. What's more, these underdog teams often hailed from smaller cities, without massive stadiums or deep-pocketed owners.

That level playing field, some fear, has disappeared entirely. With their global chains of superstores and vast array of television deals, the big clubs have become weathier, not just in absolute terms, but relative to the poorer clubs. Sales from Ronaldo and Beckham's replica jersey bring Real Madrid more income in a month than many clubs make in a year, so it's no surprise that Real so frequently rolls over its poorer foes in the Spanish game. Indeed, the results of domestic competition are virtually preordained. Manchester United or Arsenal of London has won 10 of the last 11 English Premier League titles. If you support an Italian team other than Juventus or AC Milan, you wake up every morning with a depressingly accurate sense of how the final league table will ultimately shake out.

Such lamentations, which sound a lot like the left's critiques of global free trade, are hard to resist. They have an aura of romance. But they simply don't withstand close examination. The richest clubs have always dominated their leagues. They might not be the same rich clubs; neither Liverpool nor Atlético Madrid nor Borussia Moenchengladbach dominates as they once did. Even so, the ruling elite of European and Latin American soccer has been extraordinarily constant over time. Teams like Juventus and Manchester United only fall from their thrones for brief and historically insignificant spells.

Globalization has actually added a measure of mobility to the system. Foreign investors have created new powerhouses overnight. Chelsea, funded by Russian oil money, looks poised to break the monopoly in English soccer. Parmalat has used money from its international sale of dairy products to rocket clubs in Italy and Brazil to success. Of course, it is possible to overstate the glory of the new soccer order. A few years ago, a Swedish parliamentarian named Lars Gustafson nominated the game for a Nobel Peace Prize, unleashing a fury of ridicule. And his critics have a point. Soccer doesn't deserve a prize for peace. It deserves one for economics.

—F.F.

of big games, and forgave men for lack of training so long as they spilled their guts on the field. For all their inspirational power, though, these English managers tended to lack tactical acumen. They recycled stodgy formations that encouraged the same, ineffectual mode of attack—a long ball kicked over the midfield to a lone attacker, a style that perfectly reflected stereotypes about stiff-upper-lip English resoluteness. Their lack of creativity was evident in the national trophy case. Despite England's singular place in the game's history, it has won a lone World Cup (in 1966, as the tournament's host team), and not a single European championship.

The English Football Association installed Eriksson to remedy this sorry situation. He seemed precisely the character for the job. The Swede, a reader of Tibetan verse, exudes cosmopolitanism. During his celebrated career, he has managed clubs in Lisbon, Genoa, and Rome. Whereas the English managers had tended to wear tracksuits, Eriksson dresses in impeccable Italian threads and wears a pair of tiny, chic spectacles. He speaks in far more complete and far more elegant English sentences than any of his predecessors.

Soccer clubs such as Manchester United and Real Madrid have acquired cable stations, restaurants, and megastores, catering to audiences as far away as Kuala Lumpur and Shanghai.

Never before had a man from across the channel coached England. This break with precedent wasn't just grist for pub trivia. Throughout the post-imperial decades, Britain has worried that the continent would encroach on its distinctive way of life—a repeat of the Norman and German assaults on the island. Now, that debate echoed on the populist pages of London's tabloids. A *Daily Mail* headline argued, "We've Sold Our Birthright down the Fjord." Gordon Taylor, the head of the English players' association spluttered, "I think it's a betrayal of our heritage." Some of these reactions can be chalked up to xenophobia. But the matter is psychologically deeper than that. English fans loved their old managers because they were such authentic representatives of the country, in all their faults and glories.

But the Eriksson era has taken an entirely unexpected direction: The new coach has practiced a caricature of old-fashioned, gritty English football. His system depends on goonish performances from relatively no-name defensive midfielders. Goals come from long passes to the fleet-footed forward Michael Owen. Every time Eriksson abandons the classic English formula, he gets in trouble.

Why hasn't Eriksson been able to remake English soccer in his suave continental image? The answer has to do with the deeply ingrained culture of the game. From a very early age, English players learn certain virtues—hard tackling, reckless winning of contested balls—and not others, such as fancy dribbling or short passing. Remaking these instincts isn't possible in a few seasons, let alone a few training sessions, of Swedish coaching.

The Eriksson story is archetypal. Portugal handed its squad over to a Brazilian; the Polish national team fronted a Nigerian striker who starred for a club in Greece; one of Japan's best players, Alessandro Santos (a.k.a. Alex), was born and raised in Brazil. None of these foreigners has succeeded in transforming the style and culture of national soccer teams. When Eriksson succumbed to Englishness, he upended one of the great clichés of the antiglobalization movement: that a consequence of free markets is Hollywood, Nike, and KFC steamrolling indigenous cultures. It is ironic that the defenders of indigenous cultures so often underestimate their formidable ability to withstand the market's assault.

CORRUPTION 1, INVESTORS 0

While globalization's critics have overestimated the market's destruction of local cultures, so too have its proponents. Take Brazil, for instance. How could a country with so many natural resources be so poor? How could a country with so much foreign investment remain so stymied?

Based on the stylishness of Brazil's most recent World Cup triumph in 2002—Edmilson springing backwards, catapult-like, into a poster-quality bicycle kick; Ronaldo scoring in-stride with a poke of the toe—outside observers would have no conception of the crisis in the national passion. But then, study the team's roster, and a pattern emerges. Between appearances for the national team, Edmilson contributes his stunts to a club in Lyon, France. The 27-year-old Ronaldo, now with Real Madrid, hasn't played professionally in Brazil since he was 17. Of the 23 players who wore their country's radioactive yellow jerseys in the 2002 World Cup, only 12 currently play in their home country. An estimated 5,000 Brazilians have contracts with foreign teams. While pumping out the world's greatest players, Brazil's sport couldn't be in a sorrier state. Only a handful of clubs operates anywhere in the vicinity of the black. Signs of decay are visible everywhere. Attending games in some of the country's most storied stadiums, buying their most expensive tickets, fans find themselves worrying about splinters and rusty nails protruding from the rotting wooden seats. Thousands more fans attend the average soccer game in Columbus, Ohio, and Dallas, Texas, than in the top flight of the Brazilian league.

Global capital was supposed to provide an easy fix. Foreign investors promised, implicitly at least, to wipe away the practices of corrupt elites who ran the Brazilian game and replace them with the ethic of professionalism, the science of modern marketing, and a concern for the balance sheet. In 1999, a Dallas-based investment fund called Hicks, Muse, Tate & Furst sunk tens of millions of dollars into the São Paulo club Corinthians and Cruzeiro of Belo Horizonte. ISL, a Swiss marketing company, acquired a share of the famed soccer club Flamengo in Rio de Janeiro. A few years earlier, the Italian food giant Parmalat took over Palmeiras of São Paulo. "Capitalism is winning out against the feudal attitudes that have prevailed in the sport for too long," crowed Juca Kfouri, Brazil's most venerable soccer journalist, at the height of the foreign influx. Kfouri predicted that soccer would generate 4 percent of Brazil's gross domestic product in just a few years.

Less than three years after the foreign investors arrived triumphantly in Brazil, they left in disgrace. At Corinthians, fans held demonstrations against Hicks Muse, protesting its failure to build a modern stadium. At Flamengo, ISL collapsed into bankruptcy. Foreign capital didn't turn Brazilian soccer into a commercial force like the National Basketball Association. In fact, by all objective measures, the game is now in worse shape than it was five years ago.

Why was the era of foreign investment such a debacle? The answer has to do with the men who ran the Brazilian game, perfect avatars of Latin American populism—corrupt, charismatic, and sly. When the foreign investors arrived in Brazil, they had no choice but to deal with these *cartolas*, or "top hats." But then the predictable happened. The *cartolas* siphoned funds into accounts in Bahamas and built themselves homes in Florida, expenditures documented in a congressional investigation. After they took the foreign investors' cash, the *cartolas* turned on their partners. When I visited Eurico Miranda, the president of the Rio de Janeiro club Vasco da Gama, he complained how the foreign investors brought in guys "who barely speak Portuguese." He condemned the foreign investors for selling star players to hated crosstown rivals—previously an unthinkable act. Miranda's genius was that he made these antiglobalization arguments only after he allegedly robbed his foreign investors blind. A culture of corruption, as it turns out, is not any easier to remedy in soccer than it is elsewhere in the global economy. People have an attachment to their populist leaders and politicians, not just because of their cult of personality and their ability to deliver goods. They like them because the populists paint themselves as defenders of the community against the relentless onslaught of outsiders. It's going to take more than Dallas-based pension funds and Wharton business school graduates to sweep them away.

WINNING THE PEACE

Local hostilities, even outright racism, ought to be the easiest sort of legacy for global soccer to erase. When people have a self-interested reason for getting along, they are supposed to put aside their ancient grudges and do business. But there's a massive hole in this argument: Glasgow, Scotland.

Glasgow has two teams, or rather, existential enemies. Celtic represents Irish Catholics. Its songs blame the British for the potato famine, and its games have historically provided fertile territory for Irish Republican Army (IRA) recruiters. Across town, there is Rangers, the club of Tory unionism. Banners in the stadium trumpet the Ulster Defense Forces and other Northern Irish protestant paramilitaries. Before games, fans—including respectable lawyers and businessmen—shout a song with the charming line, "We're up to our knees in Fenian blood." They sing about William of Orange, "King Billy," and his masterminding of the Protestant triumph in 1690 at the Battle of the Boyne. Until 1989, Rangers consciously forbade the hiring of Catholic players. Crosstown rivalries are, of course, a staple of sports, but the Celtic-Rangers rivalry represents something more than the enmity of proximity. It is the unfinished fight over the Protestant Reformation.

The Celtic and Rangers organizations desperately want to embrace the ethos of globalism, to convert themselves into mass entertainment conglomerates. They've done everything possible to move beyond the relatively small Scottish market—sending clothing catalogs to the Scottish and Irish diasporas in North America, and campaigning to join the bigger, better, wealthier English league.

But the Celtic and Rangers clubs don't try too hard to eliminate bigotry. Rangers, for example, continues to sell Orange jerseys. It plays songs on the stadium loudspeaker that it knows will provoke anti-Catholic lyrics. The club blares Tina Turner's "Simply the Best," which culminates in 40,000 fans screaming "Fuck the Pope!" Celtic, for its part, flies the Irish tricolor above its stadium. At Glasgow's Ibrox Park, I've watched Protestants celebrate a goal, egged on by former team captain Lorenzo Amoruso, a long-haired Italian with the look of a 1980s model. He applauds the fans. Flailing his arms, he urges them to sing their anti-Catholic songs louder. The irony is obvious: Amoruso is Catholic. Since the late 1990s, Rangers has routinely fielded more Catholics than Celtic. Its players have come from Georgia, Argentina, Germany, Norway, Portugal, and Holland, because money can buy no better ones. But ethnic hatred, it seems, makes good business sense. In fact, from the start of their rivalry, Celtic and Rangers have been nicknamed the "Old Firm," because they're seen as colluding to profit from their mutual hatreds. Even in the global market, they attract more fans because their supporters crave ethnic identification—to join a fight on behalf of their tribe.

Crosstown rivalries are, of course, a staple of sports, but the Celtic-Rangers rivalry in Scotland represents something more than the enmity of proximity. It is the unfinished fight over the Protestant Reformation.

There are plenty of economic causes for illiberal hatreds—unemployment, competition for scarce jobs, inadequate social safety nets—but none of those material conditions is especially widespread in Glasgow. Discrimination has faded. The city's unemployment problem is no better or worse than the rest of Britain. Glasgow has kept alive its tribalism, despite the logic of history, because it provides a kind of pornographic pleasure. Thousands of fans arrive each week from across the whole of Britain, in ferries from Belfast and buses from London, all aching to partake in a few hours of hate-filled tribalism. Once they release this bile from their system, they can return to their comfortable houses and good jobs.

If there were any place one would expect this sort of hostility to get messy, it would be Chelsea. During the 1980s, the club was the outfit most associated with English hooliganism. Its fans joined the xenophobic British National Party and merged with violent racist gangs like the notorious Combat 18. There are famous stories of Chelsea fans visiting Auschwitz, where they would walk around delivering *Sieg Heil* salutes to the tourists and try to climb inside the ovens. When the Holocaust denier David Irving went on trial for libel in 2001, the hooligan group Chelsea Headhunters provided security for his rallies.

Like a college alumni association, older, retired Chelsea hooligans make a point of sticking together. They stay in

[Want to Know More?]

Much of the literature on soccer veers from the vacuous to the absurdly academic. There are, however, some worthy exceptions. *Financial Times* columnist Simon Kuper charts the sport's intersection with politics through witty travelogue in ***Football Against the Enemy*** (London: Orion, 1994). Uruguayan novelist Eduardo H. Galeano provides a less journalistic, more poetic rendering of the same subject in ***Soccer in Sun and Shadow*** (New York: Verso, 1998).

In addition to these tours through the global game, several excellent case studies are available. David Winner's ***Brilliant Orange: The Neurotic Genius of Dutch Football*** (London: Bloomsbury, 2000) finds the aesthetic underpinnings for the national style in art, architecture, and politics. ***Offside: Soccer and American Exceptionalism*** (Princeton: Princeton University Press, 2001) by Andrei S. Markovits and Steven L. Hellerman explores why the United States has sat on the periphery of this most globalized phenomenon. Alex Bellos uses the game as an anthropological vehicle for understanding Brazilian race, class, and corruption in ***Futebol: The Brazilian Way of Life*** (New York: Bloomsbury, 2002). And Bill Murray covers the history of the Celtic and Rangers clubs in ***The Old Firm: Sectarianism, Sport, and Society in Scotland*** (Atlantic Highlands: Humanities Press, 1984).

Several top journalists and leading publications examine the political economy of soccer. The London-based columnist **Gabriele Marcotti** writes the **"Inside World Soccer"** column for ***Sports Illustrated*** and contributes to various European publications. See Franklin Foer's **"Glooooooooooo-balism!"** (Slate, February 12, 2001) for perspectives on soccer and globalization. Argentina's monthly *El Gráfico* remains one of Latin America's top soccer publications, while the lively English monthly ***Four Four Two*** usually contains at least one lengthy story on the social significance of a crosstown rivalry, a band of hooligans, or some other unexpected aspect of the beautiful game.

FOREIGN POLICY coverage of the intersection of culture, politics, and economics includes Joshua Fishman's **"The New Linguistic Order"** (Winter 1998–99), Theodore C. Bestor's **"How Sushi Went Global"** (November/December 2000), Mario Vargas Llosa's ***"The Culture of Liberty"*** (January/February 2001), Alberto Fuguet's ***"Magical Neoliberalism"*** (July/August 2001), Douglas McGray's **"Japan's Gross National Cool"** (May/June 2002), and Kym Anderson's **"Wine's New World"** (May/June 2003).

For links to relevant Web sites, access to the *FP* Archive, and a comprehensive index of related FOREIGN POLICY articles, go to www.foreignpolicy.com.

touch through an online message board, where they exchange war stories and debate the fortunes of their beloved club. The board makes a point of declaring, "WELCOME TO THE CHELSEA HOOLIGANS FORUM, FOR CHELSEA AND LOYAL FANS. PLEASE DONT LEAVE RACIST MESSAGES AND DONT USE THIS BOARD TO ARRANGE VIOLENCE." The warning is intended to inoculate the site against any exceptionally offensive posts, but it doesn't exactly deter the anti-Semitism. Almost immediately after oil baron Roman Abramovich, the second richest man in Russia, and a Jew, bought Chelsea, a guy calling himself West Ken Ken referred to Abramovich as a "yid," and moaned, "I like the money but the star of David will be flying down the [Stamford] bridge soon."

However, as the Abramovich era began and the new owner spent more than $150 million stocking his new team, the complaints became less apparent. And then, when Chelsea jumped to the top of the English Premier League table, the anti-Semitism vanished altogether.

Chelsea, it seems, has discovered the only effective palliative for the vestiges of localism—not global cash or global talent, but victory.

Franklin Foer is an associate editor at The New Republic. *His book on soccer and politics is forthcoming from HarperCollins.*

Croesus and Caesar

The Essential Transatlantic Symbiosis

Richard Rosecrance

A VENGEFUL United States is snubbing its German allies for opposing the war against Iraq. Prickly France has become a porcupine sitting on America's lap; Moscow and Beijing have followed Paris' lead since the United States and Britain subdued Ba'athi Iraq. Some opine, and others worry, that a belligerent United States or a reactive European Union will throw NATO into the dust bin of fast-moving history. The United States may think it does not need Europe, and Europe may reject association with the American hyperpower. Some Continentals now call more openly for European balancing against a "rogue United States."

This, however, will not happen. The EU will not seek to become another military superpower to contend with or threaten America. The defeat of Iraq will not lead to the end of the American era, with Europe as America's residuary legatee. Nor is the altercation between Europe and America properly cast as a spat between different visions of the social order—with Brussels and the European Commission representing the "paradise" of social democracy and Washington the "power" of the preponderant state. Europe, in fact, does not play Venus to America's Mars.

Amid much bluster on both sides, the paramount truth is that the two major centers are both powerful, but act in different spheres—and they desperately need each other. Neither can, or at any rate should, talk cavalierly of going it alone. Most specifically, America needs Europe's financial power, and Europe requires America's military protection.

America Needs Europe

T HE MOST appropriate metaphor for comparing Europe and America, for those who insist on them, has nothing to do with planets. It rather compares Europe's Croesus to America's Caesar, the paradigmatic example of the power of Rome. Croesus, King of Lydia in the 6th century B.C.E., accumulated vast wealth, and so does Europe today. America is a new Rome that bestrides the world politically and militarily, but rests on

shaky financial foundations. Power requires wealth, and wealth cannot protect itself without power. In Washington, Americans berate Europeans for failing to share the burdens of military spending and for free-riding on American largesse. The Bush Administration vilifies most of its NATO allies for lacking military might and having to rely on American air and sea power to get their slender forces to any theater of operations. The so-called European "rapid reaction force" is not rapid, nor will it ever be much of a force. American pundits declare that only European rearmament will remedy this deficiency and restore equality to the link with the United States.

This view is as mistaken as it is common. Military clout is not the appropriate way to measure the European contribution to NATO or to America. With little fanfare, the Europeans (and their Japanese cohorts) have shored up American power against the force of financial tides, enhancing Washington's strength and resiliency. Without the help of Europe and Japan, the United States could not have undertaken or sustained its frequent international military operations. Lacking this financial shield, U.S. foreign and security policy would have been checked and doomed to failure. So it has been for decades.

European strength has always rested on monetary foundations. In the past, Holland and then Great Britain commanded the world economy and plucked imperial spoils as a willful child eats chocolate. These nations could finance operations overseas, run commercial economies at home and still expand their territorial orbits. As Niall Ferguson points out, Britain won contest after contest because its money market was deeper and more resilient than those of its rivals.[1]

The United States inherited this role after World War I and enlarged it after World War II, when the reigning American trade surplus emptied European coffers and American credit refloated capitalist growth. Brimming with wealth, the United States could finance both its military role and expand American private capitalism worldwide while contributing to Europe's rehabilitation. But the American financial reservoir ran dry in the 1960s: since then, the United States has not been able to bankroll its military and economic sectors at the same time.

So Europe (and Japan) has. Since the late 1960s, the Europeans have been called on repeatedly to bail America out economically, and they have done so. In 1971, Europe permitted the United States to go off the gold standard, devalue its currency and levy import surcharges against its trading partners without retaliation. Europe did not punish America for its breach of faith: instead it helpfully maintained the high value of the franc and the deutschemark. In 1978-79, American inflation rates rose above the interest rate, negating any incentive to save in the United States. This led, predictably, to a run on the dollar. Spurred by German Chancellor Helmut Schmidt, Federal Reserve Chief Paul Volcker moved to collar inflation with a high interest rate policy, and the dollar regained its footing. After the dollar had climbed to prodigious heights under Ronald Reagan (fostering a huge trade deficit in the United States in the bargain), Europe helped to engineer a soft landing in both 1985 and 1987 as the dollar returned to earth. This prudent division of Western labor represents a still underappreciated reason for the West's victory in the Cold War. The Soviets, in contradistinction, had nothing comparable to it.

Nor did Europe ever take advantage of American weakness when the United States ran into financial trouble. In 1987, the Europeans stood ready to help the Federal Reserve halt the collapse of the New York stock market on October 19. On that date the Dow-Jones Average suffered its largest one-day decline in history, plunging 22 percent in a single session. When the new Fed Chief Alan Greenspan lowered interest rates and ploughed $100 billion of liquidity into the financial system—floating struggling firms and the exchange itself—the Bundesbank did not devalue the deutschemark in response. Europe and Japan continued to hold funds in New York and did not raise interest rates to counter the inflationary effect of American credit. Germany asked the United States to cut its fiscal deficit but remained ready to support America even when forced in effect to import U.S. inflation. Japan also resumed buying dollars.

Such past services from Europe were, of course, self-interested in the larger sense, but that takes nothing away from their importance. Nonetheless, these services pale beside the tasks the European Croesus will have to undertake in the future. The United States is now running unparalleled trade and fiscal deficits. The trade deficit alone stands at more than $500 billion a year, almost 5 percent of GDP. America is now borrowing 5 percent of world savings, sucking $1.5 billion a day into the United States. The budget deficit, fanned by President Bush's tax cuts, may rise to $500 billion a year, reaching $2 trillion in the next several years. If foreign capital does not enter the United States in huge amounts, the dollar could go into free fall, resulting in renewed inflation. The Fed would then have to raise interest rates, and the American economy would likely plunge into deep recession.

This crisis could occur well before 2004, and President Bush's re-election would then be up for grabs. The political outcome in the United States could thus depend upon what politicians and politically influential bankers in Brussels, Paris, Frankfurt, London and Tokyo decide to do to help—or not help, as the case may be.

It is true that investing in America is still in Europe's own economic interest. Investments yield more in New York than they do in Paris. In the short term, though, Europeans and Japanese have lost money in the United States. The New York stock market has not gone up, and European returns on U.S. investments are cancelled by currency losses. As Europe enlarges, the expansion of the European market will draw money back to Brussels. In the longer term, however, U.S. growth will return, and Europe will want to share in its benefits. As the dollar falls, Europe will substitute new foreign direct investment for trade with the United States. If one takes even a casual glance at the size of such investments in the United States, one sees immediately that their value dwarfs that of commerce. This is not going to change radically or soon. Europe will still place large funds in the American economy. But enough funds, and on what terms?

Europe Needs America

As a "peaceful power", Europe does not want, or at any rate should not want, to separate from the United States and rebuild its own major military capabilities. It needs American defense protection while integrating with a still unstable region stretching beyond east-central Europe to Central Asia. Robert Mundell, the father of the euro, believes that as many as fifty countries will eventually join the EU—even some from North Africa—and adopt the euro as their national currency. America, not Europe, will defend these expanded borders—and Europeans recognize that they must play their financial part to receive help in protecting their growing perimeter.

To be sure, the euro will probably rise to become a competitive world currency with the dollar, but this prospect ought not be feared. It will give America an opportunity to export to European markets in a major way. Washington will also then have a partner in providing liquidity to other nations through deficits in the balance of trade. Europe's economic power will probably grow, despite its demographic doldrums, equaling or even surpassing that of the United States. But this does not mean that Europe will shoulder the world's, or even its own, military burden—and the richer it gets the more it will need U.S. military protection.

Yet Europe as a pacific economic giant will not be inward-looking. It will seek to influence others to follow its peaceful mission. Other nations may well emulate its strikingly successful institutions and policies. By expanding, the new Europe will bring peace and prosperity to eastern Europe and perhaps even to Russia, but it will not police the world. The NATO link with the United States will still be essential in all of this. That is why it is both erroneous and self-defeating to claim that America does not need Europe or Europe the United States. America and Europe are essential and reciprocal counterparts to one another in the structure of world power.

Then there is the continuing threat of terrorism. This threat will not disappear now that the war with Iraq is over; terrorism may even spread, at least in the short run. Neither the technical revolution in military affairs that has occurred in the United

States nor the growing imperial reach of America's conventional forces can vanquish terrorism. For that, the United States will need cooperation and the shared intelligence of many other countries, particularly those of an enlarged European Union. There are no superpowers in the war against terrorism—every nation can be a military theater where battles may be won, and Pakistan's cooperation may be as important as Russia's. International cooperation between the United States and Europe is essential to win the long bout against terrorism.

EUROPEANS and Americans alike have been predicting the demise of NATO and the end of U.S. and European cooperation since Suez in 1956. The alliance, however, has not collapsed—and it won't. In the last thirty years, Europe has carried the primary financial burden, allowing the United States to maintain an essentially unbalanced economy while acting as the world's gendarme. Europe has not built up its military strength, but instead has done something much more important: it has created the financial conditions that have allowed the United States to act.

Americans may complain about the recent actions of Monsieur Chirac and Herr Schröder, but if they insist on remaining oblivious to Europe's critical financial role, they only contribute to an atmosphere in which that role is weakened. And Europe, which will not act militarily on its own in any serious way, should stop squawking about the use of American force as though, cases aside, there is something wrong about it in principle. The Atlantic economic and military partnership serves both Americans and Europeans, and it will become firmer as the true basis of the relationship dawns on both sides of the Atlantic. One hopes this happens sooner rather than later, for Caesar needs Croesus, and Croesus needs Caesar, now.

Notes

1. See, at the least, "True Cost of Hegemony: Huge Debt", *New York Times*, April 20, 2003.

Richard Rosecrance is professor of political science at UCLA and project director of the UCLA-Carnegie study on the effect of globalization on national self-determination..

Where the Money Went

How did developing countries wind up owing $2.5 trillion?
Hint: Western institutions haven't exactly been innocent bystanders.

By James S. Henry

By the dawn of the 21st century, after 30 years of development strategies that were designed in Washington, New York, London, Frankfurt, Paris, and Tokyo, and trillions of dollars in foreign loans, aid, and investment, more than half of the world's population still finds daily life a struggle, surviving on less than $2 a day—about the same level of real income they had 30 years ago. More than two billion people still lack access to basic amenities like electricity, clean water, sanitation, land titles, police and fire protection, and paved roads, let alone their own phones, bank accounts, or medical care.

And despite years of rhetoric about debt relief and dozens of structural adjustment plans, the real value of Third World debt has continued to grow, to more than $2.5 trillion. The cost of servicing that debt now exceeds $375 billion a year—more than all Third World spending on health or education, almost 20 times what developing countries receive each year in foreign aid, and more than twice as much as they have recently received annually in foreign direct investment.

How did 30 years of greatly expanded international lending, investment, aid, and development efforts end up producing such a fiasco? Where did all that money actually go? And what can we do, if anything, to undo all the damage that has been done?

There is no shortage of armchair analyses of the so-called Third World debt crisis, or even of the globalization crisis that succeeded it and continues to this day. The 1980s debt crisis became visible as early as August 1982, when Mexico, Argentina, and 26 other countries suddenly rescheduled their debts at once. Our disappointments with globalization have been a popular subject for economists and development policy-makers at least since the Mexico crunch of January 1995, as amplified by the East Asian and Russian crises in 1997–98, and in Turkey, Ecuador, Bolivia, Argentina, Venezuela, and other countries since then.

But there has been no detailed account of the *structural roots* of this prolonged development crisis. Among orthodox economists, the conventional wisdom is that the crisis originated in a combination of unpredictable shocks and Third World policies that were either stupid or corrupt, on top of factors like bad geographic luck. In other words, we have a slightly more sophisticated version of the same blame-the-victim ideology that ruling elites have used for centuries to explain poverty and wealth: It is either "tough luck" or "their own damn fault."

For example, the conventional view of the 1980s debt crisis is that, in response to the 1973 oil-price rise, Western banks recycled oil deposits from the Middle East back to the Third World, lending to finance oil imports and development projects. Independently, a huge tidal wave of capital flight sprang up alongside all this lending. Then an unfortunate combination of events—rising interest rates, recession, and the 1982 Falklands conflict—supposedly took everyone by surprise. In this version of the crisis, no one was responsible for it. There were no villains or victims—just innocent bystanders.

Conventional portraits of the global development crisis are fairy tales.

Similarly, in the case of more recent debacles in submerging markets like Indonesia and Thailand, the official story is that these resulted almost entirely from Third World mistakes and market imperfections; poorly designed bank regulations, faulty accounting, inexplicable failures to privatize and liberalize fast enough, or indigenous "corruption, cronyism, and collusion." The notion that there might be serious oversimplifications in the development paradigms themselves, that Western banks, investors, policy-makers, and the structure of international capital markets might have aggravated such problems—or that they might have conspired with local elites and even taught them a thing or two about "clever chicanery"—has received much less attention. And while numerous globalization critics have recently emerged, including some prominent economists and international investors as well as a rising tide of mass protesters, much of this criticism has dealt in generalities, lacking the gory investigative details needed to drive the points home and do justice to the problem's global scale.

Behind the Official Story

The fact is that conventional portraits of the global development crisis are economists' fairy tales. They leave out the blood and guts of what really happened—all the payoffs for privatizations, fraudulent loans, intentionally wasteful projects, black market "round-trip" transfers, arms deals, insider deals, and the behind-the-scenes operation of the global-haven banking network that has facilitated all this and more. They ignore the fact that in the mid-1970s, and again in the mid-1990s, repeated warnings of deep trouble were ignored: Irresponsible overlending, poorly conceived projects and privatizations, phony back-to-back loans, outright looting of central-bank reserves, and massive capital flight continued right under the noses of Western bankers and government officials who were in a position to do something about the problems but chose not to.

This begs the question of why some developing countries and banks got into so much more trouble than others, and why certain bankers, investors, and officials got rich. Standard analyses focus on uncontrollable shocks to the system, skate far too quickly over the structure of that system itself, and ignore the systematic role of specific global interests in shaping this structure.

Most importantly, the official story of shocks, surprises, and indigenous corruption ignores the pivotal role that was often played by sophisticated banks and multinationals. They were aided by their governments and the local elites in pressing countries first to overborrow, then to overservice their debts—to honor their overpriced privatization contracts—and then to liberalize and privatize far too quickly. Many of these players were also aggressively recruiting flight capital and investment deals from these very same countries and teaching their clients the basics of how to launder, plunder, and conceal.

The conventional fairy tales gloss over many key questions. What really became of all the loans and investments? Why did foreign banks lend so much money to these governments, even while their private banking arms knew full well that they were financing a massive capital-flight exodus? Who ended up owning all the juicy assets? How much did the IMF and the World Bank know about all these shenanigans, and why didn't they do more to stop them?

It is not easy to give precise answers to such questions. Studying the global underground economy is an exercise in night vision, not double-entry accounting or armchair analysis. One actually has to get up out of the armchair and do some investigative reporting. And the patterns that become visible turn out to be full of villains and victims.

Unfortunately, it has taken years to uncover the truth about such matters, and we've really only begun to scratch the surface. In the last decade, bits and pieces of this story have become more accessible. The demise of "kleptocracies" like those of Abacha, Andres Perez, Collor, Duvalier, Marcos, Mobuto, Milosevic, Salinas, Suharto, and Stroessner focused attention on the billions that such nefarious regimes managed to steal and stash abroad. The collapse of leading money-laundering banks like BCCI and BNL, and the corrupt regimes of Salinas in Mexico, Menem in Argentina, the ruling cliques in Turkey, and Suharto in Indonesia demonstrated the risks that corrupt banking poses to the entire global financial system.

However, partly because this kind of investigative research is so difficult, there are only a few rather armchairish studies of the global underground economy and the "real-economique" of underdevelopment. And these have also usually treated the subjects of irresponsible lending, wasteful projects, capital flight, corruption, money laundering, and havens separately, In fact, they go hand in hand. The rise of Third World lending in the 1970s and 1980s laid the foundations for the haven network that now shelters the wealth of the world's most venal citizens. And the corruption that this network facilitated was just a special case of a much more general phenomenon—the export of vast quantities of capital and tax-free incomes by the elites of poor countries, even as their countries were incurring vast debts and struggling to service them. Individual kleptocratic regimes and evil dictators come and go, but this sophisticated transnational system is more vibrant than ever.

A Marshall Plan in Reverse

This haven network has matured in the last 30 years, coinciding with the rise of global lending, the liberalization of capital markets, and the development crisis. Of course, "private banking" is hardly new. Except for electronic transfers and airplanes, most of its paraphernalia—secret accounts, shell companies, black-market exchanges, misinvoicing, back-to-back loans, and backdated transactions—were very important early innovations in the history of capitalism. They helped to bring "mattress money" out into the open so that it could be productively invested.

For decades, these tools were underutilized. Then, from the 1970s through the 1990s, their use expanded tremendously under the impact of the greatest torrent of loose lending in history. This torrent was driven by a coalition of influential interests that included leading Western private banks, equipment vendors, and construction companies—plus their allies among aid donors, export finance agencies, the multilateral banks, and local elites. It drove a hole in developing-country defenses against overborrowing and created a source of fundamental instability in the world's financial system that continues to this day.

It is possible to estimate the volume and composition of the flight capital that was financed by all this lending. Sag Harbor Group estimates rely on a combination of statistical methods and interviews with more than 100 private bankers and their wealthy clients. They show that at least *half* the funds borrowed by the largest "debtor" countries flowed right out the back door, usually the same year or even the same month that the loans arrived. For the developing world as a whole, this amounted to a huge Marshall Plan in reverse.

The corresponding stock of unrecorded foreign wealth owned by Third World elites is even larger than these outflows, since the outflows were typically invested in tax havens where—unlike the earnings of low-income "guest workers"—they accumulate tax-free interest, dividends, and capital gains. By the late 1980s, there was already enough anonymous Third World flight wealth on hand in Europe and the United States that the income it generated would have been able to service the *entire* Third World debt—if only this stock of "anonymous capital" had been subjected to a modest global tax. By the late 1990s, the market value of private wealth accumulated outside developing countries by their resident elites totaled at least $1.5 trillion. As one Federal Reserve official chuckled at the time, "The problem is not that these countries don't have any assets. The problem is, they're in Miami."

Most of the resulting flight wealth—defined simply as foreign capital whose true ownership is concealed—ended up in just a handful of First World havens. Their identity may surprise those who usually associate "tax havens" and money laundering with obscure nameplate banks in sultry tropical paradises. BCCI's exceptional case notwithstanding, shady banks have never been very important in the flight market—most rich people would never trust them. Rather, Third World decapitalization has taken place with the active aid of pre-eminent global financial institutions, including Citigroup, JP Morgan Chase (Chemical MHT), UBS, Barclays, Credit Suisse First Boston, ABN-AMRO Merrill Lynch, ING Bank, the Bank of New York, American Express Bank, and two dozen other leading Swiss, British, Dutch, French, German, and Austrian banks. These august institutions led the way in knowingly facilitating these perverse capital flows. In fact, despite their reputations as lenders, many of these institutions were actually net borrowers from poor countries for most of the last 30 years. International private banking—a large share of which was for Third World elites—thus became one of their most profitable lines of business.

The leading global banks were successful at profiting from Third World grief.

Because their leading global banks were so successful at profiting from Third World grief, the United States, the United Kingdom, and Switzerland were also—despite their reputations as major capital providers and aid donors—net

debtors with respect to the Third World throughout the last 30 years. It is really misleading to speak of a "Third World debt" crisis—for developing countries, it has really been an *assets* crisis, while for these key First World countries and their banks, it was an incredibly profitable *global bleed-out*.

The upshot is that ownership of Third Wealth onshore and offshore wealth is even more concentrated than it was before the 1980s debt crisis and the 1990s privatization wave. Depending on the country, the top 1 percent of households now accounts for 70 to 90 percent of all private financial wealth and real estate. We are not talking about millions of diligent middle-class savers. These are the 700 members of the Rio Country Club, the 1,000 top landowners of Argentina, El Salvador's *catorce*, the 50 top families of Caracas, the 300 top families of Mexico City, and so on. These are not people who merely observe the rules of the game. When a minister builds a dam, nationalizes a private company or its debts, privatizes a state enterprise, and floats a currency or manipulates tax provisions, these folks have the inside track. This is not to say that there are no new self-made Third World elites, or that the top tier is limited to capitalists—there are also quite a few politicians, generals, diplomats, union bosses, and even a bishop or cardinal.

Conventional explanations ignore the international community's responsibility.

But the overall system is beautifully symbiotic. On the one hand, havens provide an ideal way to launder loot. On the other, in the long run, corruption, insider deals, and the inequalities they generate encourage still more disinvestments and emigration. For individuals living in these countries, of course, the incentives for engaging in money laundering were often so powerful that it would have been quixotic not to do so. At the macroeconomic level, however, conventional explanations of capital flight, overborrowing, and mismanaged privatizations, which focus on technical policy errors and the "riskiness" of Third World markets, ignore these systemic factors. They also ignore the international community's collective responsibility for perpetuating a system that encourages noneconomic lending, tax evasion, flight-prone speculative investments, and perverse privatizations.

The Borderline of Existence

Other key ingredients in the "money trail" puzzle, in addition to capital flight, were wasteful projects and arms purchases. Hundreds of billions of Third World loans were devoted to nonproductive projects and the corruption that encouraged them. Many of these debt-financed projects also had harmful long-term consequences. In some cases, there was novel chicanery on a purely local level. But what is most striking are the global patterns—overpricing, rigged bids, endless delays, loans to front companies with close ties to the government, investments in dubious technologies and excessively capital-intensive projects, "public" projects undertaken for private motives, and private debts assumed by the state.

Over and over again, the handiwork of the very same international banks, contractors, equipment vendors, and export credit agencies grew fat while the countries grew poorer. These were not ideological errors—regimes of different ideological hues proved equally vulnerable. Nor were they due to random policy mistakes or indigenous corruption. A sophisticated *transnational system* of influential institutions contrived to produce similar mistakes over and over again, in every region of the world. Corruption has always existed, but without this global system, the abuses simply could not have been generalized on such a massive scale.

Recently, some of the leading players in the global haven industry, Citigroup and JP Morgan Chase, had also applied their "haveneering" skills to help Enron, WorldCom, and Global Crossing, using Panama and Cayman Islands shell companies to conceal billions in off-balance-sheet loans from their stockholders. When this scandal surfaced in 2001-02, the resulting bankruptcies cost investors several hundred billion dollars. Analysts who had not followed the history of these banks in the Third World were shocked ... shocked! Those who knew them better were just reminded that "character is destiny" and that "what goes around comes around." As a wise, rather ethical business colleague of mine once said, "The problem with a rat race is that, even if you win, you are still a rat."

The emphasis on these darker details is not meant to imply that every banker was a briber, or every public official a crook. But dirty money, bad banking, money laundering, and self-seeking chicanery were not incidental to the development crisis. As a governor of the Bank of International Settlements admitted privately in the late 1980s, "If Latin America's corrupt politicians simply gave back all the money they've stolen from their own countries, the debt problem could be solved. And most of that thievery simply could not have occurred without the active assistance of leading First World banks, contractors, vendors, multilateral lenders, advisors, and governments. This was not a natural catastrophe but a *manmade* one. For the developing world to overcome it, it is not only the developing world that will have to be reformed."

Today, despite decades of official development efforts and trillions in foreign loans, bonds, and investments, the vast majority of the world's residents are living on the borderline of existence. Yet the "developing countries" they inhabit are not really poor at all, in terms of natural resources, technical know-how, and raw human talent.

Some have searched for the explanation among the natural disadvantages of climate, pestilence, and topography that many of these countries experience. Some have pointed toward cultural deficiencies—for example, a purported lack of trust outside the family in some countries, or an unusual propensity for corruption at all levels of society in others. Some have invoked the *deus ex machina* of "policy errors" like overvalued exchange rates, excessive borrowing, and weak security laws, as if these were uncaused causes and as if policy was made in a disinterested vacuum, where ministers don't own bank accounts and everything would have worked out fine if only they'd been Ivy-educated. Still others have emphasized the ill winds of misfortune to which developing countries are subject—the HIV/AIDS epidemic in Africa, China, India, and Russia; Indonesia's bad luck in having Thailand as a neighbor when its currency plummeted in July 1997, Brazil's bad luck in having Russia's 1998 crisis compound Indonesia's, Argentina's in having Brazil as a neighbor, and so on.

These explanations are all the profound contribution that First World countries and their global agents have made, not only in the last 30 years but for much longer, in tolerating, contributing to, and profiting from the immiseration around them.

If one really looks objectively at why countries like the Philippines, Guatemala, Indonesia, the Congo, South Africa, Argentina, Venezuela, Brazil, Mexico, Haiti, and India have ended up as they have in the world economy, one cannot ignore the negative influence of First World corporations, governments, and financial institutions. As a rule, the closer and more unequal those relationships have been, the worse things have turned out for developing countries.

Today, the developing world is in its deepest crisis in a half-century. The First World, now so concerned about security, needs to place these security concerns and the "war on terrorism" in the context of an even more global war—underdevelopment. The world is simply getting too rickety, too interdependent, for us to ignore this other war any longer. Forest fires in the Amazon or Kalimantan set by poor people clearing land because they can't find jobs threaten the whole world's air supply. Epidemics left untreated in Africa or Russia threaten to create drug-resistant diseases that could sweep the planet. And the hatred bred by the real "weapons of mass destruction"—outrageous poverty and inequality—are only a plane or boat ride away in Port-au-Prince, Jakarta, Cairo, Kabul, and Karachi.

The first step is understanding. First Worlders, in particular, are living in a bubble—only about 20 percent of Americans have passports, just 21 percent of non-Hispanic Americans speak a second language, less than a fifth of adults have traveled abroad, and a poll taken in late 2002 showed that only 13 percent of 18-to-24-year-olds in the United States could even find Iraq on a map, 14 percent could find Israel, 31 percent could find the United Kingdom, and 42 percent could find Japan, and only two-thirds could even find the Pacific Ocean. This kind of ignorance is not just unfortunate. It is dangerous. It is especially menacing to the citizens of the Third World, because the First World now has more political, economic, and military hegemony than ever before. It is also a menace to ourselves, because if this power is not used wisely, a growing portion of the Third World will simply disappear into "the Fourth World," a vast, impoverished, hostile labor camp without visitors or investments. And that, in turn, will only heighten our insecurities, raise the drawbridge even higher, and increase hostilities.

Those who wish to alter these current trends toward immiseration and anti-development may choose to put their faith in the global economy, free trade, investment, technology, and entrepreneurship. But these market-based nostrums have not been sufficient. To go beyond them, people will need to invest in their own globalization, their own practical education about how the world really works.

Today, the developing world is in its deepest crisis in a half-century.

This kind of education is not available in university economics courses. People need to understand why political parties, the police, the military, the media, the courts, and the church are often so unresponsive to popular demands, even in nominal "democracies"; why senior officials, banks, corporations, and the elite continue to prefer monster projects to schools and clinics; why courts rarely enforce laws against people of means, let alone global companies like Freeport McMoran or Citigroup; why the radical liberalization of global capital markets and trade has taken precedence over the enforcement of tax codes, labor laws, health codes, security regulation, environmental laws, education rights, pension reform, and property rights for ordinary people; and why the poor are subject to unavoidable excise taxes while the elite are encouraged to invest, tax-free, at home and abroad.

They also need to ask why developed and developing countries alike, after 50 years of malpractice, still permit First World bankers, corporations, and investors to engage in business practices in the Third World that are grossly illegal; why the anti-foreign bribery statutes of the United States and the OECD countries are so underenforced; why undocumented capital is recruited so aggressively from developing countries while undocumented labor is increasingly harassed; and why the huge proportion of the Third World's $2.5 trillion debt that was contracted illegally and spent on failed projects and elite bank accounts deserve to be serviced at all.

But first, they may want to start with the question of where the money went.

JAMES S. HENRY is founder and managing director of the Sag Harbor Group, a strategy consulting firm, former chief economist for McKinsey & Co., and former vice president of strategy for IBM/Lotus. From The Blood Bankers: Tales From the Underground Global Economy *(Four Walls Eight Windows).* © 2003

Render Unto Caesar: Putin and the Oligarchs

MARSHALL I. GOLDMAN

"Because of the flawed manner in which the privatization process was carried out, the new owners will always lack the legitimacy necessary for a stable political climate and sustained economic investment and growth. Having constructed a faulty foundation, the builders must live with the possibility that their edifice of privatization will periodically shift and crack and may even collapse."

In 1991, a small group of Russians emerged from the collapse of the Soviet Union to claim ownership of some of the world's most valuable oil, natural gas, and metal deposits. This resulted in one of the greatest transfers of wealth ever seen. By 1997, five of these individuals, who in the 1980s had only negligible net worth, were listed by *Forbes* as among the world's richest billionaires. This year seventeen of these men were on the *Forbes* list, including two who merged their companies to create what promises to be an oil conglomerate worth $35 billion. Almost all of these self-styled oligarchs have been accused of guile, intimidation, and even murder in reaping their rewards. Not surprisingly, their rise—and for some, their fall—mirrors the travails of the Russian economy.

SHOCKING THE SYSTEM

The oligarchs gained their wealth by taking advantage of economic policies forged in the early 1990s by President Boris Yeltsin and the reformers he brought into Russia's first post-Soviet government. Why did Yeltsin and his acting prime minister, Yegor Gaidar, whom he put in charge of the Russian economy, choose the policies that they did?

The reformers were determined to transform Russia from what had been the largest republic in the centrally planned communist Soviet Union into an independent, democratic, market-based country. But, given Russia's near-impoverished condition, Gaidar realized he had to act quickly to prevent not only a complete economic collapse but also what his fellow reformer, Anatoly Chubais, feared might be an effort to reconstitute a communist, centrally planned system. It was imperative, they concluded, to resuscitate the economy so that it could pro-

vide jobs and supply the needs of the public. At the same time they wanted to create a sense of ownership and personal wealth among Russians generally so that these newly empowered stakeholders could resist any effort to reintroduce state control and ownership of factories and stores.

To do all this, Gaidar and his staff opted for a set of reforms often referred to as shock therapy. This involved embracing market mechanisms and ending price, labor, and production controls and privatizing ownership of state enterprises.

It was not a painless process. But shock therapy had proved reasonably successful in a few other countries, such as Bolivia and Poland. While higher prices meant that there would be more consumers who could no longer afford to purchase various goods, they would also mean higher profits and thus more output. Fewer consumers and more producers would soon bring an end to queues and bring about what economists call "equilibrium," which would obviate the need for rationing.

As for privatization, it was generally accepted among Western economists as well as Russia's new reformers that privately owned businesses would be run more efficiently and with higher productivity than state-owned enterprises. Private entrepreneurs were more likely to make decisions based on market imperatives, instead of the political and patronage considerations that had almost always characterized state-owned enterprises. Arranging for Russia's factory directors and their staffs to acquire up to 50 percent of the stock of the newly privatized entities, the reformers reasoned that these new "owners" were less likely to strip the enterprise of its assets or allow others to steal from "their" company.

To inculcate the same possessive attitude among the general public, the Yeltsin government arranged to issue vouchers to every citizen. These vouchers could be converted into shares of stock of one of the newly privatized companies. Approximately 49 percent of a company's stock would be set aside for outsiders. Each citizen would also become a stakeholder, a form of people's capitalism.

At first glance, this seemed a reasonable way to dismantle the communist system. However, the plan had several serious flaws. For example, the reformers assumed that once industrial ministries and the planning organizations were eliminated and state enterprises were privatized, the formerly government-controlled companies would quickly begin to think and operate like private businesses in a competitive environment. Unfortunately, nothing of the sort occurred. It was simply too much to expect that a market system would suddenly materialize out of nothing. After all, for 70 years the state and the Communist Party had done everything they could to stamp out any remnant of market behavior, including discouraging the start-up of new businesses and services and eliminating all market-oriented business codes.

Moreover, without a competitive infrastructure, the state enterprises, which for the most part were monopolies, continued to operate as monopolies—only now they were private, not state, monopolies. Thus their behavior differed little from what it had been before privatization. As with most monopolies, whenever they could, they cut back production and raised prices. The combined effect of relatively few start-ups, the dominance of state monopolies, and the sudden downsizing of the military-industrial complex as a consequence of the end of the cold war devastated the economy. GDP fell between 40 and 50 percent from 1990 to 1998 while prices soared; it took almost 1,170,000 rubles in 1998 to buy what 100 rubles did in 1990.

THE NEW OWNERS EMERGE

In the near anarchy created by the reforms, it was inevitable that a certain amount of institutional chaos would follow, especially in sorting out who would end up in control of the state-owned enterprises. The new owners can be divided into three categories: the former factory directors, the nomenklatura, and the non-nomenklatura, or upstarts.

The former factory directors formed the largest group of new owners. It was hoped that they would share control with the rest of the workers, and in some cases they did. But because the workers were usually poorly organized and because they had been told since birth that stock certificates were a fraud, they had no understanding of what stock ownership might mean. Most workers were only too happy to swap their stock for a bottle of vodka. When a vodka-for-stock swap did not work, some especially determined factory directors resorted to threats and violence to obtain the stock they wanted.

The newly constituted factory owners with their newly acquired net wealth quickly became known as the "new Russians" because of their nouveau riche ways and exhibitionism. One or two became millionaires and almost none fell below the poverty line—unlike one-third of the rest of the population. But it was the other two categories of enterprise owners that are the most interesting. Both have come to be known as oligarchs.

Initially it was generally thought that almost all the very rich oligarchs had emerged from what had been the nomenklatura. This was the official Soviet-era list of names (nomenclature) of all the senior-level people in the country. This included top echelons from the Communist Party, the KGB, the army, the ministries, and the arts. They were the big brothers. Nearly all the members of this group did not have much in the way of personal wealth but they had no need for it since the country was essentially theirs to use as they pleased. They had access to special stores as well as special restaurants, hospitals, and housing denied to the public. Because in the initial phases of privatization many of the early and blatant oligarchs had been members of the nomenklatura, it was assumed that most of the oligarchs possessed a similar background. After all, they were the officials who had been in control of the country's assets, and it seemed like a simple matter for them to take over and privatize the industries they had controlled.

Three of the most prominent nomenklatura oligarchs are Vagit Alekperov, Viktor Chernomyrdin, and Rem Vyakirev. Alekperov had been minister of the petroleum ministry, so it was relatively simple for him to set aside three producing fields—Langepaz, Urengoi, and Kogalym—and combine them into Lukoil, making it the largest of the privatized petroleum companies. Even more brazen, Chernomyrdin and Vyakirev, the minister and deputy minister, respectively, of the gas industry, transformed not selected parts, but the entire ministry into a private company they called Gazprom. While the state continued to hold on to 38 percent of the stock, Vyakirev, after Chernomyrdin became prime minister, ended up controlling the largest number of Gazprom's shares and ran the company as if it were his personal fiefdom.

A fourth member of the nomenklatura, Vladimir Potanin, was an official of the Foreign Trade Organization, which traded nonferrous metals in the Ministry of Foreign Trade. Potanin in effect privatized the FTO into a trading company and built it into a commercial bank and financial holding company. Eventually his holding company acquired control and ownership of Norilsk Nickel, Russia's primary producer of nickel and precious metals such as platinum and palladium. All four men, Alekperov, Chernomyrdin, Potanin and Vyakirev, made the *Forbes* list in 1997.

It is the third category of oligarchs, the non-nomenklatura, that is most intriguing. Unlike those who were members of the nomenklatura, these upstarts were anything but part of the Communist establishment. Many were regarded as living on the margins of society, at best on the fringe of it and in some cases beyond. Members of minority groups for the most part, they were considered untrustworthy and therefore excluded from any responsible official positions in the Soviet era, including the Communist Party, the military, the KGB, and the ministerial bureaucracies.

If Russia is to attract investment from both inside as well as outside the country, some effort has to be made to curb the arbitrary use of state power and the intimidation of enemies and opponents.

Excluded from positions within the power elite, many of those with extra energy or initiative chose instead to go into private business. This meant dealing in the black market for jeans, exchanging rubles, or engaging in construction work and other services on a private contractual basis. There are many examples. Alexander Smolensky was caught selling Bibles that he produced after-hours on a state printing press. Vladimir Gusinsky used his private car as a cab and Mikhail Fridman washed windows. All of them provided construction services. They became adroit at finding and providing goods and services in short supply. Since the economic planning system seldom worked effectively, the opportunities were considerable.

Unfortunately, the state treated all private business activities as economic crimes. Since economic crimes were often punished by imprisonment (reportedly Smolensky was arrested for his activities) and even death, there was a bit of a downside. Offsetting that, the possibility of punishment discouraged competition, which in turn made it all the more lucrative for those daring and determined enough to take this risk.

To those dealers and speculators who persevered, the decision by Soviet leader Mikhail Gorbachev in 1987 to authorize the establishment of cooperative and private businesses meant that what they had been doing had suddenly become legitimate overnight. Not only could they go about their businesses openly without fear of arrest, they actually had an advantage over members of the nomenklatura. In contrast to the upstarts, the nomenklatura had heretofore played by the rules and obtained and supplied goods according to state plans through state agencies. They had virtually no experience in securing goods in short supply or dealing in markets.

This was a serious handicap. In January 1992, with the end of central planning and the shift to the market, the supply system all but disintegrated. Store shelves were empty, and shortages abundant. This was ideal for the upstarts. They knew how to obtain goods in short supply; those in the nomenklatura did not. Suddenly the men who had been social outcasts and on the fringe of economic life were now at its very center.

With their ability to scout out scarce supplies, the former upstarts were able to respond to the hunger that Russian consumers had for goods long denied to them, especially foreign products. In no time they became extremely wealthy. It helped that there was little or no competition from the former state sector and little or no regulation by the now disorganized and disoriented state.

After they began to amass all this new wealth, most of it in cash, these businessmen began to establish their own banks. This too became a possibility after 1987 and for a time all it took was as little as $75,000 in capital. These banks also proved to be an important tool in providing the financing needed for the vouchers and shares of stock that many of the would-be oligarchs began to accumulate. Before long, Boris Berezovsky was able to boast that seven of the oligarchs, six of them "upstarts," had come to control 50 percent of the economy. That was an exaggeration, but if Alekperov and Vyakirev were included, it would not be too far from the truth.

THE BACKLASH

The accumulation of so much wealth by so few at a time of hyperinflation and economic depression inevitably gave rise to political resentment and social dysfunction. Criminal gangs began to appear throughout Russia. These mafia groups, as the Russians called them, were said to control as much as 70 percent of the Russian economy. Equally troublesome, the breakdown in discipline emboldened government officials seeking to exploit their positions. Many bureaucrats justified the breakdown in morality and the growing ubiquity of bribes with the rationalization that their government salaries had fallen far behind the rate of hyperinflation. Given the political and economic tumult, most anyone with money sent a substantial portion of it outside the country. The result was massive capital flight, estimated to exceed $1 billion per month. This more than exceeded the funds provided by international financial organizations in an effort to help finance Russia's economic and investment recovery.

Various forms of capital flight occurred. The most common technique was for an oil or raw materials company to keep in the West the payments for its exports. In other cases, enterprise managers would strip assets from their companies (usually where the state remained a part owner of the company), then transfer the proceeds to another company located offshore and wholly owned by the enterprise manager or his relatives. ITERA, for a time Russia's second-largest producer of natural gas, is a prime example. Located in Jacksonville, Florida, ITERA has received enormous gifts from Gazprom, including control of natural gas fields and access to natural gas supplies. Worth billions of dollars, these assets were often provided free of charge; if there was a payment, it was usually far below the market value.

As members of the public became aware of the magnitude of the theft and the misuse of what once belonged to the state, they began to blame the government and Boris Yeltsin in particular. Harnessing that anger, in December 1995, the Communist Party scored a major victory in the parliamentary elections for the Duma, the lower house of parliament. A month later, in January 1996, Yeltsin's approval rating in opinion polls had dropped to less than 10 percent. This was far below that of his chief opponent in the June 1996 presidential race, Gennady Zyuganov, the head of the Communist Party. Chubais, one of the original architects of the economic reforms, was right: it did look as if the Communist Party might return to power. Not because of the failure to privatize, but because the privatization process was seen to have been implemented unfairly.

The Oligarchs

Roman Abramovich
Major stockholder in Sibneft Oil as well as Millhouse Capital and Russian Aluminum (RUSAL, which controls 70 percent of Russian aluminum production). Elected to the Duma in December 1999 and governor of the Chukotka region in 2000.

Vagit Alekperov
President of Lukoil.

Petr Aven
President of Alfa Bank (see Mikhail Fridman).

Boris Berezovsky
Held effective operating control of ORT (Russia's largest television network), Sibneft Oil, Logovaz, a bank, a large auto dealership, and the newspaper *Nezavisimaia Gazeta.* Dismissed by Yeltsin from leadership of the Commonwealth of Independent States. Elected to Duma but resigned and went into exile. Wanted for questioning. Sold interest in ORT.

Vladimir Bogdanov
President of Surgutneftgas.

Viktor Chernomyrdin
Minister of gas industry and chairman of Gazprom. Former prime minister.

Anatoly Chubais
President of United Energy System.

Oleg Deripaska
General Director of RUSAL and Base Element. Major holdings in Sibneft and RusProm Auto, which in turn controls the GAZ automobile works as well as bus and truck and diesel engine manufacturing plants.

Mikhail Fridman
Chairman, Alfa Group (controls Alfa Bank and Tyumen Oil).

Vladimir Gusinsky
Founder, MOST-Bank and Media-MOST (controls NTV, second-largest TV network) and *Segodnya* newspaper. MOST-Bank collapsed; switched interest to Media-MOST. Arrested and jailed for three days. Fled to Spain. Borrowed from Gazprom, which then took over ownership of Media-MOST.

Mikhail Khodorkovsky
Founder, Menatep Bank and Rosprom (controls Yukos Oil). Being sued for sleight-of-hand tricks at Yukos Oil and subsidiaries. Now says he wants to be honest. State raided offices in 2003.

Vitaly Malkin
President, Russian Credit Bank. Russian Credit Bank's assets shifted to IMPEX Bank.

Vladimir Potanin
President, Oneximbank and Interros Trading Company. Oneximbank assets shifted to Rosbank. Owner of Norelsk Nickel.

Alexander Smolensky
Chairman of SBS AGRO Bank. Assets shifted to Soyuz and First Mutual Trust Society with subsidy from central bank. In hiding in Vienna.

Rem Vyakhirev
President of Gazprom. Stripped assets and was fired, May 2001.

M. I. G.

Concerned that a Zyuganov victory might mean the loss not only of their acquisitions but of their lives, seven of the oligarchs agreed to put aside whatever feuds they may have had with each other and unite to work for Yeltsin's victory. (According to one report, Berezovsky had earlier ordered a contract killing of Gusinsky.) Since these seven oligarchs controlled the major television networks and newspapers, they thought they could engineer Yeltsin's election, but it was not easy. Yeltsin was often drunk and had serious health problems; he had suffered several heart attacks. To erase these images, the oligarchs resorted to selective editing. Whenever Yeltsin appeared in the media, he was shown in a favorable light. In contrast Zyuganov was always depicted negatively. One of the more notable scenes from television at the time was a sequence of Yeltsin doing the twist with two young women in miniskirts. Shortly thereafter, Yeltsin suffered a heart seizure that somehow was omitted from any current news reports.

Ultimately, the oligarchs managed to shift public opinion and Yeltsin won. In reward for their help, Yeltsin agreed to repay his backers. Gusinsky was permitted to acquire more television outlets, allowing him to create a nationwide network. Potanin and Berezovsky were appointed to senior government positions, and Yeltsin agreed to expand a "loans for shares" program that Potanin had proposed in 1995. This proved to be the biggest asset giveaway of all.

Ostensibly, loans for shares was a generous offer by several banks to assist the government with its ever-growing budget deficit. To help the government pay its bills, several of the oligarchs agreed that their banks would lend the government the money it needed. The banks asked the government to use as collateral for the loans government-owned stock it still held in several companies. The government refrained from completely selling off several of the energy and raw materials businesses (which eventually turned out to be the most valuable portions of its portfolio). According to the plan, once the government repaid its loans, the collateral would be returned. In the event the government found itself unable to repay its debt, the banks would organize a series of auctions, the stock would be sold at a good price, and the government would collect the proceeds.

Of course, everyone understood that the government would not be able to repay the loans: revenues would remain inadequate because tax evasion was widespread. What was not anticipated, however, was that the bankers would rig the bids so that in almost every instance, the bank conducting the auction would end up as the only bidder (disguised but in fact a "straw" for the bank). As a result, the price was rarely much above the minimum price that had been determined in advance.

No one was mugged in the holdup, but the state found its pockets picked of hundreds of millions, even billions, of dollars in the process. As an example, Mikhail Khodorkovsky paid not much more than $300 million for Yukos Oil, which by 2003 was worth between $20 billion and $25 billion. Similarly, Berezovsky paid only slightly more than $100 million for Sibneft Oil, which in 2003, when it agreed to merge with Yukos, was valued at over $10 billion.

RENDER UNTO CAESAR, OR ELSE

As good as the early years of the Yeltsin presidency had been for the oligarchs, the meltdown of the economy on August 17, 1998, brought about the weakening, if not downfall, of most of them, especially those who had neglected to include energy and raw materials companies in their portfolios. Almost all their banks except for one closed down. After many years, there was some restitution, but the public was outraged when it was discovered that several of the banks, including Khodorkovsky's, had managed to funnel most of their remaining good assets into newly created entities. As a result, these funds were put beyond the reach of the banks' depositors and those who had lent money to or invested in these banks.

The economic climate began to improve in 1999 and Russia began to show the first signs of a prolonged economic revival. Some attributed this growth to Vladimir Putin, who had been appointed prime minister in August that year. A few of the oligarchs, especially Berezovsky, were among those who had urged Yeltsin to appoint Putin as prime minister. Initially they welcomed what they saw as Putin's firm hold over the economy.

Putin's stable presence, both as prime minister and, after the March 2000 elections, as president, stood in sharp contrast to Yeltsin's unstable ways. But the turnaround in the economy had little to do with anything Putin did. Putin was appointed as prime minister in August 1999; the turnaround in industrial production began several months earlier, in March 1999.

Two changes made the difference. The first was a threefold jump in world oil prices from about $11 a barrel to almost $30 a barrel. The second was the extreme deval-

uation of the ruble that followed in the wake of the August 17, 1998, financial collapse. On August 1, only slightly more than 6 rubles would buy one dollar; a few months later, more than 24 rubles were required. Since Russia was the world's second-largest exporter of oil, the increase in prices provided a windfall not only for the oil oligarchs, but for the government, which collected an export tax on oil. Of course, the price increase was a consequence of world economic forces and not anything that Putin or Russia had done. The cheaper ruble was also a windfall for Russian manufacturers, who moved into markets heretofore dominated by foreign imports. Almost immediately, imports shrank by 50 percent.

The combined effect of all this in 2000 was to produce a 10 percent increase in GDP. Since then economic growth has fallen to a range of 4 to 6 percent annually. That is still robust compared to the rest of the world, but is generally below the 7 percent rate that Putin needs if the Russian GDP is to double in 10 years—his proclaimed goal.

For Russia to grow that fast, it will take more than exhortation. Part of the problem is Russia's traditional overreliance on raw materials such as oil, gas, and ferrous and nonferrous metals. The dollar revenues that the oil exports, for example, generate are converted into rubles. This increased demand for rubles pushes the price of the ruble up relative to the dollar, which gradually erodes the advantage that Russian manufacturers enjoyed after the August 1998 devaluation. As a result, domestic manufacturers, once again, find themselves facing stiff foreign competition.

To generate the growth he wants, Putin must restructure the country's economy and institutions to make investing in Russia more attractive to foreign investors and Russians, including the oligarchs. To his credit, once installed as president, Putin did try to facilitate economic growth by initiating some needed structural changes.

Following the advice of his economic advisers, German Gref and Andrei Illarionov, Putin introduced legislation eliminating high marginal personal income tax rates and instead created a flat tax of 13 percent. Many economists in the West oppose a flat tax because it imposes a disproportionate burden on the poor. In the Russian context, however, it may not be such a bad idea because so few of the Russian rich actually paid the 35 percent rate they were supposed to pay under the old tax system. For Russia, a flat tax is better than no tax.

Russian chinovniks (bureaucrats) have survived the czar and the Soviets; they seem determined to outlive current reform attempts.

Putin also restructured business taxes, arranged for the private ownership of land, rationalized the pension system, and tried to shrink or otherwise curb the power of the bureaucracy. These were all steps in the right direction. It remains to be seen, however, if adopting the laws will lead to their implementation, especially where the laws call for a reduction in the power of the bureaucracy.

As for his treatment of the oligarchs, Putin initially tried to reassure the public as well as the oligarchs who survived the 1998 crash that he would attempt to maintain an evenhanded policy. Admittedly it was a bit awkward that he owed his appointment in part to Berezovsky, but he insisted that he would treat the oligarchs no differently from anyone else. As he put it, "It is asked what then should be the relationship with the so-called oligarchs? The same as with anybody else. The same as with the owner of a small bakery or a shoe repair shop…." This implied a tough policy aiming to ensure equal competitive opportunities for everyone, as well as a lack of distinctive privileges for specific businessmen. Although it sounds very nice, there have been no reports of Putin meeting with bakers or cobblers; he has, however, met regularly with oligarchs.

It is true that on occasion some of the oligarchs have found these meetings rather ominous. And not all the oligarchs have liked what they have heard from Putin. Moreover, despite his stated disavowal, Putin has played favorites. He has allowed some former KGB officers to expand their private business empires and build communications monopolies. Nor has Putin been too concerned about the type of person he deals with. For example, he allowed Roman Abramovitch and Oleg Deripaska to establish RUSAL, a company that now controls 70 percent of Russia's aluminum production. The United States refuses to allow Deripaska into the country because of reports that when he decides to "eliminate" competition, he really does.

Putin did eventually move against some of the oligarchs. His decisions to fire Rem Vyakirev and Viktor Gerashchenko were the least controversial. Vyakirev had been looting Gazprom and diverting billions of dollars worth of assets to entities such as ITERA, from which he personally profited. Gerashchenko, the chairman of the Russian Central Bank, had been accused of lying to the International Monetary Fund. He also seemed more interested in using the bank to dole out favors than in conducting a measured monetary policy.

The arrest of Gusinsky in June 2000 and the exile of Berezovsky, as well as this summer's hounding and interrogation of the billionaire executives running Yukos Oil, are more suspect. Whatever the official explanation, the real reason for their harassment was that they had strayed beyond the boundaries of their businesses into the realm of politics.

In 2000, after he took over as president, Putin signaled to the oligarchs that if they stayed out of politics, he would not go back and reexamine their credentials and how they came to gain control of what once was state property. When Gusinsky and Berezovsky allowed or encouraged the television networks they controlled to criticize Putin for his handling of the Chechen conflict and the sinking of the submarine *Kursk*, it was considered a lapse

into politics. Similarly, as Putin saw it, Khodorkovsky's open acknowledgement that he planned to spend as much as $100 million to support political parties (reportedly even the Communists) that were opposed to Putin constituted a blatant challenge that had to be addressed.

To the outsider, it is hard to know whom to support. Even though several of the oligarchs, including Gusinsky, Berezovsky, and Khodorkovsky, have tried in various ways to atone for whatever past sins they may have committed, none are self-made men, and even more important, all have violated numerous codes of ethics, not to mention laws. Almost all of their wealth derives from taking over ownership of what had been state resources. (Gusinsky to some extent is a partial exception because he created his own media network.) None of this wealth is the result of new inventions or the creation of new manufacturing processes.

FAULTY FOUNDATIONS

The episodic harassment of one oligarch or another by the present and probably future presidents of Russia hangs over the Russian economy and potential investors, both domestic and foreign. In a sense it also represents a rebuke to the architects of reform such as Yeltsin's finance minister, Anatoly Chubais, and his Western advisers: Jeffrey Sachs, Andre Shleifer, and Anders Aslund. They felt strongly that the most important step was to transform state enterprise into private enterprise as quickly as possible. Any shortcomings would be resolved with time.

But that has not been the case. Because of the flawed manner in which the privatization process was carried out, the new owners will always lack the legitimacy necessary for a stable political climate and sustained economic investment and growth. Having constructed a faulty foundation, the builders must live with the possibility that their edifice of privatization will periodically shift and crack and may even collapse. Even with a statute of limitations that limits judicial redress, the owners of privatized Russian businesses live with constant uncertainty, knowing that sooner or later they may be charged with having used illegal methods to gain control of their enterprises.

If Russia is to attract investment from both inside as well as outside the country, some effort has to be made to curb the arbitrary use of state power and the intimidation of enemies and opponents and move instead to adhere to the rule of law. Resort to state harassment by the police and tax authorities, experienced by Gusinsky, Berezovsky, and Khodorkovsky, is a throwback to the pervasive use of intimidation and fear that was the hallmark of the Soviet era.

It is hardly reassuring that the KGB agents associated with Putin seem to have come to terms with the collapse of communism and now want to share in the spoils of the market system. Several of those in Putin's circle appear to be using state power not only to seize assets not fully privatized but also to take over assets already claimed by private owners. This behavior, as well as Putin's difficulty in shrinking the bureaucracy, underscores how deeply embedded is the presumption among senior Russian officials that those in authority can act without restraint. It also suggests that at best change will come very slowly. Those surrounding Putin as well as Putin himself seem instinctively to embrace the traditional bureaucratic modes of behavior satirized by Gogol's short story "The Inspector General." Russian *chinovniks* (bureaucrats) have survived the czar and the Soviets; they seem determined to outlive current reform attempts as well.

MARSHALL I. GOLDMAN, a Current History *contributing editor, is associate director of the Davis Center for Russian and Eurasian Studies at Harvard University and author of* The Piratization of Russia: Russian Reform Goes Awry *(Routledge, 2003).*

Is Chile a Neoliberal Success?

Chile is often heralded as the global South's best case for free-trade economic policies, but the facts tell a different story.

BY JAMES M. CYPHER

Chile is commonly portrayed as the great exception to Latin America's long and difficult struggle to overcome economic backwardness and instability. In 1982, conservative economist Milton Friedman of the University of Chicago pronounced the market-driven policies of Gen. Augusto Pinochet's military dictatorship "an economic miracle." Friedman was hardly an impartial observer. He and other Chicago economists had trained many of the dictatorship's ultra-free-market economic advisors, a group of Chilean economists who became known as the "Chicago Boys." Other prominent U.S. economists, however, also tout Chile's "economic miracle." In 2000, Harvard economist Robert Barro asserted in *Business Week* that Chile's "outstanding performance derived from the free-market reforms instituted by … Pinochet." Even Nobel laureate Joseph Stiglitz, a strong critic of the Chicago School, described Chile in his 2002 book *Globalization and its Discontents* as an exception to the failure of unregulated free markets and free trade policies in developing nations.

Neoliberalism, a term first employed in Latin America, describes the experiment in unregulated capitalism that the Pinochet dictatorship embraced in the years following the 1973 coup that toppled the elected government of Socialist President Salvador Allende. Chile has seen three elected governments since Pinochet's fall in 1990. None, however, including the present Socialist-led government, has broken sharply with the neoliberal economic model instituted by the dictatorship. For years, these post-Pinochet *Concertación* governments (a coalition of the Christian Democratic and Socialist parties) were content to administer the economic boom that had begun in the latter years of the dictatorship.

But the boom turned to stagnation in 1997: average per capita income rose only 0.7% per year between 1998 and 2002, while unemployment stayed above 9% through 2003. Export growth, widely viewed as the engine of the Chilean "miracle," stagnated, with total exports barely rising from $17 billion in 1997 to $17.4 billion in 2002.

While both the *Concertación* economists and those of the far right sought to blame Chile's woes on outside factors—the Asian crisis of 1997, the Argentine implosion of 2000, the U.S. slump of 2001, and so on—a few dissident economists had predicted all along that the boom would inevitably reach an impasse. One, economist Graciela Moguillansky of the U.N. Economic Commission for Latin America and the Caribbean, argued that the large Chilean finance/resource-processing conglomerates which dominate the economy had exhausted the easy resource-processing opportunities handed to them by the government through programs created decades ago. The "Chilean miracle" had reached its own self-imposed limits. Along similar lines, Orlando Caputo, director of the Santiago-based Centro de Estudios sobre Transnacionalización, Economia y Sociedad, argued that the underlying cause of the crisis was Chile's breakneck overproduction of copper. The price of copper fell so sharply between 1995 and 2002 that dollars received for copper exports actually declined over this period even while total output increased by 85%.

Rather than grapple with the need to realign economic policy and adopt a new model of development, however, the leaders of the *Concertatión* have decided to intensify the free-trade, export-oriented model—for example, by signing as many free-trade agreements as possible. In 2003 Chile was able to sign agreements with the two largest trading areas of the world, the European Union (EU) and the United States. These agreements would allow Chile's exports to surge, *Concertación* politicians and economists argued, pulling the rest of the economy along. The Chilean ruling elite is once again basking in self-satisfaction. But can they expect another lengthy period of export-fueled growth?

The chief target of the Chicago Boys (and other right-wing economists), the military dictatorship, and the business class was not state intervention in economic life, but the organized power of Chilean working class.

A review of recent Chilean economic history suggests that it's not likely. Despite the claims of free marketeers, Chile's economic performance has been mixed, and its successes owe more to state intervention than to the invisible hand of the free market. In fact, it would be hard to find any major sector of the economy that did not owe much of its existence to state intervention—intervention which continued in a variety of forms under the nominally neoliberal Pinochet dictatorship.

THE REAL TARGET: LABOR'S POWER

While the Chicago School is known for its devotion to free-market policies and its hostility to government regulation, the chief target of the Chicago Boys (and other right-wing economists), along with the military dictatorship and the business class, was not state intervention in economic life, but rather the organized power of the Chilean working class.

Through protracted struggles over many decades, Chilean unions had grown, spreading from the mining sector into manufacturing and eventually into agriculture. The 1973 coup destroyed the power of the working class in the political and economic systems. Its union and party leaders were tortured, assassinated, imprisoned, or exiled. Political parties were banned and unions made virtually illegal. The dictatorship introduced a "flexible" labor system that left workers with the formal right of individual contract, but stripped them of any right to organize and bargain collectively. Power shifted to the employers in the mines, factories, fields, and ports. The dictatorship spared no form of state intervention to crush workers' power.

The core political and economic strength of the unions when Pinochet came to power was in the industrial sector. Given the dictatorship's anti-labor aims, its economic development policy thus had to be an anti-industry policy. This meant abandoning the longtime basis of Chilean development policy, import substitution industrialization (ISI). When the Great Depression of the 1930s hit Chile harder than any other nation in the world, advocates of ISI argued that Chile had to stop exporting natural resources (e.g., nitrates, copper, and other minerals), at falling prices, and importing machinery and consumer products, at rising prices. ISI advocates claimed that nations such as Chile would be better off and more stable if they developed domestic industries and an internal market. To sustain industrialization, infant industries needed to develop behind a tariff wall so firms could learn, adapt, and grow. A developed internal market requires mass-based consumption (rather than just elite consumption of imported luxuries),

trained workers, and the diffusion of production knowledge. Unions thrived in this environment: high wages provided the mass consumption base, while unions helped maintain morale and skill levels which facilitated higher productivity.

The economists associated with the military dictatorship realized that an economic development strategy based on industrialization could not exclude a real voice for labor. One that drew upon Chile's vast treasure of untapped natural resources, however—the unceasing ocean, the endless forests of the south, the exceedingly fertile farm lands—could do just that. (Peter Winn's *Victims of the Chilean Miracle* brilliantly documents the pivotal anti-labor focus, particularly in the agricultural, forestry, and the fisheries sectors.) Thus, the creation of a "nontraditional" export sector became the keystone of the dictatorship's economic strategy.

The plan worked—for a time. Exports soared from 14.5% of GDP in 1974 to 31.4% in 1999. Copper fell to 40.5% of exports by 1995, while nontraditional exports such as salmon and resource-based "manufactured" products such as wood pulp, paper, cardboard, disposable diapers, and processed woods boomed. By 2002 all of the new resource-based exports combined—fresh produce and processed foods, forestry products, wine, and fishing—totaled $6.7 billion, less than total mining exports of $7.3 billion, but ahead of copper at $6.3 billion.

It was not the invisible hand of the market that caused resource-based exports to boom. Surprisingly, most of the strategies and many of the personnel involved in the new Chilean "miracle" were products of the old, and much derided, state "interventionism."

The export boom undeniably fueled the country's economic growth. After enduring a deep recession from 1982 to 1985 (unemployment reached 20% in 1982), Chile's economy more than bounced back. Between 1987 and 1998, per capita income grew by 88%.

MARKET OR STATE?

But it was not the invisible hand of the market that caused the new boom in resource-based exports. Most of the credit belongs to the state: most of the strategies—such as new product development, risk capital, technical training/advising, marketing, quality control—and many of the personnel involved in the new Chilean "miracle" were products of the old, and much derided, state interventionism of the ISI era.

How could the core changes generating the "miracle" come from the detested state sector? One part of the answer is that many military officers in the upper ranks of the dictatorship were "developmentalists"—believing that the economic growth of Chile was partly a by-product of an agile and creative state. (When the nationalized copper giant CODELCO fell into the military's control, it *remained* a state-owned corporation. The military reluctantly permitted the privatization of the electric grid and the telephone system, but not copper, the state-owned oil corporation, or several other key state entities.) Another answer is that nations cannot quickly change their economic structure. The Chilean economy was put on a particular development path in 1939 with the creation of the CORFO, the state agency mandated to carry out the ISI strategy and build the national production base. CORFO's many new public and mixed public/private firms accounted for the great bulk of Chile's industrial growth from 1940 to 1974: a 1993 study pointed out that of the 20 top private exporting companies, at least 13 had been *created* by CORFO.

For a while under the dictatorship, it seemed that CORFO's mission was nothing more than to sell off all the state-owned firms, then disappear. But the agency still exists, and after the 1982-1985 recession it became more active in the funding and development of new resource-sector firms.

In one key instance, CORFO was responsible for the funding and creation of the forestry sector—a strategy that it had advocated and supported for decades prior to the coup. The Chicago School economists have portrayed the boom in forestry products, now the largest export sector after mining, as a result of good policy and private initiative. The real story is that, while private capital lacked the initiative and foresight to develop the forestry sector, CORFO introduced forest management techniques, provided credits and subsidies, financed projects for technological development and labor training, and fostered the development of the allied paper, cardboard, and wood industries. CORFO also created an affiliate, the Forestry Institute, which launched a marketing and information campaign designed to promote forestry exports, while carrying on massive reforestation programs and introducing new tree varieties.

The same basic story holds for the fishing industry, as well as for most of the developments in fresh produce and processed food. Rather than the invisible hand moving through market forces, the visible (but largely ignored) hand of *Fundación Chile*—a public-private agency designed to develop firms in new areas where private capital would not invest, then sell them to the private sector—was responsible for most of this diversification. *Fundación Chile* began in 1976, with the assistance of a prominent economist, Raul Saez, who had headed CORFO for many years. Like many of the military officers in the highest ranks of the dictatorship, Saez was contemptuous of the pretensions and ignorance of the Chicago School neoliberals. With CORFO under attack from this quarter, Saez moved laterally and gathered a group of experts who achieved major changes in the productive apparatus of the Chilean economy. (Incubating institutions similar to *Fundación Chile* have played a key role in the recent history of economic development in Korea, Taiwan, and much of East Asia—but these nations have consciously avoided neoliberal free

trade policies. Instead of *accepting* the dictates of the market, they have sought to *govern* the market.)

Likewise, the dictatorship created ProChile in 1974 to assist the private sector in locating and selling to foreign markets. Today, many of the activities of ProChile are coordinated with support programs fostered by CORFO. Through its Export Promotion Fund, ProChile has co-financed export projects, providing up to 50% of the necessary capital—often using funds obtained from or through CORFO.

In yet another instance of state intervention, the government facilitated a boom in the private mining sector from the 1980s by allowing the mines to operate essentially tax-free. The tax rate was an all but invisible 0.8% of sales in 2002. From 1992 to 2002, eight of the top 10 private mining companies paid no taxes, in spite of the fact that Chile has the most profitable copper mining companies in the world. (Meanwhile, the poor pay a value-added tax of 19% on their consumption, including food and medicine.)

What all this adds up to should not be too surprising. In spite of a brutal military dictatorship that sought the total restructuring of the economy and the elimination of the state's guiding role in it, the state sector was a crucial ingredient in Chile's efforts to build an export-led economy in the Pinochet years and beyond. Thus, although neoliberals occasionally imposed their free-market ideas in the financial sector, the restructuring of the economy was led by a *stealth* government development policy. (Even in the financial sphere, the Chicago Boys were forced by real circumstances to retreat, imposing protective tariffs in the 1980s and accepting capital controls on imported short-term "hot" money.) While Chile is nearly always portrayed as a *neoliberal* success story, the reality is that Chile's transformation was *not* neoliberal at its core—that is, within the system of production.

END OF AN ERA?

Why has the Chilean "miracle" stagnated? CORFO, the *Fundación Chile*, and ProChile—the core triangle of state institutions responsible for the stealth development policy—are no longer receiving the funding to create new export sectors. In theory, the large forestry companies and others involved in resource processing *could* expand and upgrade their exports, but Moguillansky's work has demonstrated that these corporations are unwilling to take risks and plow their profits into new economic activities. They have no intention of making the long-term investments in machinery and equipment, personnel and technology, and marketing that would be necessary to develop, for example, a strong, dynamic furniture-making sector. The same criticism was raised by the vice-president of Chile's Institute of Mining Engineers and by the Association of Metallurgy Industries, who argued that Chile needed an industrial policy (i.e., state intervention directing investment to strategic new sectors) to develop copper processing and the manufacture of copper-intensive products. Currently, only 1%

of the copper mined in Chile is processed or turned into manufactured products in the country.

As Moguillansky stresses, Chile's financial/industrial groups are not interested in technological modernization. In a study of 15 similar nations, the United Nations ranked Chile next to last in its index of technological capabilities, and 13th in terms of expenditures on research and development by private firms. Yet, in spite of all this, and because of the role of state intervention in the past, Chile does have a somewhat competitive manufacturing sector: metalworking exports in the manufacturing sector in 2002 were larger than processed food exports and nearly equal to fresh produce exports. Other manufactured exports include plastics, containers, and textiles. With large investments and a new government policy fostering technological research and development and massive labor training, Chile could develop high value-added manufacturing. High wages flowing from a strongly unionized manufacturing sector could, in turn, enlarge the internal market for industrial products.

Both forestry and mining have great potential in terms of expansion into clusters of high value-added industries. Both would demand the massive participation of a trained, skilled industrial work force. Support for these sectors is logical from the standpoint of development economics; however, it would create more favorable conditions for unionization and, because independent unions in Chile have always been political as well as economic organizations, this would help bring the working class back into the political arena. Such a development would threaten to revive fundamental struggles not only over the distribution of income, but also over the institutional organization of the economy—the worst fear of Chile's economic, political, military, and religious elites.

THE FREE TRADE OPTION?

For now, the elite is hoping that the new free-trade agreements will cure the country's economic stagnation. Exports to Europe are up, along with copper prices, and Chile now awaits a further opening to the U.S. market. The new trade agreements are expected to increase foreign direct investment, but no one has attempted to quantify it. Chile has been so wide open to foreign investment, and foreign investors have enjoyed the assurance of a pro-investor climate for so long, that it is difficult to imagine much of a surge of new investment from either Europe or the United States.

Thus, if the new free-trade agreements are to be the next short-term fix for Chile's unraveling economic model, the impact will have to come through more export opportunities. Chile has an edge in trade with the EU and the United States because its fresh produce is largely grown and harvested during the summer in the southern hemisphere, which coincides with the winter months in Europe and the United States. This edge is tenuous, though, since other southern nations can also grow and transport these crops.

Boom for Whom?

The Chilean economy unquestionably enjoyed a boom in the late 1980s and 1990s. Some Chileans enjoyed the boom more than others, however. Even as Chile's average per capita income nearly doubled between 1987 and 1998, workers' average wages increased by only 53%, and because of wage losses over the 22 years between 1970 and 1992, real wages in 1998 were only 29.5% higher than in 1970. Chile now has the third most unequal income distribution in Latin America (behind Brazil and Guatemala). According to official government data, the top 20% received 57.5% of the national income in 2000; the top 10% received 42.5%. Tax evasion is widespread, with an estimated 23% of total income going unreported, virtually all of it flowing to the top 20%. So in reality, income distribution is even more unequal than official figures acknowledge. The skewed distribution of income, more unequal than it was in the 1960s, is a deliberate result of government policy. One element of the policy is to keep income taxes low—the average income tax rate on the top 10% is a mere 2.5%. Another is to keep wages low in the export sectors to keep them internationally competitive. But this also means that wages must be low throughout the economy.

Redistribution policies instituted by the Concertación have brought the official poverty rate from 45% in 1987 to 21% in 2000. Certainly this is a laudable accomplishment, but it fails to address the inequality and the near-poverty status of workers that result from the neoliberal strategy. Those in power may simply have no interest in addressing these problems on a systemic level. For the top 20% (and this includes the political class—right, left, and center), Chile is a great country full of expensive imported SUVs, cheap servants, spiffy private schools, marvelous skiing, and exquisite weekend beach houses. No matter that monthly tuition in one of the private schools exceeds the entire monthly wage of the average worker, or that one day of skiing would cost that worker three to four days' income. Unlike in most of Latin America, the poor are kept out of sight of the comfortable, thanks to homogeneous neighborhoods on the U.S. model.

Chile may be a few years ahead of its competitors, but it has nothing unique to offer. Since 1996, reforestation has stagnated and Chile's two largest forest-product companies have shifted investment to Argentina and Brazil. Another successful niche has been aquaculture, particularly farmed salmon. Salmon production and prices are up for the moment and Chile has a huge coastline well adapted to fishing. But other countries can become competitors in this sector as well. In short, although Chile has developed some important niche markets like high-end produce, fish, wine, and wood products, there is no reason to believe that these nontraditional exports will long be immune to global competition. The result, as always: overproduction and falling prices, a problem that has cursed Latin America for centuries.

There is, of course, another side to free-trade agreements. The United States did not sign the 2003 agreement with Chile out of goodwill. While Chile's pundits foresee huge export growth, they are virtually silent about surging U.S. imports. Which parts of the production system will be knocked out by competition from U.S. firms? The Chilean government has conducted some relevant studies, which predict that the agreement's overall results will be positive for Chile, but small. Between 2004 and 2010, when the stimulus of the trade agreement is expected to end, Chile's GDP growth might rise between one-half and one percentage point per year. Since 1997, Chile's growth has slowed to 2.3% per year. Thus, in the most optimistic scenario, the U.S. trade agreement could lift the growth rate for a six year period to 3.3 %. The EU free-trade agreement may have a similar impact, since the EU market is of similar size. If so, Chile could project, at best, a 4.3% growth rate. Chilean government economists claim U.S. imports will essentially *complement* Chile's expansion, providing machinery and equipment that will permit more exports. But competition from U.S. wheat, potato, corn, sugar beet, and dairy producers will probably destroy much of the farm sector.

Still, all this is just toying with economic models. Back in the real world, if Chile manages to keep an edge in its non-mineral exports, it will only be by keeping wages down. If growth does pick up modestly for a few years, the benefits of that growth will not flow to the mass of Chileans.

Unfortunately for them, the two new free-trade agreements will probably prolong the life of the export-led model a bit longer. Rather than facing the inherent limitations and injustices of this development model, the *Concertación* government has sidestepped the real question: After export-led policies fail, what next?

In the absence of a critical dialogue, it is hard to imagine Chile turning away from its free-market, free-trade orientation. Even a devastating economic crisis might not spur change if no critical vision can be put forward. Unfortunately, the legacy of the dictatorship still lingers over Chilean public opinion and political discourse: The economics de-partments (with one or two small exceptions) all speak with a single, free-trade voice, the independent research centers are silent, and the government and the press laud the idea of more and greater export possibilities. In this climate, as a Chilean colleague said: "*You* can be critical, but if *we* say these things we will be committing economic suicide—our careers will be destroyed." Less than 15 years removed from the end of the military dictatorship, in Chile dissent is still largely an unpracticed art.

SOURCES

Rafael Agacino, "Reestructuración Productiva, flexiblidad y empleo" (mimeo 2003); Carlos Alvarez, "La CORFO y la transformación de la industria manufacturera" en *La transformación de la producción en Chile,* CEPAL (Santiago: Estudios y informes de la CEPAL no. 84, 1993); Orlando Caputo and Juan Radrigán, "Economia Chilena: agotamiento relativo del modelo (octubre 2003) www.cetes.cl; Graciela Moguillansky, *La inversión en Chile: ¿el fin de un ciclo en expansion?* (Mexico: Fondo de Cultura, 1999); Peter Winn (ed.), *Victims of the Chilean Miracle: Workers and Neoliberalism in the Pinochet Era, 1973–2002* (Durham, N.C.: Duke University Press, 2004).

James Cypher teaches economics at California State University, Fresno. In 2003 he was visiting research professor at the Facultad Latinoamericana de Ciencias Sociales (FLACSO-Chile). He is a *Dollars & Sense* associate and co-author of *The Process of Economic Development,* 2nd edition (Routledge, 2004).

From *Dollars & Sense,* September/October 2004, pp. 30-35. Reprinted by permission of Dollars & Sense, a progressive economics magazine. www.dollarsandsense.org

The Fall of the House of Saud

Americans have long considered Saudi Arabia the one constant in the Arab Middle East—a source of cheap oil, political stability, and lucrative business relationships. But the country is run by an increasingly dysfunctional royal family that has been funding militant Islamic movements abroad in an attempt to protect itself from them at home. A former CIA operative argues, in an article drawn from his new book, Sleeping With the Devil, *that today's Saudi Arabia can't last much longer—and the social and economic fallout of its demise could be calamitous.*

Robert Baer

In the decades after World War II the United States and the rest of the industrialized world developed a deep and irrevocable dependence on oil from Saudi Arabia, the world's largest and most important producer. But by the mid-1980s—with the Iran-Iraq war raging, and the OPEC oil embargo a recent and traumatic memory—the supply, which had until that embargo been taken for granted, suddenly seemed at risk. Disaster planners in and out of government began to ask uncomfortable questions. What points of the Saudi oil infrastructure were most vulnerable to terrorist attack? And by what means? What sorts of disruption to the flow of oil, short-term and long-term, could be expected? These were critical concerns. Underlying them all was the fear that a major attack on the Saudi system could cause the global economy to collapse.

The Saudi system seemed—and still seems—frighteningly vulnerable to attack. Although Saudi Arabia has more than eighty active oil and natural-gas fields, and more than a thousand working wells, half its proven oil reserves are contained in only eight fields—including Ghawar, the world's largest onshore oil field, and Safaniya, the world's largest offshore oil field. Various confidential scenarios have suggested that if terrorists were simultaneously to hit only a few sensitive points "downstream" in the oil system from these eight fields—points that control more than 10,000 miles of pipe, both onshore and offshore, in which oil moves from wells to refineries and from refineries to ports, within the kingdom and without—they could effectively put the Saudis out of the oil business for about two years. And it just would not be that hard to do.

The most vulnerable point and the most spectacular target in the Saudi oil system is the Abqaiq complex—the world's largest oil-processing facility, which sits about twenty-four miles inland from the northern end of the Gulf of Bahrain. All petroleum originating in the south is pumped to Abqaiq for pro-

cessing. For the first two months after a moderate to severe attack on Abqaiq, production there would slow from an average of 6.8 million barrels a day to one million barrels, a loss equivalent to one third of America's daily consumption of crude oil. For seven months following the attack, daily production would remain as much as four million barrels below normal—a reduction roughly equal to what *all* of the OPEC partners were able to effect during their 1973 embargo.

Oil is pumped from Abqaiq to loading terminals at Ras Tanura and Ju'aymah, both on Saudi Arabia's east coast. Ras Tanura moves only slightly more oil than Ju'aymah does (4.5 million barrels per day as opposed to 4.3 million barrels), but it offers a greater variety of targets and more avenues of attack. Nearly all of Ras Tanura's export oil is handled by an offshore facility known as The Sea Island, and the facility's Platform No. 4 handles half of that. A commando attack on Platform 4 by surface boat or even by a Kilo-class submarine—available in the global arms bazaar—would be devastating. Such an attack would also be easy, as was made abundantly clear in 2000 by the attack on the USS *Cole,* carded out with lethal effectiveness by suicide bombers piloting nothing more than a Zodiac loaded with plastic explosives.

Another point of vulnerability is Pump Station No. 1, the station closest to Abqaiq, which sends oil uphill, into the Aramah Mountains, so that it can begin its long journey across the peninsula to the Red Sea port of Yanbu. If Pump No. 1 were taken out, the 900,000 barrels of Arabian light and superlight crude that are pumped daily to Yanbu would suddenly stop arriving, and Yanbu would be out of business.

Even the short pipe run from Abqaiq to the Gulf terminals at Ju'aymah and Ras Tanura is not without opportunity. If heavy damage were inflicted on the Qatif Junction manifold complex, which directs the flow of oil for all of eastern Saudi Arabia, the flow would be stopped for months. The pipes that connect the terminals and processing facilities can be replaced off the shelf, but those at Qatif require custom fabrication.

Promoters of Alaskan, Mexican Gulf, Caspian, and Siberian oil all like to point out that the United States has been weaning itself from Saudi Arabian oil, for protection against the effects of just such an attack on the Saudi oil system. Saudi Arabia may sit on 25 percent of the world's known oil reserves, they argue, but it provides somewhere around 18 percent of the crude oil consumed by the United States—and that is down from 28 percent in only a decade. What these people fail to mention is that Saudi Arabia has the world's only important surplus production capacity—two million barrels a day. This keeps the world market liquid. Not only that, but because the Saudis more or less determine the price of oil globally by deciding how much oil to produce, even countries that don't buy Saudi oil would be vulnerable if the flow of that oil were disrupted.

The Saudis have repeatedly used their surplus production capacity to stabilize the international oil market. They used it to break the OPEC embargo (but not before they had enriched themselves by tens of billions of dollars), in 1974. They used it again during the protracted Iran-Iraq war, to keep oil flowing to the industrialized West. They used it during the Gulf War, in 1990-1991; with help from a couple of other Gulf states, they produced an extra five million barrels a day, making up for the loss of Iraqi and Kuwait oil.

And they used it again on September 12, 2001. Less than twenty-four hours after the attacks on the World Trade Center and the Pentagon, the Saudis decided to send nine million barrels of oil to the United States over the next two weeks. The result was that the United States experienced only a slight inflation spike in the wake of the most devastating terrorist attack in history. Had that same surplus capacity been taken out of play with twenty pounds of Semtex, all bets would have been off. The U.S. Strategic Petroleum Reserve can support the domestic market for only about seventy days. And if Saudi Arabia's contribution to the world's oil supply were cut off, crude petroleum could quite realistically rise from around $40 a barrel today to as much as $150 a barrel. It wouldn't take long for other economic and social calamities to follow.

Americans have long considered Saudi Arabia the one constant in the Arab Middle East. The Saudis banked our oil under their sand, and losing Saudi Arabia would be like losing the Federal Reserve. Even if the Saudi rulers one day did turn anti-American, the argument went, they would never stop pumping oil, because that would mean cutting their own throats. This, at any rate, is the way we looked at the matter before fifteen Saudis and four other terrorists launched their suicide attacks on September 11; before Osama bin Laden suddenly became for the Arab world the most popular Saudi in history; before *USA Today* reported last summer that nearly four out of five hits on a clandestine al Qaeda Web site came from inside Saudi Arabia; and before a recent report commissioned by the UN Security Council indicated that Saudi Arabia has transferred $500 million to al Qaeda over the past decade.

Five extended families in the Middle East own about 60 percent of the world's oil. The Saud family, which rules Saudi Arabia, controls more than a third of that amount. This is the fulcrum on which the global economy teeters, and the House of Saud knows what the West is only beginning to learn: that it presides over a kingdom dangerously at war with itself. In the air in Riyadh and Jidda is the conviction that oil money has corrupted the ruling family beyond redemption, even as the general population has grown and gotten poorer; that the country's leaders have failed to protect fellow Muslims in Palestine and elsewhere; and that the House of Saud has let Islam be humiliated—that, in short, the country needs a radical "purification."

We can try to wish this away all we want. But the reality is getting harder and harder to ignore. Per capita income in Saudi Arabia fell from $28,600 in 1981 to $6,800 in 2001. The country's birth rate has soared, becoming one of the highest in the world. Its police force is corrupt, and the rule of law is a sham. Saudi Arabia almost certainly leads the world in public beheadings, the venue for which is often a Riyadh plaza popularly known as Chop-Chop Square. Illegal arms routinely flow into and out of the country. Taking into account its murky "off-budget" defense spending, Saudi Arabia may spend more per capita on defense than any other country in the world (some estimates put the figure at 50 percent of its total revenues), and the House of Saud believes this is necessary for its personal protection. The regime is threatened by increasingly hostile neighbors—and by determined enemies within the country's borders. Popular preachers all over Saudi Arabia call openly for a *jihad* against the West—a designation that clearly includes the royal family itself—in terms as vitriolic as anything heard in Iran at the height of the Islamic revolution there. The kingdom's mosque schools have become a breeding ground for militant Islam. Recent attacks in Bali, Bosnia, Chechnya, Kenya, and the United States, not to mention those against U.S. military personnel within Saudi Arabia, all point back to these schools—and to the House of Saud itself, which, terrified at the prospect of a militant uprising against it, shovels protection money at the fundamentalists and tries to divert their attention abroad.

Recent examples of Saudi support for the fundamentalists abound. In 1997 a high-ranking member of the royal family coordinated a $100 million aid package for the Taliban. In Los Angeles two of the 9/11 hijackers met with a Saudi working for a company contracted to the Ministry of Defense. A raid on the Hamburg apartment of a suspected accomplice of the hijackers turned up the business card of a Saudi diplomat attached to the religious-affairs section of the embassy in Berlin. Most of the more than 650 al Qaeda prisoners being held at the Guantanamo Bay Naval Base in Cuba—"the worst of the worst," according to Secretary of Defense Donald Rumsfeld—are rumored to be Saudis.

I served for twenty-one years with the CIA's Directorate of Operations in the Middle East, and during all my years there I accepted on faith my government's easy assumption that the money the House of Saud was dumping into weaponry and national security meant that the family's armed forces and bodyguards could keep its members—and their oil—safe. "The royal family is like the fingers of a hand," my colleagues at the State Department liked to say. "Threaten it, and they become a fist." I no longer believe this. Saudi Arabia is more and more a breathtakingly irrational state. For a surprising number of Saudis, including some members of the royal family, taking the kingdom's oil off the world market—even for years, and at the risk of destroying their own economy—is an acceptable alternative to the status quo.

Saudi Arabia has existed as a formal nation only since 1932, when the tribal leader Abdul Aziz ibn Saud gained control of much of the Arabian Peninsula, named the territory after his clan, and proclaimed himself king. But the House of Saud had been powerful in the region ever since the eighteenth century, when the radical cleric Muhammad ibn Abdul Wahhab, the founder of the puritanical Wahhabi movement, wandered into Dar'iya, near present-day Riyadh, and made a bargain with its ruler, Muhammad ibn Saud. The Saud family would provide the generals, and the Wahhabis would provide the foot soldiers. Until recently it was a marriage made in heaven.

An attack on Saudi Arabia's largest oil-processing facility could create a reduction in the flow of oil roughly equal to what *all* of OPEC effected in its 1973 embargo.

If I had to pick a single moment when the House of Saud truly began to fall apart, it would be when Abdul Aziz ibn Saud's son Fahd, who has been king since 1982, suffered a near fatal stroke, in 1995. As soon as the royal family heard about Fahd's stroke, it went on high alert. From all over Riyadh came the *thump-thump* of helicopters and the sirens of convoys converging on the hospital where Fahd had been taken.

Among the first to arrive were Jawhara al-Ibrahim, Fahd's fourth and favorite wife, and their spoiled, megalomaniac twenty-nine-year-old son Abdul Aziz—or "Azouzi" ("Dearie"), as Fahd called him. Anyone who knew how Fahd's court ran knew the extent to which Fahd had come to depend on Jawhara, who helped him with everything from remembering his medicine to handling intricate problems of foreign policy. If a prince wanted a matter immediately brought to Fahd's attention, he called Jawhara. As for Abdul Aziz, he was the youngest of Fahd's children and the apple of his father's eye. Fahd indulged him in everything. Stories circulated widely about Abdul Aziz's riding a Harley-Davidson inside his father's palace, chasing servants and smashing furniture. Most of the royal family found the king's indulgence strange. Abdul Aziz was pimply, craven, a bit slow. Apparently, though, he was regarded as the king's good-luck charm. Fahd's favorite soothsayer had once told him that as long as Abdul Aziz was by his side, the king would have a long, fulfilling life. So Fahd did not complain when Abdul Aziz spent $4.6 billion on a sprawling palace and theme park outside Riyadh, because Abdul Aziz was "interested" in history. The property includes a scale model of old Mecca, with actors attending mosque and chanting prayers twenty-four hours a day, and also replicas of the Alhambra, Medina, and half a dozen other Islamic landmarks.

Next to arrive at the hospital, in a great show of solidarity, were Fahd's full brothers—Sultan, the Defense Minister; Nayef, the Interior Minister; and Salman, the governor of Riyadh province. To outsiders, they were a tight bunch. Their mother, from the Sudayri clan, had taught them from an early age to stick together or risk being elbowed out by the forty-odd other children of their father.

Other princes—the children and grandchildren of Ibn Saud's children—hurried to the hospital too, from all over the kingdom and the rest of the world. Private executive jets were lined up wing to wing at Riyadh's airport. These princes couldn't get anywhere near Fahd, but by being close at hand they could pick up more-reliable news and, just as important, demonstrate their fealty. Most of them lived off his largesse—royal stipends, which ran from $800 to $270,000 a month. The princes knew they were breaking the treasury—all told, their brethren numbered 10,000 to 12,000. Would Crown Prince Abdullah—Fahd's half brother, a seventy-one-year-old reformer who was next in line for the throne—cut back on their stipends, or even eliminate them if Fahd died? They had to stick around to find out.

A recent report commissioned by the UN Security Council indicated that Saudi Arabia has transferred $500 million to al Qaeda over the past decade.

At this point Fahd's brothers were calling doctors in the United States and Europe. They wanted to know not whether Fahd would ever recover his mental capacities, or what kind of life he would be able to live, but what it would take to keep his heart beating and his body warm. Money, of course, wasn't a problem. They told the doctors they were prepared to lease as many Boeing 747 cargo jets as needed to bring in mobile hospitals and medical teams. The doctors couldn't understand the reasoning behind the questions—but only because they didn't understand the politics of the kingdom. What the family knew and the doctors didn't was that Crown Prince Abdullah had long been eager to take power. The only way to keep him at bay was to keep Fahd alive—God willing, until Abdullah died.

Abdullah had always been the odd prince out. To begin with, his mother was from the Rashid tribe, traditional enemies of the Saud. Ibn Saud had married her to cement a truce with the Rashid, and although the Rashid were now loyal subjects, Abdullah was still mistrusted by Fahd's full brothers. Almost alone among the top members of the royal family, Abdullah had chosen the way of the desert, turning his back on the luxuries of Riyadh, Jidda, and Ta'if. He never vacationed lavishly in Europe, unlike King Fahd and his entourage, who typically spent $5 million a day during visits to the palace at Marbella, on the Spanish Riviera. Abdullah preferred to spend his time in a tent, drinking camel's milk and eating dates. He interspersed his conversation with Bedouin aphorisms and turns of phrase. All his children were raised according to the customs of the desert. It is Abdullah who has recently called publicly for democratic reforms, the reining in of the conservative clergy, and military disengagement from the United States.

The royal family hated being reminded that they had abandoned their Bedouin roots, but they hated still more that Abdullah was trying to cut back royal corruption and entitlements. Aping the senior members of the family, the lesser princes had fantastic financial expectations, and their stipends didn't suffice. The third-generation princes were getting only about $19,000 a month—a fraction of what they needed for the lifestyles they sought. To keep even a modest yacht on the French Riviera requires a million dollars a year. What were they supposed to do? In order to make ends meet they had been getting into nastier and nastier business, taking bribes from construction firms (mostly the bin Laden family's) seeking government contracts, getting involved in arms deals, expropriating property from commoners, and selling Saudi visas to guest workers. Another trick they'd discovered was borrowing money from private banks and simply refusing to pay it back. It wasn't as if the larger family could somehow discipline or shame them. There were so many princes that they didn't even all know one another.

Abdullah had made no secret of his intention to put an end to the thievery when he became king—and for a while it looked as if he might get his way even before becoming king. In the mid-1990s, as Saudi Arabia was facing increasingly dire financial difficulties, he persuaded King Fahd to appoint a handful of reformist ministers. Abdullah first had them zero in on expropriations. The practice had become so widespread among the lesser princes that it was completely alienating Saudi Arabia's traditional merchant class and fledgling middle class. A prince might walk into a restaurant, see that it was doing well, and write out a check to buy the place, usually well below market price. There was nothing the owner could do. He knew that if he resisted, he'd end up in jail on trumped-up charges.

The senior princes used their government positions to do the same thing, but on a much grander scale. One of them would pick out a valuable piece of property—maybe a particularly good location for a shopping mall or a new road—and then order a court to condemn it in the name of the state, which would clear the way for the king to award it to him. The money to be earned was staggering, and senior princes had started to rely on the practice to maintain their ever more bloated personal budgets. In Abdullah's view, however, crooked property deals and the like were only a small part of the problem. The off-budget deals were a much bigger part. In off-budget spending, revenue from oil sales goes directly to special accounts, bypassing the Saudi treasury altogether. The money is then used to pay for pet projects, from defense procurement to construction, with no government audits or accountability of any sort. Commissions and bribes are enormous.

As a reformer, Abdullah was kept out of the tight circle that gathered around Fahd after his stroke. Bitterness against Abdullah within the family was so deep that he was in fact blamed for the stroke. One version had it that Fahd and Abdullah had been on the telephone, arguing about who would attend a meeting of the Gulf Cooperation Council in Oman. It was a fundamentally unimportant decision, but relations between the two men had become so toxic, it was said, that Fahd's anger brought on the event. Another rumor in circulation held that Fahd and Abdullah had been arguing about what they always argued about—looming financial collapse. There were even whispers that Abdullah had intentionally provoked Fahd, knowing his health wouldn't withstand a shouting match.

It eventually became clear that Fahd would live, but the extent of his impairment also became clear—embarrassingly so when, during a therapy session not long after the stroke, Fahd defecated in his pool in front of his family. His mind was affected too. Those close to him knew that he would never truly rule again, though he is still led out for ceremonial appearances.

A year and a half after Fahd's stroke Sultan had come to so despise Abdullah that he stopped attending cabinet meetings chaired by him. For Abdullah, the feeling was mutual. In July of 1997 he simply bypassed the Council of Ministers, which was heavily stacked in favor of the Sudayri, and tried to get Fahd to sign off on decrees and laws he thought needed passing. Jawhara and Abdul Aziz teamed up to thwart him.

Mind you, it is not as if the rest of the Fahd clan is united. Sultan, Salman, and Nayef may have arrived at the hospital together in a show of solidarity, but they got a rude shock once they pushed through the front doors. Jawhara and Abdul Aziz blocked them from seeing their brother. The two had set up camp outside Fahd's hospital room and were deciding who and what would or would not get in. That included ministers, senior princes, and doctors, along with petitions, decrees, and everything else.

Saudi succession doesn't operate according to primogeniture. By tradition, senior princes come to a consensus on succession, usually choosing one from their ranks who is thought to have the necessary experience and wisdom. So far the system had served the royal family well, even though Abdullah had become a gadfly, but now Fahd's brothers were afraid that Abdul Aziz was trying to circumvent custom and place himself higher in the line of succession. For one thing, he had started getting more and more involved in national security, from foreign affairs to intelligence. Even the Americans noticed it. When the commander of U.S. forces in the Middle East, General J. H. Binford Peay, came to Riyadh to meet with Fahd, in July of 1997, he was surprised to find Abdul Aziz at Fahd's side, whispering in his father's ear. Where was Abdullah? What had become of Sultan? Peay had to meet with Abdullah separately, and even then Abdullah didn't talk about the issues at hand.

What really worried some members of his family was that Abdul Aziz was funding radical Wahhabi causes and was gaining strength and popularity as a result. They had little doubt that money was going to clerics and causes that were associated with Osama bin Laden. Abdul Aziz hadn't rediscovered his faith, of course: he was courting favor with the Wahhabis because he knew he would need their support to become king. In September of 1997 he helped to coordinate that $100 million aid package for the Taliban, even though the Taliban were protecting bin Laden—a man who not only had vowed to overthrow the House of Saud but also seemed increasingly capable of doing so. Abdul Aziz was buying support wherever he could find it. In December of 1993 Abdul Aziz authorized $100,000 for a Kansas City mosque. On September 15, 1995, he opened the King Fahd Academy, in Bonn, and two days later he dedicated a new mosque there. Nine days after that he invited the head of the Islamic Society of Spain, Mansur Abdul Salam, to Riyadh. In May of 1996 he and Jawhara arranged for King Fahd to release Muhammad al Fasi from prison. Al Fasi had been imprisoned for opposing the Gulf War and the presence of U.S. troops in Saudi Arabia; in other words, he shared some of bin Laden's chief grievances. In December of 1999 the press finally caught wind of Abdul Aziz's penchant for backing radical Islamic causes. One regional account made available by U.S. translation services noted that he was believed to have been funding an associate of bin Laden's, Sa'd al Burayk, who in turn was giving the money to Islamic groups dedicated to killing Russian soldiers and civilians in Chechnya. Nayef promised to put a stop to Abdul Aziz and bring his charity back under control—but he appears to have done nothing.

All the while, throughout the 1990s, the royal family kept growing and growing. A prince might sire forty to seventy children during a lifetime of healthy copulation; however, the resources to support the growing population of the entitled were shrinking, not just in relative terms but in absolute ones. Young royals were pushing up from below, chafing at leaders who were slipping into their late seventies and eighties. The incapacitated King Fahd will turn eighty this year; Crown Prince Abdullah will turn seventy-nine. Many of the most active court intriguers are also in their seventies.

The Saudi royals number 30,000 and may grow to 60,000 in a generation. What would oil have to cost to support even the most basic privileges they now enjoy?

The House of Saud currently has some 30,000 members. The number will be 60,000 in a generation, maybe much higher. According to reliable sources, anecdotal evidence, and the Saudi gossip machine, the royal family is obsessed with gambling, alcohol, prostitution, and parties. And the commissions and other outlays to fund their vices are constant. What would the price of oil have to be in 2025 to support even the most basic privileges—for example, free air travel anywhere in the world on Saudia, the Saudi national airline—that the Saudi royals have come to enjoy? Once the family numbers 60,000, or 100,000, will there even be a spare seat for a mere commoner who wants to fly out of Riyadh or Jidda? Reformers among the royal family talk about cutting back the perks, but that's a hard package to sell.

Saudi Arabia operates the world's most advanced welfare state, a kind of anti-Marxian non-workers paradise. Saudis get free health care and interest-free home and business loans. College education is free within the kingdom, and heavily subsidized for those who study abroad. In one of the world's driest spots water is almost free. Electricity, domestic air travel, gasoline, and telephone service are available at far below cost. Many of the kingdom's best and brightest—the most well-educated

and in theory, the best prepared for the work world—have little motivation to do any work at all.

About a quarter of Saudi Arabia's population, and more than a third of all residents aged fifteen to sixty-four, are foreign nationals, allowed into the kingdom to do the dirty work in the oil fields and to provide domestic help, but also to program the computers and manage the refineries. Seventy percent of all jobs in Saudi Arabia—and close to 90 percent of all private-sector jobs—are filled by foreigners.

Among men, at least, the Saudis have an admirably high literacy rate, especially for a place that only three generations back was inhabited mostly by nomadic tribesmen. About 85 percent of Saudi men aged fifteen and older can read and write, as opposed to less than 70 percent of Saudi women of the same age. But because in recent years the Saudi education system has been largely entrusted to Wahhabi fundamentalists, as a form of appeasement that many in the royal family hope will direct the fundamentalists' animus at foreign targets, its products are generally ill prepared to compete in a technological age or a global economy. Today two out of every three Ph.D.s earned in Saudi Arabia are in Islamic studies. Doctorates are only very rarely granted in computer sciences, engineering, and other worldly vocations. Younger Saudis are being educated to take part in a world that will exist only if the Wahhabi *jihad*ists succeed in turning back the clock not just a few decades but a few centuries.

Then there's the demographic problem. Saudi Arabia has one of the highest birth rates in the world outside Africa—37.25 births for every 1,000 citizens last year, compared with 14.5 per 1,000 in the United States. Ninety-seven percent of all Saudis are sixty-four or younger, and half the population is under eighteen. The simple presence of so many people of working age, and especially so many just now ready to enter the work force, places enormous pressure on an economy—particularly one designed less to accommodate those who want to work than to provide sustenance for those who would rather contemplate original intent in the Koran. A middle class stabilizes society. Saudi Arabia's middle class is imploding.

The functioning of the world's most advanced welfare state is influenced overwhelmingly by fluctuations in the price of oil. In 1981, when the entire kingdom was in effect being put on the dole, oil was selling at nearly $40 a barrel, and the annual per capita income was $28,600. A decade later, just before Iraq invaded Kuwait, refiners were able to buy oil for about $15 a barrel. The Gulf War sent prices back up to about $36 a barrel before they quickly fell. Today a barrel of oil once again fetches around $40, but twenty years' worth of inflation, combined with a population explosion, has brought per capita income down to below $7,000. Because roughly 85 percent of Saudi Arabia's total revenues are oil-based, every dollar increase in the price of a barrel of oil means a gain of about $3 billion to the Saudi treasury. In the early 1980s the kingdom boasted cash reserves on the order of $120 billion; today the figure is estimated to be $21 billion.

Given all these threatening forces, one might think that every map in official Washington would have a red flag sticking out of Riyadh, as a reminder that Saudi Arabia is on life support. The truth is quite the opposite. Before 9/11 the United States never issued an advisory indicating the obvious security problems for Americans traveling to Saudi Arabia. Dependents of U.S. citizens residing there were never advised to leave. According to official Washington, even today the country is stable: its government is in undisputed control of its borders; its police force and army are efficient and loyal; its people are well clothed, well fed, and well educated.

Consider the way the State Department has handled visas for Saudi nationals. Until 9/11, Saudis were not even required to appear at the U.S. embassy in Riyadh or the consulate in Jidda for a visa interview. Under a system called Visa Express a Saudi had only to send his passport, an application, and the application fee to a travel agent. The Saudi travel agent, in other words, stood in for the U.S. government. Just about any Saudi who had the money could book a flight to New York after a mere twenty-hour wait. Until recently Saudis were exempt from the new anti-terrorism entry regulations that apply to citizens of other Middle Eastern countries, despite the fact that most of the 9/11 terrorists were Saudis.

"The Saudi Arabian Government, at all levels, continued to reaffirm its commitment to combating terrorism," the State Department's 1999 report *Patterns of Global Terrorism* soberly asserted. The report went on to state, "The Government of Saudi Arabia continued to investigate the bombing in June 1996 of the Khobar Towers." This was false; Prince Nayef, Saudi Arabia's grim Interior Minister, had been stalling the investigation for years. Nayef told the kingdom's other senior princes that he was reluctant to help the United States with the Khobar investigation. In one heated meeting Nayef ignored Defense Minister Sultan when Sultan warned that stonewalling the FBI would end up causing a rift with the United States. To make his point Nayef went out of his way to avoid meeting the FBI's director, Louis Freeh, when Freeh showed up in Saudi Arabia to see what he could do to get the Khobar investigation going. Nayef put himself out of reach—on his yacht, anchored off the coast near Jidda, in the Red Sea—and turned the chore over to two low-ranking officials in the internal-security service, neither of whom knew anything about the Khobar investigation.

Even after the 1998 attacks on the U.S. embassies in Kenya and Tanzania, which were organized by Osama bin Laden from his bases in Afghanistan, the Saudi royals continued to aid the Taliban and its main supporter in the region, Pakistan. This was hardly a secret: in July of 2000 *Petroleum Intelligence Weekly*, which calls itself the "bible" of the international petroleum industry, reported that Saudi Arabia was sending as many as 150,000 barrels of oil a day to Afghanistan and Pakistan in off-budget foreign aid that had a value of something like $2 million a day. Furthermore, the United States had known since 1994 that the Saudis were supporting Pakistan's nuclear development program, ultimately contributing upwards of a billion dollars. More recently, because Saudi law does not allow foreign agencies to directly question Saudi citizens, the FBI has not been allowed to interview Saudi suspects, including the families of the fifteen Saudi hijackers, about the 9/11 attacks. For more than a year after September 11 Saudi Arabia refused to provide advance manifests for flights coming into the United States—which could have led to a basic and potentially fatal breach of

security. Although there are plenty of possible al Qaeda members awaiting trial, as of this writing there hasn't been a single Saudi arrest related to 9/11—not even of a material witness.

As for the CIA, the Agency let the State Department take the lead and decided simply to ignore Saudi Arabia. The CIA recruited no Saudi diplomats to tell us, for instance, what the religious-affairs sections of Saudi embassies were up to. The CIA's Directorate of Intelligence avoided writing national intelligence estimates—appraisals, drawn from various U.S. intelligence services, about areas of potential crisis—on Saudi Arabia, knowing that such estimates, especially when negative, have a tendency to find their way onto the front pages of U.S. newspapers, where they might have an undesired effect on public opinion. The CIA's line became the same as State's: There's no need to worry about Saudi Arabia and its oil reserves.

No need to worry, of course, means business as usual—and for decades now that's meant that almost every Washington figure worth mentioning has been involved with companies doing major deals with Saudi Arabia. Spending a lot of money was a tacit part of the U.S.-Saudi relationship practically from the very beginning: the Americans would buy Saudi Arabia's oil and would provide the Saudis with protection and security; the Saudis would buy American weapons, construction services, communications systems, and drilling rigs. In the global-economics game this is known as "recycling," and in this case it worked well: two-way trade between Saudi Arabia and the United States grew from $56.2 million in 1950 to $19.3 billion in 2000.

Consider the case of the Carlyle Group—a private investment company, founded in 1987, that almost since its inception has been conducting immensely profitable business with Saudi Arabia. From 1993 to 2002 the chairman of Carlyle was Frank Carlucci, who served first as Ronald Reagan's National Security Adviser and then as his Secretary of Defense. Carlyle's senior counselor is James Baker, who served as Secretary of State under George H.W. Bush—who in his post-presidency also happens to be a Carlyle adviser. Others who hang their hats at Carlyle include Arthur Levitt, the head of the Securities and Exchange Commission under Bill Clinton, and now Carlyle's senior adviser; John Major, a former Prime Minister of Great Britain and the current chairman of Carlyle Europe; William Kennard, who chaired the Federal Communications Commission during the second Clinton Administration; Afsaneh Mashayekhi Beschloss, a former treasurer and chief investment officer of the World Bank; and Richard Darman, who ran the Office of Management and Budget under the first President Bush and also served as deputy secretary of the treasury under Reagan.

Carlyle isn't the only company in this business. Halliburton, run by Dick Cheney between his stints as Secretary of Defense under the first George Bush and Vice President under the second, has been a frequent beneficiary of Saudi money. In late 2001 Halliburton landed a $140 million contract to develop a new Saudi oil field. For many years Condoleezza Rice, now President Bush's National Security Adviser, served on the board of Chevron, which merged in 2001 with Texaco. The new corporation, ChevronTexaco, is a partner with Saudi Aramco in several ventures and has recently joined forces with Nimir Petroleum to develop oil fields in Kazakhstan. Currently on the board of ChevronTexaco are Carla Hills, who served as the Secretary of Housing and Urban Development under Gerald Ford and as a U.S. trade representative under George H.W. Bush; the former Louisiana senator J. Bennett Johnston, who made a specialty of energy issues while in Congress; and the former Georgia senator Sam Nunn, who served most notably as head of the Senate Armed Services Committee.

Elsewhere, Nicholas Brady, the Secretary of the Treasury under the first President Bush, and Edith Holiday, a former assistant to the first President Bush, serve on the board of Amerada Hess, which has teamed with some of Saudi Arabia's most powerful royal-family members to exploit the rich oil resources of Azerbaijan. In 1998 Amerada Hess formed a joint venture, Delta Hess, with the Saudi-owned Delta Oil to explore and exploit petroleum resources in Azerbaijan. The Houston-based Frontera Resources Corporation joined the Azerbaijan hunt the same year, teaming with the newly created Delta Hess. Among the members of Frontera's board of advisers: the former Texas senator former Secretary of the Treasury, and 1988 Democratic vice-presidential candidate Lloyd Bentsen; and John Deutch, a former CIA director.

Just to make sure that no one upsets the workings of this system, perhaps by meddling in internal Saudi affairs, Saudi Arabia now keeps possibly as much as a trillion dollars on deposit in U.S. banks—an agreement worked out in the early eighties by the Reagan Administration, in an effort to get the Saudis to offset U.S. government budget deficits. The Saudis hold another trillion dollars or so in the U.S. stock market. This gives them a remarkable degree of leverage in Washington. If they were suddenly to withdraw all their holdings in this country, the effect, though perhaps not as catastrophic as having a major source of oil shut down, would still be devastating.

The U.S.-Saudi relationship would not be as cozy as it is without there being someone well connected on both sides who can move comfortably between them. That someone is the fifty-four-year-old Prince Bandar. Although he ranks low on the royal bloodline (his father is King Fahd's brother Sultan, the Saudi Defense Minister, but his mother was a house servant), Prince Bandar has been the Saudi ambassador to the United States since 1983. He is the only foreign ambassador to have a security detail assigned to him by the State Department. A daredevil fighter pilot in his younger years, a Muslim with a taste for single-malt Scotch, and an envoy with a perpetually open wallet, Bandar has proved adept at working both the public and the private sides of diplomacy. As the Saudi military attache to the United States, he scored a stunning coup in 1981 by persuading Congress to approve the sale of AWACS air-defense technology to his country, over the objections of AIPAC, the pro-Israeli Washington lobby. Later, as ambassador, Bandar conveyed the kingdom's thanks by secretly placing $10 million in a Vatican City bank, as reported last year in *The Washington Post*; the money, deposited at the request of William Casey, then the director of the CIA, was to be used by Italy's Christian Democratic Party in a campaign against Italian Communists. Later still, in June of 1984, Bandar started paying out $30 million from the royal family so that Lieutenant Colonel Oliver North could buy arms for the Nicaraguan contras.

Leading U.S. corporations hire and rehire known Saudi crooks and known financiers of terrorism, so that they can land deals back in Saudi Arabia.

It is on the personal front, however, where Bandar shines. A visit in the early nineties to the summer home of George H.W. Bush, in Kennebunkport, Maine, earned the prince the affectionate family sobriquet "Bandar Bush." Bandar reciprocated by inviting Bush to hunt pheasant on his estate in England. For good measure he also contributed a million dollars to the construction of the Bush Presidential Library, in College Station, Texas. King Fahd sent another million to Barbara Bush's campaign against illiteracy. (He had donated a million dollars to Nancy Reagan's "Just Say No" campaign against drugs four years earlier.) Bandar was once Colin Powell's racquetball partner.

Press accounts portrayed Bandar as largely on the outside during the Clinton years, passing melancholy weeks at his mountain compound in Aspen, Colorado (more than 50,000 square feet, thirty-two rooms, sixteen bathrooms). If Bandar was less physically present, however, he was his usual useful self. In 1992 he persuaded King Fahd to donate $20 million to the University of Arkansas's new Center for Middle Eastern Studies, a gesture of respect for the Arkansas governor who had just been elected President. He is said to have played a role in persuading the Libyans, in 1999, to turn over two intelligence operatives suspected in the 1988 bombing of Pan Am Flight 103, over Lockerbie, Scotland. As he reportedly does at the end of every administration, whether he is perceived as friend or foe, Bandar also invited each of the Clinton Cabinet members out to dinner, at a restaurant of their choice, in a private room or a public one, depending on their willingness to be seen with him.

Prince Bandar once told associates that he is very careful to look after U.S. government officials when they return to private life. "If the reputation then builds that the Saudis take care of friends when they leave office," Bandar has observed, according to a source cited in *The Washington Post*, "you'd be surprised how much better friends you have who are just coming into office." Practically every deal with the Saudis eventually becomes hard to trace, lost in some desert sandstorm back near the wellheads where the money sprang from in the first place. Many of Washington's lobbyists, PR firms, and lawyers live off Saudi money. Just about every Washington think tank has taken it. So have the John F. Kennedy Center for the Performing Arts, the Children's National Medical Center, and every presidential library built in the past thirty years.

Bandar hurried back to prominence after the election of George W. Bush, occupying a spot somewhere between ambassador and permanently enthroned visiting head of state. But after 9/11 he began to experience some difficulty in maintaining a positive Saudi image. In March of last year agents of the Treasury Department raided the northern-Virginia offices of four Saudi-based charities: the SAAR Foundation, the Safa Trust, the International Institute for Islamic Thought (IIIT), and the International Islamic Relief Organization (IIRO). Also raided was the local headquarters for the Muslim World League, an umbrella group funded by the Saudi government. All five organizations are located only a few miles from Bandar's mansion overlooking the Potomac River. The organizations can point to a long list of genuinely humanitarian causes they have aided and supported; but they also have a long list of alarming associations. Testifying before Congress in August of 2002, Matthew Levitt, a senior fellow with the Washington Institute for Near East Policy, noted that Tarik Hamdi, an IIIT employee, had personally provided Osama bin Laden with batteries for his satellite phone—a critical link in the stateless world that bin Laden inhabits. IIIT and the SAAR Foundation are suspected of helping to finance Hamas and the Palestinian Islamic Jihad, the sponsors of some of the most lethal suicide bombers in the Middle East. From 1986 to 1994 Muhammad Jamal Khalifa, a brother-in-law to Osama bin Laden, ran the IIRO's Philippine office, from which he channeled funds to al Qaeda. Only excellent work by the Indian police prevented another IIRO employee, Sayed Abu Nasir, from bombing the U.S. consulates in Calcutta and Madras.

In mid-2002 word leaked to the press that the semi-official Defense Policy Board, chaired by the notorious cold warrior Richard Perle, had sponsored a report declaring Saudi Arabia to be part of the problem of international terrorism rather than part of the solution. Saudi Arabia, the report stated, was "central to the self-destruction of the Arab world and the chief vector of the Arab crisis and its outwardly-directed aggression." It went on to say, "The Saudis are active at every level of the terror chain, from planners to financiers, from cadre to foot-soldier, from ideologist to cheerleader." Within hours Colin Powell was on the phone to the Saudi Foreign Minister, assuring him—and through him, the royal family—that such apostasy was not and never would be the official stance of the Bush Administration. To reinforce the message, President Bush invited Bandar down to the family ranch at Crawford, Texas.

And yet the image problems have continued. In October of 2001, NATO forces raided the offices of the Saudi High Commission for Aid to Bosnia, founded by Prince Salman, and discovered, among other items, photos of the U.S. embassies in Kenya and Tanzania, before and after they were bombed; photos of the World Trade Center and the USS *Cole*; information on the use of crop-duster planes; and materials for forging U.S. State Department badges. His job wasn't made any easier when, in the fall of last year, Bandar found himself having to explain away the fact that about $130,000 in charitable contributions from his wife, Princess Haifa, might have ended up with two of the 9/11 hijackers.

In the wake of these revelations a U.S. delegation headed by Alan Larson, President Bush's undersecretary of state for economic affairs, traveled to Riyadh last November, ostensibly to prod the Saudis toward increasing the surveillance of their charities and financial networks. But U.S. and Saudi sources say that one of the main reasons for Larson's trip was to ensure that if the United States invaded Iraq, the Saudis would guarantee the flow of extra oil into the World market. The U.S. embrace of the House of Saud was as tight as ever.

Washington's answer for Saudi Arabia—apart from repeating that nothing is wrong—is to suggest that a little democracy will cure everything. Talk the royal family into ceding at least part of its authority; support the reform-minded princes; set up a model parliament; co-opt the firebrands with a cabinet position or two, a minor political party, and some outright bribery; send Jimmy Carter in to monitor the first election; and in a few generations Riyadh will be Ankara, maybe even London. The governmental mechanism may be faulty, the Washington view maintains, but the people who administer the government are for the most part committed to rooting out corruption, rounding up terrorists, and recognizing the right of the people to self-government.

It's utter nonsense, of course. If an election were held in Saudi Arabia today, if anyone who wanted to could run for the office of president, and if people could vote their hearts without fear of having their heads cut off afterward in Chop-Chop Square, Osama bin Laden would be elected in a landslide—not because the Saudi people want to wash their hands in the blood of the dead of September 11, but simply because bin Laden has dared to do what even the mighty United States of America won't do: stand up to the thieves who rule the country.

Saudi Arabia today is a mess, and it is our mess. We made it the private storage tank for our oil reserves. We reaped the benefits of a steady petroleum supply at a discounted price, and we grabbed at every available Saudi petrodollar. We taught the Saudis exactly what was expected of them. We cannot walk away morally from the consequences of this behavior—and we *really* can't walk away economically. So we crow about democracy and talk about someday weaning ourselves from our dependence on foreign oil, despite the fact that as long as America has been dependent on foreign oil there has never been an honest, sustained effort at the senior governmental level to reduce long-term U.S. petroleum consumption.

Not all the wishing in the world will change the basic reality of the situation.

- Saudi Arabia controls the largest share of the world's oil and serves as the market regulator for the global petroleum industry.
- No country consumes more oil, and is more dependent on Saudi oil, than the United States.
- The United States and the rest of the industrialized world are therefore absolutely dependent on Saudi Arabia's oil reserves, and will be for decades to come.
- If the Saudi oil spigot is shut off, by terrorism or by political revolution, the effect on the global economy, and particularly on the economy of the United States, will be devastating.
- Saudi oil is controlled by an increasingly bankrupt, criminal, dysfunctional, and out-of-touch royal family that is hated by the people it rules and by the nations that surround its kingdom.

Signs of impending disaster are everywhere, but the House of Saud has chosen to pray that the moment of reckoning will not come soon—and the United States has chosen to look away. So nothing changes: the royal family continues to exhaust the Saudi treasury, buying more and more arms and funneling more and more "charity" money to the *jihad*ists, all in a desperate and self-destructive effort to protect itself.

The fact is that the West, especially the United States, has left the Saudis little choice. Leading U.S. corporations hire and re-hire known Saudi crooks and known financiers of terrorism to represent their interests, so that they can land the deals that will pay the commissions back in Saudi Arabia—commissions that will further erode the budget and thus further divide the ruling class from everyone else. Former CIA directors serve on boards whose members have to hold their noses to cut deals with Saudi companies—because that's business, that's the price of entry, that's the way it's done. Ex-Presidents, former prime ministers, onetime senators and congressmen, and Cabinet members walk around with their hands out, acting as if they're doing something else but rarely slowing down, because most of them know it's an endgame too. But sometime soon, one way or another, the House of Saud is coming down.

Robert Baer served for twenty-one years with the CIA, primarily as a field officer in the Middle East. He resigned from the agency in 1997 and was awarded its Career Intelligence Medal in 1998. This article is adapted from the forthcoming book Sleeping With the Devil *(Crown Publishers), to be published in June.*

From *The Atlantic Monthly*, May 2003, pp. 53-62. © 2003 by Robert Baer. Reprinted by permission of the author.

Thirty Years of Petro-Politics

Daniel Yergin

The 1973 oil embargo was the unsheathing of "the Arab Oil weapon." Coming the day after a decision to double prices, it was meant to be an act of retribution—punishing the United States for resupplying arms to Israel, which at that moment was still reeling from the surprise attack on Yom Kippur, several days earlier. Beyond that, it was also meant to pressure the entire Western world, whose oil consumption had been rising rapidly, into supporting the Arab side in the Arab-Israeli confrontation.

Though hardly anyone anticipated such cuts, there had been warnings. Indeed, during the 1967 Arab-Israeli war, some of the Arab exporters had tried to impose an embargo. It failed because there was plenty of extra production capacity that could be called into service. But by 1973 the world market had changed: Every well in the world, it seemed, was producing flat-out, at full capacity. The reason was the torrid growth in demand. The United States, its domestic production having flattened, had turned to the world market, and in a few short years went from minor importer to the world's largest importer. And this time there was no place to go for extra oil.

Are we less vulnerable today than in the 1970s? Oil is still the preeminent strategic commodity. And in a tight market, even without the political motives that brought on the 1973 embargo, the oil market is vulnerable to shocks and disruptions. True, markets are flexible and can adjust more quickly, but extensive turmoil in the Middle East—or other major oil-producing regions—could send new shocks throughout the world. After the last OPEC meeting, oil prices are again hovering around $30 a barrel. All this has had a significant impact on the world economy.

Thirty years ago today, a group of oil ministers gathered in Kuwait City and made a decision that shook world politics and ignited the energy crisis of the 1970s. It continues to shape the way we see energy issues today, when tight oil and gas markets and high prices are a key factor in our economy—and when the strategic significance of oil is once again evident.

The embargo created a massive global panic as buyers competed furiously with each other to get what they could, pushing the price up further. In the United States, it hit home for most consumers in infuriating gas lines—long waits for limited amounts of gasoline. (The lines were in fact largely self-in-flicted, a result of government controls that prevented flexibility and accentuated shortages in the marketplace.)

The whole international order seemed to have been transformed. Now politics was also about economics. On the day the embargo was announced, President Nixon told his advisers, "No one is more keenly aware of the stakes: oil and our strategic position." The vast flood of "petro-dollars" to the exporters turned "petro-power" into a central fact of international politics.

Prices went up fourfold in the crisis; then, a few years later, with the Iranian revolution, they doubled again. The oil crisis marked the end of the postwar economic boom and did much to turn the 1970s into the worst decade, in economic terms, since the Great Depression. Moreover, there were fears that the crisis portended a permanent shortage of oil. People speculated that the price of oil might go to $100 a barrel.

That's not how things turned out, of course. Within less than a decade, the "permanent shortage" turned into a glut, triggering a price collapse that, among other things, hastened the end of the Soviet Union, which had been depending on its oil exports as the lifeline to keep its economy alive.

There are many lessons here. Nations that had taken their energy supplies for granted suddenly realized how important reliable, reasonably priced supplies were to their well-being. Oil became high politics, and energy became part of public policy.

One of the less obvious but lasting lessons is that markets work, even in circumstances as dramatic as these were. Supply and demand adjusted. The United States and other industrial countries have since become much more efficient in the use of oil. Today—SUVs notwithstanding—the United States uses only half as much oil per unit of GDP as it did in the 1970s. New, non-OPEC sources of oil, led by Alaska and the North Sea, came on stream quickly. And the world switched from oil to other energy sources.

It is also now clear that the starting point for energy security is diversification of supplies—that is, production coming from many sources. The United States now imports oil from a large number of countries; it has a Strategic Petroleum Reserve as a supply source of last resort and can coordinate with other countries through the International Energy Agency.

The response to the crisis also demonstrates the power of technology. Technological advances have brought both greater efficiency in oil consumption and greater range in production. The oil industry is able to accomplish feats—such as drilling beneath the ultra-deep waters of the Gulf of Mexico—that were simply inconceivable in the 1970s.

Exporters learned their own powerful lesson: Customers matter. When prices skyrocketed, the exporters found that they were losing market share; customers had choices. After the price collapse, most of the exporters set out to rebuild their credibility as reliable suppliers, which meant making oil nonpolitical again. Consumers worry about security of supply, but producers—who often depend overwhelmingly on petroleum earnings for their national budgets—learned they need to worry about security of demand.

In November 1973, a few weeks after the embargo went into effect, President Nixon announced Project Independence to make the United States self-reliant in energy. Thirty years of rhetoric later, we are no closer to that goal—indeed, we're farther away. At the time of the embargo, the United States was importing a third of its oil; today it's almost 60 percent. We will continue to hear much about energy independence, but the real challenge is how to manage our dependence through diversification, efficiency, technological advances and the stability of relations with a wide range of suppliers.

The writer is chairman of Cambridge Energy Research Associates and author of "The Prize: the Epic Quest for Oil, Money, and Power."

India's Hype, Hope, and Hazards

Optimistic signs of economic development obscure the disparities of economic growth and opportunities in different regions, states, and segments of India's population.

By Mahmood H. Butt

The largest democracy in the world has recently concluded a nationwide electoral campaign to choose the leaders and policies that will guide the destiny of India and its people in the rapidly globalizing world of the early twenty-first century. Ideally, political campaigns give the contestants opportunities to present to the electorate their past record of accomplishments, long-term vision of the future, and agenda for the next five years, the maximum constitutional term of office for India's National Assembly, the Lok Sabha.

A fresh mandate to govern and guide a nation of over one billion people is the result of the current campaign and will determine the direction of this vastly complex nation, full of promise yet mired in paradox and privations. Those in power and their spin doctors are portraying India as finally on the verge of breaking the shackles of poverty, ignorance, disease, and scarcity; in their view, it is well on its way to becoming the third-largest economy of the world, behind China and the United States, in the next three decades. There is some truth to that projection. They point out the long-term optimistic signs—significantly increased economic growth (8 percent in 2004, up from the average growth rate of about 4 percent in the past), a sizable foreign exchange re-serve of over $100 billion, and growth in export of manufactured goods, software, and IT services—as indicators supporting their bullish attitude.

Yet these signs of economic development also obscure the disparities of uneven growth and opportunities in different regions, states, and segments of the Indian population. The rural sector, where 70 percent of Indians still live on small landholdings eking out existence at bare subsistence levels, constitutes a large backwater of deprivation and degradation. The Indian Union consists of 28 states and six union territories. Some of these states are as large as medium-sized countries. Each state has its unique linguistic, demographic, geographic, and economic characteristics. The uneven progress in human resource development, gender equity, and economic growth among and within states is rooted in cultural characteristics.

State-level policies are often controlled by the old, caste-based ruling elites, whose inertial tendencies and preference for slow, incremental change are partly responsible for keeping the majority in their "proper place" and subordinate state. National-level aggregation of data regarding human development and economic growth hides the regional disparities and gives a partial and often erroneous glimpse of progress. Rural India, where most Indians still live, has yet to experience clean water, electricity, or transparent democratic government.

The scale of rural deprivation can be measured by the stark facts. For example, 63 percent of all rural households lack electricity. Those who have it can't count on having it all of the time. Forty percent of India's more than 600,000 villages have no all-weather road connection to markets and social services. Local government in rural areas is still controlled by the higher-caste elites. In an article in the Daily Tribune entitled "India Shining, Dimly," L.H. Naqvi wrote, "The rural republic is where the real India still lives. It has yet to reap the benefits of the amazing advances that the country has made in the fields of science, technology and medicine. It is an impoverished, brutalized and criminalized republic. The flyovers that do not lead to rural India or the six-lane expressway that bypasses the villages will be of little use in converting India into a robust republic."

Social (r)evolution

In the social sector, the positive signs include the emergence of a well-educated and professionally competent urban middle class. On these 250 million young and upwardly mobile Indians rest the hopes of creating a brighter nation, one that is both full of confidence and caring and competent enough to deliver the promise of a modern secular democracy. This quarter of the population is the product of the quality sector of the educational system, which is increasingly available to males and females in India's rapidly expanding urban centers. This large middle class provides both a competent workforce and the growing market for consumer goods and industries. These are the twin pillars of social and economic change mentioned in the recent "India Shining" slogans.

The rate of population growth has slowed from 2.14 percent between 1991 and 2001 to the current annual growth of 1.8 percent and is projected to decline further over the next few years. This trend has a positive statistical impact on per capita income and reduced needs for social services. The emergence of the large middle class and declining population growth rates have also resulted in a decline in the poverty ratio. Yet, India remains desperately poor. All indicators of human poverty and deprivation—including illiteracy, infant mortality, life expectancy, and the availability or lack of such basic services as clean water, roads, and health care—point to the daunting challenges facing the country and its leaders.

There are still 400 million Indians, most of them female, who do not know how to read and write in their native language. Another 390 million live on less than a dollar a day. Ninety-five out of every 1,000 children born in India do not reach their fifth birthday, and more than one-third of the country's 200 million school-age children are still not in school. So long as this sizable segment of the population is not included in India's prosperity, the government will continue to be of the few, by the few, and for the few. The policy agenda will be driven disproportionately by the 20 million who work directly for the government at the national, state, and local levels, as well as the 30 percent ur-

banites who are the prime beneficiaries of these policies. Their claims on public expenditures, investment priorities, and privatization options in social sectors have a higher priority, while the majority continue to live in urban squalor or villages.

In a survey entitled "India's Shining Hopes," the *London Economist* examined the bases of the prevailing optimism in the country and concluded that "much of it is overdone or at least based on hope rather than achievement." Despite the favorable external and internal factors, India is not shining yet, but it can, provided needed policy reforms addressing unemployment, underinvestment, and growing polarization between the rich and poor parts of the country are implemented.

The International Monetary Fund recently issued an optimistic forecast of 7.6 percent growth in India's GDP for 2003-04 but warned that drastic reforms were needed to push the rate higher to the desired 8-10 percent. If India is to achieve sustained growth at that level, it must overcome the vagaries of the monsoon rains by investing in multipurpose irrigation schemes to ensure availability of freshwater for crops, flood control, and hydropower generation. IMF chief economist Raghuram Rajan said that "for India to really shine, more aggressive infrastructure-sector and public investment needs to be made." India also must expedite reforms in five critical areas of public policy: improving the climate for investment, both foreign and domestic; initiating wide-ranging labor reforms; striving for better corporate governance; privatizing state-owned businesses; and achieving full capital account convertibility.

Improving all the people

While India is poised for an economic takeoff, achieving its full potential will require a serious and sustained effort. It must eradicate poverty by developing not some but *all* its human resources, renew strong commitment to its founding goal of creating a politically and

morally just social order, and ensure that the rights of all its citizens are guaranteed irrespective of caste, color, or creed. It must embark on a path of peaceful coexistence with all its neighbors in the South Asian region. These three difficult challenges have to be met in an enlightened and deliberate manner, building on the progress made so far. India has lots of promises to keep and miles to go.

Though the nation has made significant gains in human resource development, the challenges it faces are still huge. According to the UNDP Human Development Index, India is still ranked among the less-developed countries of the world. Article 45 of the Indian constitution calls for the state to "endeavor to provide, within a period of 10 years from the commencement of the constitution, for free and compulsory education for all children until they complete the age of 14 years." A constitutional amendment has made elementary education an inalienable right of all children in order to achieve the goal of universal primary education. These constitutional requirements have not been met. They have been constrained through planned and actual investments in basic primary education.

So long as India invests less than 4 percent of its GDP in human resource development, millions of children ages 6 to 10 will not attend primary schools and almost 37 percent of those who enter will leave before completing fifth grade. Built on such a weak foundation, the rest of the educational edifice cannot be strong. This problem is reflected in a secondary school enrollment rate that is half that of the primary level. In a country where one-third of the population is under 15, these figures tell a sorry tale. Victims of deprivation and disenfranchisement, a significant segment of young Indians can only watch the procession go by.

The cumulative result of this meager investment is a dubious distinction: India has the largest number of illiterates of any country in the world. Regional disparities further complicate matters. While Kerala has achieved universal literacy, the traditionally backward states like Rajasthan, Bihar, Uttar Pradesh, and Madhya Pradesh are still struggling to reach even the national literacy level of 60 percent. The ambitious goals of the eighth, ninth, and tenth five-year development plans to provide universal primary education, eradicate illiteracy through traditional and alternate measures, and improve the quality of educational opportunities have not yet been achieved. As long as these goals remain unattained, the cancer of poverty will continue to spread at disproportionate rates in different parts of India.

The higher-education sector, available to only 6 percent of the population, has diversified its curricular range and quality. The crown jewels of this sector have been the Indian Institutes of Technology and a variety of research centers. They were established in the late fifties and have matured into centers of excellence whose research and productivity are as good as any in the world. They have not only produced world-class statisticians, mathematicians, computer scientists, and engineers but have created a culture of discovery and dissemination of truth that Indian higher education previously lacked.

Another far-reaching contribution made by these new institutions is the transparency in competitive admission, retention, and graduation based on demonstrated quality of performance and affirmation of the principle of merit in all operational aspects. These crown jewels are in marked contrast to the rest of the higher educational system, which is producing barely credentialed and not intellectually creative individuals. In this system, merit is recognized as a guiding principle but is often overridden by other factors (who you know rather than what you know). India has to increase the accessibility, quality, and range of its educational enterprise; this cannot be done without a grad-

ual and significant increase in investment. Indian statisticians have worked out computer models showing the impact of increasing federal, state, and district expenditures on education from the current 4-7 percent of GDP. Using these models, planners can set ambitious yet attainable goals that will make the nation really shine. What India lacks is the concerted political will to make hard choices when setting its domestic and foreign policy agenda.

Foreign policy challenges

Indian policy planners in defense and foreign affairs have been mainly concerned with developing reactive policies vis-à-vis two potential adversaries, China and Pakistan. Since the conflicts with China in 1962 and Pakistan in 1965, India's defense posture has been to achieve credible military deterrence through self-reliance. This policy has resulted in huge expenditures in the defense sector, which is consuming 35 percent of the country's fiscal resources.

More recently, India has embarked upon even more ambitious plans to project its power as the preeminent regional player. Acquiring a credible naval force and a technologically advanced, integrated defense system using land, air, and sea forces has meant increased expenditures. While evolving geopolitical realities in South Asia and its environs are often cited as the reason for developing this national defense policy, it is imperative for India, at this stage of its development, to avoid excessive entanglements in policing the region. Initial diplomatic steps are being taken to pave the path of peaceful coexistence with China and Pakistan. Much more needs to be done.

Kashmir has been a bone of contention for the last half century. Though India and Pakistan have fought three wars to settle the issue, this has proved to be a costly and futile effort. By arming themselves with nuclear weapons and increas-

ingly sophisticated delivery systems, both countries have raised the stakes. As pointed out by Jonah Blank in an article published in *Foreign Affairs*, "Every border skirmish between India and Pakistan carries the potential, however remote, for catastrophic escalation .... There will be no safety for either state without stability in Kashmir and there will be no stability in Kashmir without the cooperation of its people."

Kashmir is not just a territorial dispute between two states. It will not be settled by a divvying-up of the real estate through military or even diplomatic means. The people of Kashmir, as the primary party in the dispute, have to decide their future, and their factional leaders have been insisting on having a place at the table. So far, they have paid a heavy price, losing 70,000 lives over a 10-year struggle for their right for self-determination. It has been a costly venture for India and Pakistan as well. A peaceful resolution of this dispute could pay a tremendous dividend, one that both countries could use to develop their human resources and make a dent in poverty. Continuation of the dispute might lead to increased fundamentalist fervor in both India and Pakistan. The longer Kashmir is controlled through sheer brute force, the dimmer the chances that India will shine as a multiethnic, religiously diverse, secular society.

Hope or hype?

The election manifesto of the ruling National Democratic Alliance (NDA), released on April 7, 2004, calls for making India an economic superpower in five years by creating 10 million new jobs each year. It promises to electrify India's 600,000 villages by 2009. The manifesto pledges to carry forward the peace initiative with Pakistan. It calls for reduction in communal and caste violence and creation of a riot-free India. Many thoughtful Indians consider this to be either election hype or a list of pious hopes at best.

The Indian finance minister, Jaswant Singh, while presenting the 2003-2004 budget to the parliament, identified five conceptual priorities that were to drive the budget. These included "poverty reduction, addressing the 'lifetime concerns' of citizens, covering health, housing, education and employment; infrastructure development; fiscal consolidation through tax reforms; agriculture and irrigation; and enhancing manufacturing-sector efficiency including promotion of exports." He concluded his speech by recognizing that "there is palpable impatience in the country for progress and growth. The nation cannot afford the luxury of prolonged periods of reflection, or a leisurely implementation schedule. The world will otherwise pass us by."

More recently, India has embarked upon even more ambitious plans to project its power as the preeminent regional player.

The NDA election manifesto and Singh's speech identify the challenges facing India and the policy options available to its leaders. If India is to shine, it will not be through the luster of a segment of its population but its collective will to leave no Indian behind. India's progress should be measured not only by its growth rate but by how the GDP is used to produce a morally upright and peaceful country, where citizens of many races, castes, creeds, and colors live in harmony. Indians have to realize that economic growth, worthwhile as it is, is only a means to a larger end: meeting the need of all its citizens to enjoy a decent quality of life in a tolerant, secular, and democratic India. The promises made in the political manifestos of contending parties will remain hazardous slogans if they are not implemented. India's glitter is not yet gold.

Mahmood H. Butt is chairman of the secondary education department at Eastern Illinois University in Charleston.

How Nike Figured Out China

The China market is finally for real. To the country's new consumers, Western products mean one thing: status. They can't get enough of those Air Jordans

By MATTHEW FORNEY DAREN FONDA/ BEAVERTON; NEIL GOUGH/GUANGZHOU

Nike swung into action even before most Chinese knew they had a new hero. The moment hurdler Liu Xiang became the country's first Olympic medalist in a short-distance speed event—he claimed the gold with a new Olympic record in the 110-m hurdles on Aug. 28—Nike launched a television advertisement in China showing Liu destroying the field and superimposed a series of questions designed to set nationalistic teeth on edge. "Asians lack muscle?" asked one. "Asians lack the will to win?" Then came the kicker, as Liu raised his arms above the trademark Swoosh on his shoulder: "Stereotypes are made to be broken." It was an instant success. "Nike understands why Chinese are proud," says Li Yao, a weekend player at Swoosh-bedecked basketball courts near Beijing's Tiananmen Square.

Such clever marketing tactics have helped make Nike the icon for the new China. According to a recent Hill & Knowlton survey, Chinese consider Nike the Middle Kingdom's "coolest brand." Just as a new Flying Pigeon bicycle defined success when reforms began in the 1980s and a washing machine that could also scrub potatoes became the status symbol a decade later, so the Air Jordan—or any number of Nike products turned out in factories across Asia—has become the symbol of success for China's new middle class. Sales rose 66% last year, to an estimated $300 million, and Nike is opening an average of 1.5 new stores a day in China. Yes, a day. The goal is to migrate inland from China's richer east-coast towns in time for the outpouring of interest in sports that will accompany the 2008 Summer Olympics in Beijing. How did Nike build such a booming business? For starters, the company promoted the right sports and launched a series of inspired ad campaigns. But the story of how Nike cracked the China code has as much to do with the rise of China's new middle class, which is hungry for Western gear and individualism, and Nike's ability to tap into that hunger.

Americans have dreamed of penetrating the elusive China market since traders began peddling opium to Chi-

nese addicts in exchange for tea and spices in the 19th century. War and communism conspired to keep the Chinese poor and Westerners out. But with the rise of a newly affluent class and the rapid growth of the country's economy, the China market has become the fastest growing for almost any American company you can think of. Although Washington runs a huge trade deficit with Beijing, exports to China have risen 76% in the past three years. According to a survey by the American Chamber of Commerce, 3 out of 4 U.S. companies say their China operations are profitable; most say their margins are higher in China than elsewhere in the world. "For companies selling consumer items, a presence here is essential," says Jim Gradoville, chairman of the American Chamber in China.

The Chinese government may have a love-hate relationship with the West—eager for Western technology yet threatened by democracy—but for Chinese consumers, Western goods mean one thing: status. Chinese-made Lenovo (formerly Legend) computers used to outsell foreign competitors 2 to 1; now more expensive Dells are closing the gap. Foreign-made refrigerators are displacing Haier as the favorite in China's kitchens. Chinese dress in their baggiest jeans to sit at Starbucks, which has opened 100 outlets and plans hundreds more. China's biggest seller of athletic shoes, Li Ning, recently surrendered its top position to Nike, even though Nike's shoes—upwards of $100 a pair—cost twice as much. The new middle class "seeks Western culture," says Zhang Wanli, a social scientist at the Chinese Academy of Social Sciences. "Nike was smart because it didn't enter China selling usefulness, but selling status."

The quest for cool hooked Zhang Han early. An art student in a loose Donald Duck T shirt and Carhartt work pants, Zhang, 20, has gone from occasional basketball player to All-Star consumer. He pries open his bedroom closet to reveal 19 pairs of Air Jordans, a full line of Dunks and signature shoes of NBA stars like Vince Carter—

more than 60 pairs costing $6,000. Zhang began gathering Nikes in the 1990s after a cousin sent some from Japan; his businessman father bankrolls his acquisitions. "Most Chinese can't afford this stuff," Zhang says, "but I know people with hundreds of pairs." Then he climbs into his jeep to drive his girlfriend to McDonald's.

Zhang hadn't yet been born when Nike founder Phil Knight first traveled to China in 1980, before Beijing could even ship to U.S. ports; the country was just emerging from the turmoil of the Cultural Revolution. By the mid-'80s, Knight had moved much of his production to China from South Korea and Taiwan. But he saw China as more than a workshop. "There are 2 billion feet out there," former Nike executives recall his saying. "Go get them!"

Phase 1, getting the Swoosh recognized, proved relatively easy. Nike outfitted top Chinese athletes and sponsored all the teams in China's new pro basketball league in 1995. But the company had its share of horror stories too, struggling with production problems (gray sneakers instead of white), rampant knock-offs, then criticism that it was exploiting Chinese labor. Cracking the market in a big way seemed impossible. Why would the Chinese consumer spend so much—twice the average monthly salary back in the late 1990s—on a pair of sneakers?

Sports simply wasn't a factor in a country where, since the days of Confucius, education levels and test scores dictated success. So Nike executives set themselves a potentially quixotic challenge: to change China's culture. Recalls Terry Rhoads, then director of sports marketing for Nike in China: "We thought, 'We won't get anything if they don't play sports.'" A Chinese speaker, Rhoads saw basketball as Nike's ticket. He donated equipment to Shanghai's high schools and paid them to open their basketball courts to the public after hours. He put together three-on-three tournaments and founded the city's first high school basketball league, the Nike League, which has spread to 17 cities. At games, Rhoads blasted the recorded sound of cheering to encourage straitlaced fans to loosen up, and he arranged for the state-run television network to broadcast the finals nationally. The Chinese responded: sales through the 1990s picked up 60% a year. "Our goal was to hook kids into Nike early and hold them for life," says Rhoads, who now runs a Shanghai-based sports marketing company, Zou Marketing. Nike also hitched its wagon to the NBA (which had begun televising games in China), bringing players like Michael Jordan for visits. Slowly but surely, in-the-know Chinese came to call sneakers "Nai-ke."

And those sneakers brought with them a lot more than just basketball. Nike gambled that the new middle class, now some 40 million people who make an average of $8,500 a year for a family of three, was developing a whole new set of values, centered on individualism. Nike unabashedly made American culture its selling point, with ads that challenge China's traditional, group-oriented ethos. This year the company released Internet teaser clips showing a faceless but Asian-look-

ing high school basketball player shaking-and-baking his way through a defense. It was timed to coincide with Nike tournaments around the country and concluded with the question, "Is this you?" The viral advertisement drew 5 million e-mails. Nike then aired TV spots contrasting Chinese-style team-oriented play with a more individualistic American style, complete with a theme song blending traditional Chinese music and hip-hop.

Starting in 2001, Nike coined a new phrase for its China marketing, borrowing from American black street culture: "Hip Hoop." The idea is to "connect Nike with a creative lifestyle," says Frank Pan, Nike's current director of sports marketing for China. The company's Chinese website even encourages rap-style trash talk. "Shanghai rubbish, you lose again!" reads a typical posting for a Nike League high school game. The hip-hop message "connects the disparate elements of black cool culture and associates it with Nike," says Edward Bell, director of planning for Ogilvy & Mather in Hong Kong. "But black culture can be aggressive, and Nike softens it to make it more acceptable" to Chinese. At a recent store opening in Shanghai, Nike flew in a streetball team from Beijing. The visitors humiliated their opponents while speakers blasted rapper 50 Cent as he informed the Chinese audience that he is a P-I-M-P with impure designs on their mothers.

Thanks in part to Nike's promotions, urban hip-hop culture is all the rage among young Chinese. One of Beijing's leading DJs, Gu Yu, credits Nike with "making me the person I am." Handsome and tall under a mop of shoulder-length hair, Gu got hooked on hip-hop after hearing rapper Black Rob rhyme praises to Nike in a television ad. Gu learned more on Nike's Internet page and persuaded overseas friends to send him music. Now they send something else too: limited-edition Nikes unavailable in China. Gu and his partner sell them in their shop, Upward, to Beijing's several hundred "sneaker friends" and wear them while spinning tunes in Beijing's top clubs. To them, scoring rare soles and playing banned music are part of the same rebellious experience. "Because of the government, Chinese aren't allowed access to a lot of these things," says Gu's partner, Ji Ming, "but with our shop and Nike-style music, they can get what they want."

The Nike phenomenon is challenging Confucian-style deference to elders too. At the Nike shop in a ritzy Shanghai shopping mall, Zhen Zhiye, 22, a dental hygienist in a miniskirt, persuades her elderly aunt, who has worn only cheap sneakers that she says "make my feet stink," to drop $60 on a new pair. Zhen explains the "fragrant possibilities" of higher-quality shoes and chides her aunt for her dowdy ways. Her aunt settles on a cross trainer. For most of China's history, this exchange would have been unthinkable. "In our tradition, elders pass culture to youth," says researcher Zhang. "Now it's a great reversal, with parents and grandparents eating and clothing themselves like children."

Success aside, Nike has had its stumbles. When it began outfitting Chinese professional soccer teams in the mid-1990s, its ill-fitting cleats caused heel sores so painful that Nike had to let its athletes wear Adidas (with black tape over the trademark). In 1997, Nike ramped up production just before the Asian banking crisis killed demand, then flooded the market with cheap shoes, undercutting its own retailers and driving many into the arms of Adidas. Two years later, the company created a $15 Swoosh-bearing canvas sneaker designed for poor Chinese. The "World Shoe" flopped so badly that Nike killed it.

Yet all that amounts to a frayed shoelace compared with losing China's most famous living human. Yao Ming had worn Nike since Rhoads discovered him as a skinny kid with a sweet jumper—and brought him some size 18s made for NBA All-Star Alonzo Mourning. In 1999 he signed Yao to a four-year contract worth $200,000. But Nike let his contract expire last year. Yao defected to Reebok for an estimated $100 million. The failure leaves Nike executives visibly dejected. "The only thing I know is, we lost Yao Ming," says a Shanghai executive who negotiated with the star.

Nike is determined not to repeat the mistake. It has already signed China's next NBA prospect, the 7-ft. Yi Jianlian, 18, who plays for the Guangdong Tigers. And the company has resolved problems that dogged it a few years ago. Nike has cleaned up its shop floors. It cut its footwear suppliers in China from 40 to 16, and 15 of those sell only to Nike, allowing the company to monitor conditions more easily. At Shoetown in the southern city of Guangzhou, 10,000 mostly female laborers work legal hours stitching shoes for $95 a month—more than minimum wage. "They've made huge progress," says Li Qiang, director of New York City-based China Labor Watch.

In China, Nike is hardly viewed as the ugly imperialist. In fact, the company's celebration of American culture is totally in synch with the Chinese as they hurtle into a chaotic, freer time. In July, at a Nike three-on-three competition in the capital, a Chinese DJ named Jo Eli played songs like I'll Be Damned off his Dell computer. "Nike says play hip-hop because that's what blacks listen to," he says. "The government doesn't exactly promote these things. But we can all expose ourselves to something new." That sounds pretty close to a Chinese translation of "Just Do It."

What's Wrong With This Picture?

THE RISE OF THE MEDIA CARTEL HAS BEEN A LONG TIME COMING. THE CULTURAL EFFECTS ARE NOT NEW IN KIND, BUT THE PROBLEM HAS BECOME CONSIDERABLY LARGER.

MARK CRISPIN MILLER

For all their economic clout and cultural sway, the ten great multinationals—AOL Time Warner, Disney, General Electric, News Corporation, Viacom, Vivendi, Sony, Bertelsmann, AT&T and Liberty Media—rule the cosmos only at the moment. The media cartel that keeps us fully entertained and permanently half-informed is always growing here and shriveling there, with certain of its members bulking up while others slowly fall apart or get digested whole. But while the players tend to come and go—always with a few exceptions—the overall Leviathan itself keeps getting bigger, louder, brighter, forever taking up more time and space, in every street, in countless homes, in every other head.

The rise of the cartel has been a long time coming (and it still has some way to go). It represents the grand convergence of the previously disparate US culture industries—many of them vertically monopolized already—into one global superindustry providing most of our imaginary "content." The movie business had been largely dominated by the major studios in Hollywood; TV, like radio before it, by the triune axis of the networks headquartered in New York; magazines, primarily by Henry Luce (with many independent others on the scene); and music, from the 1960s, mostly by the major record labels. Now all those separate fields are one, the whole terrain divided up among the giants—which, in league with Barnes & Noble, Borders and the big distributors, also control the book business. (Even with its leading houses, book publishing was once a cottage industry at both the editorial and retail levels.) For all the democratic promise of the Internet, moreover, much of cyberspace has now been occupied, its erstwhile wildernesses swiftly paved and lighted over by the same colossi. The only industry not yet absorbed into this new world order is the newsprint sector of the Fourth Estate—a business that was heavily shadowed to begin with by the likes of Hearst and other, regional grandees, flush with the ill-gotten gains of oil, mining and utilities—and such absorption is, as we shall see, about to happen.

Thus what we have today is not a problem wholly new in kind but rather the disastrous upshot of an evolutionary process whereby that old problem has become considerably larger—and that great quantitative change, with just a few huge players now co-directing all the nation's media, has brought about enormous qualitative changes. For one thing, the cartel's rise has made extremely rare the sort of marvelous exception that has always popped up, unexpectedly, to startle and revivify the culture—the genuine independents among record labels, radio stations, movie theaters, newspapers, book publishers and so on. Those that don't fail nowadays are so remarkable that they inspire not emulation but amazement. Otherwise, the monoculture, endlessly and noisily triumphant, offers, by and large, a lot of nothing, whether packaged as "the news" or "entertainment."

Of all the cartel's dangerous consequences for American society and culture, the worst is its corrosive influence on journalism. Under AOL Time Warner, GE, Viacom et al., the news is, with a few exceptions, yet another version of the entertainment that the cartel also vends nonstop. This is also nothing new—consider the newsreels of yesteryear—but the gigantic scale and thoroughness of the corporate concentration has made a world of difference, and so has made this world a very different place.

Let us start to grasp the situation by comparing this new centerfold with our first outline of the National Entertainment State, published in the spring of 1996. Back then, the national TV news appeared to be a tidy tetrarchy: two network news divisions owned by large appliance makers/weapons manufacturers (CBS by Westinghouse, NBC by General Electric), and the other two bought lately by the nation's top purveyors of Big Fun (ABC by Disney, CNN by Time Warner). Cable was still relatively immature, so that, of its many enterprises, only CNN competed with the broadcast networks' short-staffed newsrooms; and its buccaneering founder, Ted Turner, still seemed to call the shots from his new aerie at Time Warner headquarters.

Today the telejournalistic firmament includes the meteoric Fox News Channel, as well as twenty-six television stations owned outright by Rupert Murdoch's News Corporation (which holds majority ownership in a further seven). Although ultimately thwarted in his bid to buy DirecTV and thereby dominate the US satellite television market, Murdoch wields a pervasive influence on the news—and not just in New York, where he has two TV stations, a major daily (the faltering *New York Post*) *and* the Fox News Channel, whose inexhaustible platoons of shouting heads attracts a fierce plurality of cable-viewers. Meanwhile, Time Warner has now merged with AOL—so as to own the cyberworks through which to market its floodtide of movies, ball games, TV shows, rock videos, cartoons, standup routines and (not least) bits from CNN, CNN Headline News, CNNfn (devised to counter GE's CNBC) and CNN/Sports Illustrated (a would-be rival to Disney's ESPN franchise). While busily cloning CNN, the parent company has also taken quiet steps to make it more like Fox, with Walter Isaacson, the new head honcho, even visiting the Capitol to seek advice from certain rightist pols on how, presumably, to make the network even shallower and more obnoxious. (He also courted Rush Himself.) All this has occurred since the abrupt defenestration of Ted Turner, who now belatedly laments the overconcentration of the cable business: "It's sad we're losing so much diversity of thought," he confesses, sounding vaguely like a writer for this magazine.

Whereas five years ago the clueless Westinghouse owned CBS, today the network is a property of the voracious Viacom—matchless cable occupier (UPN, MTV, MTV2, VH1, Nickelodeon, the Movie Channel, TNN, CMT, BET, 50 percent of Comedy Central, etc.), radio colossus (its Infinity Broadcasting—home to Howard Stern and Don Imus—owns 184 stations), movie titan (Paramount Pictures), copious publisher (Simon & Schuster, Free Press, Scribner), a big deal on the web and one of the largest US outdoor advertising firms. Under Viacom, CBS News has been obliged to help sell Viacom's product—in 2000, for example, devoting epic stretches of *The Early Show* to what lately happened on *Survivor* (CBS). Of course, such synergistic bilge is commonplace, as is the tendency to dummy up on any topic that the parent company (or any of its advertisers) might want stifled. These journalistic sins have been as frequent under "longtime" owners Disney and GE as under Viacom and Fox [see Janine Jaquet, "The Wages of Synergy"]. They may also abound beneath Vivendi, whose recent purchase of the film and TV units of USA Networks and new stake in the satellite TV giant EchoStar—could soon mean lots of oblique self-promotion on *USAM News*, in *L'Express* and *L'Expansion*, and through whatever other news-machines the parent buys.

Such is the telejournalistic landscape at the moment—and soon it will mutate again, if Bush's FCC delivers for its giant clients. On September 13, when the minds of the American people were on something else, the commission's GOP majority voted to "review" the last few rules preventing perfect oligopoly. They thus prepared the ground for allowing a single outfit to own both a daily paper and a TV station in the same market—an advantage that was outlawed in 1975. (Even then, pre-existing cases of such ownership were grandfathered in, and any would-be owner could get that rule waived.) That furtive FCC "review" also portended the elimination of the cap on the percentage of US households that a single owner might reach through its TV stations. Since the passage of the Telecommunications Act of 1996, the limit had been 35 percent. Although that most indulgent bill was dictated by the media giants themselves, its restrictions are too heavy for this FCC, whose chairman, Michael Powell, has called regulation per se "the oppressor."

And so, unless there's some effective opposition, the several-headed vendor that now sells us nearly all our movies, TV, radio, magazines, books, music and web services will soon be selling us our daily papers, too—for the major dailies have, collectively, been lobbying energetically for that big waiver, which stands to make their owners even richer (an expectation that has no doubt had a sweetening effect on coverage of the Bush Administration). Thus the largest US newspaper conglomerates—the New York Times, the Washington Post, Gannett, Knight-Ridder and the Tribune Co.—will soon be formal partners with, say, GE, Murdoch, Disney and/or AT&T; and then the lesser nationwide chains (and the last few independents) will be ingested, too, going the way of most US radio stations. America's cities could turn into informational "company towns," with one behemoth owning all the local print organs—daily paper(s), alternative weekly, city magazine—as well as the TV and radio stations, the multiplexes and the cable system. (Recently a federal appeals court told the FCC to drop its rule preventing any one company from serving more than 30 percent of US cable subscribers; and in December, the Supreme Court refused to hear the case.) While such a setup may make economic sense, as anticompetitive arrangements tend to do, it has no place in a democracy, where the people have to know more than their masters want to tell them.

That imperative demands reaffirmation at this risky moment, when much of what the media cartel purveys to us is propaganda, commercial or political, while no one in authority makes mention of "the public interest"—except to laugh it off. "I have no idea," Powell cheerily replied at his first press conference as chairman, when asked for his own definition of that crucial concept. "It's an empty vessel in which people pour in whatever their preconceived views or biases are." Such blithe obtuseness has marked all his public musings on the subject. In a speech before the American Bar Association in April 1998, Powell offered an ironic little riff about how thoroughly he doesn't get it: "The night after I was sworn in [as a commissioner], I waited for a visit from the angel of the public interest. I waited all night, but she did not come." On the other hand, Powell has never sounded glib about his sacred obligation to the corporate interest. Of his decision to move forward with the FCC vote just two days after 9/11, Powell spoke as if that sneaky move had been a gesture in the spirit of Patrick Henry: "The flame of the American ideal may flicker, but it will never be extinguished. We will do our small part and press on with our business, solemnly, but resolutely."

Certainly the FCC has never been a democratic force, whichever party has been dominant. Bill Clinton championed the disastrous Telecom Act of 1996 and otherwise did almost nothing to impede the drift toward oligopoly. (As *Newsweek* reported in 2000, Al Gore was Rupert Murdoch's personal choice for President. The mogul apparently sensed that Gore would happily play ball with him, and also thought—correctly—that the Democrat would win.)

What is unique to Michael Powell, however, is the showy superciliousness with which he treats his civic obligation to address the needs of people other than the very rich. That spirit has shone forth many times—as when the chairman genially compared the "digital divide" between the information haves and have-nots to a "Mercedes divide" between the lucky few who can afford great cars and those (like him) who can't. In the intensity of his pro-business bias, Powell recalls Mark Fowler, head of Reagan's FCC, who famously denied his social obligations by asserting that TV is merely "an appliance," "a toaster with pictures." And yet such Reaganite *bons mots*, fraught with the anti-Communist fanaticism of the late cold war, evinced a deadly earnestness that's less apparent in General Powell's son. He is a blithe, postmodern sort of ideologue, attuned to the complacent smirk of Bush the Younger—and, of course, just perfect for the cool and snickering culture of TV.

Although such flippancies are hard to take, they're also easy to refute, for there is no rationale for such an attitude. Take "the public interest"—an ideal that really isn't hard to understand. A media system that enlightens us, that tells us everything we need to know pertaining to our lives and liberty and happiness, would be a system dedicated to the public interest. Such a system would not be controlled by a cartel of giant corporations, because those entities are ultimately hostile to the welfare of the people. Whereas we need to know the truth about such corporations, they often have an interest in suppressing it (as do their advertisers). And while it takes much time and money to find out the truth, the parent companies prefer to cut the necessary costs of journalism, much preferring the sort of lurid fare that can drive endless hours of agitated jabbering. (Prior to 9/11, it was Monica, then *Survivor* and Chandra Levy, whereas, since the fatal day, we have had mostly anthrax, plus much heroic footage from the Pentagon.) The cartel's favored audience, moreover, is that stratum of the population most desirable to advertisers—which has meant the media's complete abandonment of working people and the poor. And while the press must help protect us against those who would abuse the powers of government, the oligopoly is far too cozy with the White House and the Pentagon, whose faults, and crimes, it is unwilling to expose. The media's big bosses want big favors from the state, while the reporters are afraid to risk annoying their best sources. Because of such politeness (and, of course, the current panic in the air), the US coverage of this government is just a bit more edifying than the local newscasts in Riyadh.

Against the daily combination of those corporate tendencies—conflict of interest, endless cutbacks, endless trivial pursuits, class bias, deference to the king and all his men—the

public interest doesn't stand a chance. Despite the stubborn fiction of their "liberal" prejudice, the corporate media have helped deliver a stupendous one-two punch to this democracy. (That double whammy followed their uncritical participation in the long, irrelevant *jihad* against those moderate Republicans, the Clintons.) Last year, they helped subvert the presidential race, first by prematurely calling it for Bush, regardless of the vote—a move begun by Fox, then seconded by NBC, at the personal insistence of Jack Welch, CEO of General Electric. Since the coup, the corporate media have hidden or misrepresented the true story of the theft of that election.

And having justified Bush/Cheney's coup, the media continue to betray American democracy. Media devoted to the public interest would investigate the poor performance by the CIA, the FBI, the FAA and the CDC, so that those agencies might be improved for our protection—but the news teams (just like Congress) haven't bothered to look into it. So, too, in the public interest, should the media report on all the current threats to our security—including those far-rightists targeting abortion clinics and, apparently, conducting bioterrorism; but the telejournalists are unconcerned (just like John Ashcroft). So should the media highlight, not play down, this government's attack on civil liberties—the mass detentions, secret evidence, increased surveillance, suspension of attorney-client privilege, the encouragements to spy, the warnings not to disagree, the censored images, sequestered public papers, unexpected visits from the Secret Service and so on. And so should the media not parrot what the Pentagon says about the current war, because such prettified accounts make us complacent and preserve us in our fatal ignorance of what people really think of us—and why—beyond our borders. And there's much more—about the stunning exploitation of the tragedy, especially by the Republicans; about the links between the Bush and the bin Laden families; about the ongoing shenanigans in Florida—that the media would let the people know, if they were not (like Michael Powell) indifferent to the public interest.

In short, the news divisions of the media cartel appear to work *against* the public interest—and *for* their parent companies, their advertisers and the Bush Administration. The situation is completely un-American. It is the purpose of the press to help us run the state, and not the other way around. As citizens of a democracy, we have the right and obligation to be well aware of what is happening, both in "the homeland" and the wider world. Without such knowledge we cannot be both secure and free. We therefore must take steps to liberate the media from oligopoly, so as to make the government our own.

Mark Crispin Miller is a professor of media studies at New York University, where he directs the Project on Media Ownership. He is the author of The Bush Dyslexicon: Observations on a National Disorder *(Norton).*

UNIT 5
Conflict

Unit Selections

Key Points to Consider

- Are violent conflicts and warfare increasing or decreasing today?

- What changes have taken place in recent years in the types of conflicts and who participates?

- How is military doctrine changing to reflect new political realities?

- How is the nature of terrorism different than conventional warfare? What new threats do terrorists pose?

- What are the motivations and attitudes of those who use terror as a political tool?

- What challenges does nuclear proliferation pose to the U.S.?

- How is the national security policy of the United States likely to change? What about Russia, India, and China?

DUSHKIN ONLINE **Links: www.dushkin.com/online/**
These sites are annotated in the World Wide Web pages.

DefenseLINK
http://www.defenselink.mil
Federation of American Scientists (FAS)
http://www.fas.org
ISN International Relations and Security Network
http://www.isn.ethz.ch
The NATO Integrated Data Service (NIDS)
http://www.nato.int/structur/nids/nids.htm

$\mathbf{D}$o you lock your doors at night? Do you secure your personal property to avoid theft? These are basic questions that have to do with your sense of personal security. Most individuals take steps to protect what they have, including their lives. The same is true for groups of people, including countries.

In the international arena, governments frequently pursue their national interest by entering into mutually agreeable "deals" with other governments. Social scientists call these types of arrangements "exchanges" (i.e., each side gives up something it values in order to gain something in return that it values even more). On an economic level, it functions like this: "I have the oil that you need and am willing to sell it. In return I want to buy from you the agricultural products that I lack." Whether on the governmental level or the personal level ("If you help me with my homework, then I will drive you home this weekend"), exchanges are the process used by most individuals and groups to obtain and protect what is of value. The exchange process, however, can break down. When threats and punishments replace mutual exchanges, conflict ensues. Neither side benefits and there are costs to both. Furthermore, each may use threats with the expectation that the other will capitulate. But if efforts at intimidation and coercion fail, the conflict may escalate into violent confrontation.

With the end of the cold war, issues of national security and the nature of international conflict have changed. In the late 1980s agreements between the former Soviet Union and the United States led to the elimination of superpower support for participants in low-intensity conflicts in Central America, Africa, and Southeast Asia. Fighting the cold war by proxy is now a thing of the past. In addition, cold-war military alliances have ei-

ther collapsed or have been significantly redefined. Despite these historic changes, there is no shortage of conflicts in the world today.

Many experts initially predicted that the collapse of the Soviet Union would decrease the arms race and diminish the threat of nuclear war. However, some analysts now believe that the threat of nuclear war has in fact increased as control of nuclear weapons has become less centralized and the command structure less reliable. In addition, the proliferation of nuclear weapons into North Korea and South Asia (India and Pakistan) is a growing security issue. Further, there are concerns about both dictatorial governments and terrorist organizations obtaining weapons of mass destruction. What these changing circumstances mean for U.S. policy is a topic of considerable debate.

The unit begins with two articles on contemporary U.S. national security, including the fundamental redefinition of U.S. security in the post 9/11 era (including the potential of nuclear terrorism). This is followed with two different perspectives on terrorism and the conflict in the Middle East. The unit concludes with a series of articles that present specific case studies, including the challenges created by nuclear proliferation.

As in the case of the other global issues described in this anthology, international conflict is a dynamic problem. It is important to understand that conflicts are not random events, but follow patterns and trends. Forty-five years of cold war established discernable patterns of international conflict as the superpowers deterred each other with vast expenditures of money and technological know-how. The consequence of this stalemate was often a shift to the developing world for conflict by superpower proxy.

The changing circumstances of the post-cold war era generate a series of important new policy questions: Will there be more nuclear proliferation? Is there an increased danger of so-called "rogue" states destabilizing the international arena? Is the threat of terror a temporary or permanent feature of world affairs? Will there be a growing emphasis on low-intensity conflicts related to the interdiction of drugs, or will some other unforeseen issue determine the world's hot spots? Will the United States and its European allies lose interest in security issues that do not directly involve their economic interests and simply look the other way? For example, as age-old ethnic conflicts become brutally violent? Can the international community develop viable institutions to mediate and resolve disputes before they become violent? The answers to these and related questions will determine the patterns of conflict in the twenty-first century.

The Transformation of National Security

Five Redefinitions

Philip Zelikow

PERHAPS THE key question of international politics and U.S. national security policy today is whether a genuinely new era has dawned since the end of the Cold War. It has. The attacks of September 11, 2001 did not create the new era, but they were a catalytic moment in our recognition of it. Like previous shocks to the United States in June 1940, December 1941 or June 1950, this shock gave emerging trends a form, brought them into mass consciousness, and forced upon us the task of defining a comprehensive national response.

Such a definition appears in the Bush Administration's recently published *National Security Strategy of the United States*.[1] This essay draws out some of the ideas that appear to undergird the administration's emerging strategy. It focuses on five essential redefinitions of what national security means for the United States in the 21st century—but first a note about the rhetoric of empire that has come to dominate much current discussion.

The Distraction of Empire

ALL NATIONAL security strategies start with a mental image of the world. The image of the new era is properly that of a modern and truly pluralistic international system. In the traditional world, populations, governance, commerce, culture and habits of life evolved slowly and changed gradually. The break point between this traditional world and that of modernity arrives, at different times for different societies, when social and technological change severs the links that had defined the relationships between humanity and the physical resources of the earth; shatters the hitherto ageless ceilings on productivity enforced by the physical limits of humans, animals, wind and water; and transforms our ability to communicate across distances, communities and nations.

These transformations acquired momentum in the 18th century and spread in the 19th until, by the year 1900, the modern world extended to Europe, North America and to their limited veins of settlement and commerce stretched out over the rest of the planet. The 20th century saw the further, nearly global extension of modernity and wrenching, worldwide efforts to adjust to its impact. Indeed, such has been the shock to many societies that the last century has been dominated by great con-

tests over how to conceive, organize and provide moral justification and political order for modern industrial nations.

Those great contests have now subsided. The world is no longer so broken and divided. The militant utopias of class, nation and race have been defeated and discredited. Modernity means that change itself becomes the constant, and the challenge of change has not disappeared. Its pressures have instead been internalized within every society trying to adapt to a quickening pace of change. In other words, today's battle lines are less *international* and more *transnational*.

In dealing with these pressures, some see the United States as a new source of world order—fearing or welcoming that prospect. This accounts in part for the current popularity of "empire" as a reigning metaphor for America's ambitions. The metaphor is seductive, yet vicious.

What is an empire? Once upon a time, "empire" was casually applied as a positive expression, as with Jefferson's "empire of liberty." In recent years "empire" has been used to describe—often with an edge—any circumstance where a powerful country exerts influence over lesser powers, whether direct or indirect, physical, cultural or commercial. This shallow equation of all sorts of economic and cultural influences with "imperialism" was first popularized about a hundred years ago by writers reacting to Britain's war against the Boers in South Africa. John Atkinson Hobson, for example, who had an immense influence on Vladimir Lenin, condemned economic imperialism as a ruling force in the world. Hobson also worked a strong element of his anti-Semitism into his theories, seeing a cabal of Jewish bankers and merchants lurking behind Britain's excesses.

The same distasteful mixture of bad thinking was prominent after World War I. Many writers in Europe, especially in the defeated nations, decried the new "Anglo-Saxon empire" that they accused Britain and America—and the Jews, of course—of having created.[2] The same fetid intellectual waters seep from this old gutter into much anti-American commentary today.

Sadly, even some American conservatives have joined the new anti-imperialists in seeing the United States as the metropole of a world empire. The appeal is understandable. After all, there is a certain gratification in imagining that one is the successor to emperors and proconsuls of past ages, a certain pleasure in opening such a dusty, venerable chest of ideas about how to sort out the affairs of faraway peoples.

But these imperial metaphors, of whatever provenance, do not enrich our understanding; they impoverish it. They use a metaphor of how to rule others when the problem is how to persuade and lead them. Real imperial power is sovereign power. Sovereigns rule, and a ruler is not just the most powerful among diverse interest groups. Sovereignty means a *direct* monopoly control over the organization and use of armed might. It means *direct* control over the administration of justice and the definition thereof. It means control over what is bought and sold, the terms of trade and the permission to trade, to the limit of the ruler's desires and capacities.

In the modern, pluralistic world of the 21st century, the United States does not have anything like such direct authority over other countries, nor does it seek it. Even its informal influence in the political economy of neighboring Mexico, for instance, is far more modest than, say, the influence the British could exert over Argentina a hundred years ago.

The purveyors of imperial metaphors suffer from a lack of imagination, and more, from a lack of appreciation for the new conditions under which we now live. It is easier in many respects to communicate images in a cybernetic world, so that a very powerful United States does exert a range of influences that is quite striking. But this does not negate the proliferating pluralism of global society nor does it suggest a will to imperial power in Washington. The proliferation of loose empire metaphors thus distorts into banal nonsense the only precise meaning of the term imperialism that we have.

The United States is central in world politics today, not omnipotent. Nor is the U.S. Federal government organized in such a fashion that would allow it to wield durable imperial power around the world—it has trouble enough fashioning coherent policies within the fifty United States. Rather than exhibiting a confident will to power, we instinctively tend, as David Brooks has put it, to "enter every conflict with the might of a muscleman and the mentality of a wimp." We must speak of American power and of responsible ways to wield it; let us stop talking of American empire, for there is and there will be no such thing.

Five Redefinitions

THE UNITED States does have unique responsibilities as the greatest power in this pluralistic world, however. Those responsibilities have moved the Bush Administration to rethink the meaning of America's national security and it is a process of thinking that transcends yesterday's partisan differences. Both conservative and liberal orthodoxies are being challenged. To attain lasting influence, these new ideas must pass into the vocabulary and assumptions of many in *both* parties—just as happened with strategies of "containment" in the late 1940s and early 1950s—even as political conflict continues around the edges of this new vision.

- This vision is redefining the geography of national security.
- It is redefining the nexus between principles and power.

- It is redefining the structure of international security.
- It is redefining multilateralism.
- And it is redefining national security threats in the dimension of time.

The New Geography of National Security

IN THE PAST, the geography of national security was defined by foreign frontiers. Dangerous enemies had to possess mass and scale as they first accumulated armies, navies or air forces and then deployed them. Today the frontiers of national security can be everywhere. The point is so obvious in the case of mass-casualty terrorism that it needs no elaboration.

Less obvious is the way the Bush Administration, following on but surpassing the Clinton Administration, has consistently identified poverty, pandemic disease, biological and genetic dangers, and environmental degradation as significant national security threats. All of these dangers have a transnational character, their social origins resting within the turmoil of modern and modernizing societies. In other words, national security threats in the new era are defined more by the fault-lines within societies than by the territorial borders between them. The decisive clashes in this phase of history are not therefore between civilizations but inside them.

The implications of this redefinition touch every major institution of national security in the U.S. government. We are witness to the largest Executive Branch transformation in half a century with the new Department of Homeland Security, the largest reorganization of the FBI in a generation, significant overhauling of the U.S. intelligence community, the creation of an entirely new unified command and other dramatic restructuring in the Department of Defense. Fundamental relationships between Federal, state, regional and local agencies—and the private sector—must change, and are changing, as well. These transformations are just getting under way, but they are already breaking down the core paradigm of the American national security system created in the 1940s and 1950s.

The geographical redefinition of national security profoundly affects the entire U.S. foreign policy agenda. Whether dealing with terrorism or public health, the division of security policy into domestic and foreign compartments is breaking down. U.S. foreign and security policies must delve into societies, into problems from law enforcement to medical care, in novel ways—challenging international institutions and the principles that define them to adapt. More fundamentally, if the United States is to develop national security policies that are aimed at the fault-lines within societies, those policies must transcend physical and material dimensions. They must include positions about fundamental principles.

The New Centrality Of Moral Principles

DURING AND after World War II a conventional image of American interests developed that posed a dichotomy between realism and idealism. Practically every thinker on foreign policy alive today has grown up under the influence of this dichotomy. Realism was usually identified as a cold-eyed focus on calculations of power. Idealism embraced a pre-eminent concern for human rights, global poverty or other facets of human welfare. These two images became a convenient shorthand for labeling political factions or individual leaders. But those stereotypes, overly simple to begin with, no longer even remotely suit our times. They do not capture the nature of the controversies within the present administration, nor do they comprehend the new fusion of power and principle that is now guiding U.S. policy.

The administration's concept of a "balance of power that favors freedom" —to note the marquee concept of the *National Security Strategy*—applies calculations of power to the worldwide capacity to support beneficial principles affecting both relations among states and conditions within them. The administration thus emphasizes both power and a readiness to distinguish good from bad, right from wrong. The administration takes to heart the observation of John Courtney Murray that

> policy is the meeting-place of the world of power and the world of morality, in which there takes place the concrete reconciliation of the duty of success that rests upon the statesman, and the duty of justice that rests upon the civilized nation that he serves.[3]

The administration drives power and principle together around a remarkably straightforward statement of the "nonnegotiable demands of human dignity." Seven of these demands, all originally appearing in the President's January 2002 State of the Union message, are listed in the strategy document: the rule of law; limits on the power of the state; respect for women; private property; free speech; equal justice; and religious tolerance. All seven focus on the relationship of individuals to the state. None deals directly with the form or processes of government to produce these relationships: There is no mention here of democracy. Far from being an assertion of American exceptionalism, therefore, or a call for others to emulate the example of our "city on a hill", both the strategy document and President Bush have stressed that these are universal principles that apply everywhere. This is not an affirmation of the Scottish Enlightenment, but of human civilization itself.

Such a stance is critical in an age that seems to contradict or qualify every universal truth, preferring instead to cultivate irony as the essential human sensibility. The Bush Administration has defined one constant as the essential complement to modernity, what the *National Security Strategy* calls a "single, sustainable model for national success." That model features a linked conception of respect for human dignity and regard for liberating human potential. If modernity implies constant change, then greater personal and economic freedom is the perpetual safety valve, the constant source of adaptation and thus the very source of structural resilience that enables different societies to organize themselves in different successful forms. Influenced by his experiences in the Balkan crises, Tony Blair put the matter well in his October 1, 2002 speech to the Labour Party conference in Blackpool:

> Our values aren't Western values. They're human values and anywhere, anytime people are given the chance, they embrace them. Around these values, we build our global partnership. Europe and America to-

gether. Russia treated as a friend and equal. China and India seeking not rivalry but cooperation and for all nations the basis of our partnership—not power alone but a common will based on common values, applied in an even-handed way.

The Bush Administration's similar emphasis on moral reasoning, on uncomfortable (to some) judgments about good and evil, challenges some conservatives and liberals alike. The challenge to conservatives flows from the fact that traditional conservatism is founded on a cultural pessimism, on an abiding skepticism about human improvement. But President Bush (and many of his key advisors) are cultural optimists who, in their own lifetimes, have witnessed great and *positive* upheaval—not least the successful end to the Cold War. Indeed, President Bush's whole education reform agenda ("Leave no child behind!") epitomizes the break with traditional conservatism, and it is an attitude that carries over to world affairs. His administration's efforts on HIV/AIDS, for example, have not bothered with the Clinton Administration's convoluted effort to justify attention to global disease in traditional national security terms. The danger is not dressed up as a threat to "stability" but is portrayed for what it really is: a moral obligation to act when tens of millions face preventable death.

The administration's rhetoric is broader still: "Including all of the world's poor in an expanding circle of development—and opportunity", the *National Security Strategy* states, "is a moral imperative...one of the top priorities of U.S. international policy." Hence the proposal for a 50 percent increase in development assistance, the largest proposed by any administration since that of John F. Kennedy. President Bush is gambling that a notoriously broken foreign aid system can be reformed to make such investments politically saleable, and perhaps thereby legitimize an even greater effort in the future.

The challenge to liberals has been more interesting, if somewhat depressing in its manifestations. The American Left in its various hues once defined itself by its fierce belief in values, a readiness to judge and to act on those judgments. It is now too often defined by dedication to a consensus that disclaims any right to make moral judgments at all—except those that condemn the U.S. government. The Left's "knowing" attitude about America's own failings has evolved into the chronic suspicion of any expression of patriotism, and into the instinctive assumption that anyone who uses words like "evil" must be some Bible-thumping rube pronouncing naive moral judgments. And to think that such a person is President of the United States!

But many liberal Democrats are now wisely pressing for a return to the ancient virtues. What, then, is the net result? One old contrarian of the Left, Christopher Hitchens wrote of it in the October 20, 2002 *Washington Post*:

> As someone who has done a good deal of marching and public speaking about Vietnam, Chile, South Africa, Palestine, and East Timor in his time (and would do it all again), I can only hint at how much I despise a Left that thinks of Osama Bin Laden as a slightly misguided anti-imperialist, or a Left that can think of Milosevic and Saddam as victims. Instead of interna-

123

tionalism, we find among the Left now a sort of affect-less, neutralist, smirking isolationism. In this moral universe, the views of the corrupt and conservative Jacques Chirac—who built Saddam Hussein a nuclear reactor, knowing what he wanted it for—carry more weight than that of persecuted Iraqi democrats.

Of course, the Bush Administration's emphasis on moral judgment is vulnerable to criticisms. One is simply aesthetic. The image of the preachy American abroad, short on talent and long on sanctimony, inspires enough resentment already, and few foreign diplomats want to encourage more such Americans to sally forth. But there are more substantive problems, too. One concerns the so-called slippery slope—and it *is* slippery. Yet it is hard to explain why, because one cannot right wrongs everywhere, one should not try to right them anywhere. If the United States simply deals and trades with all countries as we find them, then how, in an age when the frontiers of national security are increasingly defined by issues inside societies, can the United States take stands on the battle lines that matter most?

Then there is the problem of hypocrisy and double standards. (Saudi Arabia is of late the usual friendly villain hauled into the dock for illustration.) Here the administrations goals and rhetoric do bring with them a real burden. Whether it chooses to be reticent or outspoken, whether its reflections are offered in private or in public, any administration that stresses moral judgment is obliged to display a basic honesty about the character of other governments. And it assumes the burden of a consistent regard for the non-negotiable demands of human dignity, no matter the pull of other interests or the nation s accounts. Sometimes other strategic equities will win pride of place, and the administration should not lie to itself about the tradeoffs—either their difficulty or their durability—that will have to be made.

Confronting a messy world full of frustrating choices, President Bush might draw comfort from the remarks of another wartime president, Franklin Roosevelt:

> I am everlastingly angry only at those who assert vociferously that the four freedoms and the Atlantic Charter are nonsense because they are unattainable. If those people had lived a century and a half ago they would have sneered and said that the Declaration of Independence was utter piffle. If they had lived nearly a thousand years ago they would have laughed uproariously at the ideals of Magna Carta. And if they had lived several thousand years ago they would have derided Moses when he came from the Mountain with the Ten Commandments.
>
> We concede that these great teachings are not perfectly lived up to today, but I would rather be a builder than a wrecker, hoping always that the structure of life is growing—not dying.[4]

The New Structure Of World Politics

ANOTHER reason for thinking hard about the centrality of moral judgment in redefining American national security is because of the way the Bush Administration, following on the work of its predecessors, is trying to integrate universal principles into great power politics. For centuries the structure of world politics has been defined by the rivalry of great powers. It is now possible instead for the United States to form active agendas of cooperation with every major center of global power, founded on imperfect yet extraordinary degrees of agreement about underlying principles in the organization of society.

A circumstance of unprecedented great power harmony is neither an immutable fact nor a deterministic prediction. It is a contingency over which we have limited but not inconsiderable influence. Great power rivalry could return to the foreground, so present circumstances ought to be viewed as an opportunity. It is an opportunity that can be lost, however, if the Bush Administration and its successors allow serious agendas of cooperation to drift into routinized gestures and petty fractiousness. These agendas, moreover, will never be built entirely out of the bilateral relationships between the United States and other great powers. Instead, the United States must challenge its present and future partners to join in common tasks that transcend narrow concerns, offering the networks of American allies in Europe and Asia real opportunities to share the responsibilities of global leadership.[5]

In this regard, the United States and its traditional allies have already given considerable attention to Russia and China as potential great powers in transition. India, set to become not only the most populous country in the world but also its largest and most diverse democracy, deserves comparable notice. If India can balance these attributes with sustained growth and stability, the world must bring that nation, and the distinct and ancient civilization it represents, into every inner circle of global power. That would include at the least permanent representation on the United Nations Security Council and in the Group of Eight.

Varieties Of Multilateralism

THE CARTOON version of America's international policy dilemma poses a choice between unilateralism versus multilateralism, the wild cowboy versus the cooperative diplomat. This depiction is false.

Everything that America does in the world is done multilaterally. That emphatically includes the policies the Bush Administration considers most important, and even those that are the most "military" in character. The global war against terrorism is being conducted through an elaborate, often hidden, network of multilateral cooperation among scores of governments. A large number of players are interacting on intelligence, law enforcement, military action, air transportation, shipping, financial controls and more. Ongoing military operations in Afghanistan involve several countries, and were multilateral even at the height of American military activity, as the United States relied heavily on relationships with Pakistan, Russia, three Central Asian governments and a variety of Afghan factions.

The caricature of the administration's unilateralism usually rests on the recitation of a by now standard list of diplomatic actions that some other governments did not like (Kyoto, the International Criminal Court and so on). Some of these disagreements were handled in a style and manner that seemed insensitive or simply maladroit. Unfortunately, too, the caricature of the administration's unilateralism is willingly fed by some U.S. officials and unofficial advisers who relish the chance to play the role of the truth teller lancing foreign obfuscations. Sometimes they overplay the part, sensing the license they get from working for a plain-spoken president.

President Bush, however, is more sensitive to foreign opinion than some who act in his name. He knows that, to much of the world, "I'm the toxic Texan, right?" He recognizes that

> if you want to hear resentment, just listen to the word unilateralism. I mean, that's resentment. If somebody wants to try to say something ugly about us, 'Bush is a unilateralist, America is unilateral.' You know, which I find amusing.

He finds it amusing because he meets and works with foreign dignitaries almost every day, and sees himself as a "pretty good diplomat" —though he smilingly concedes that "nobody else does." From the time Bush first started making decisions about the Iraq issue, for example, he has worked at every turn through international coalitions, noting how much he had enjoyed building one for the war in Afghanistan.[6]

Europeans, of course, rush to play the part of the cosmopolitan professional who rolls his eyes and offers wittily barbed observations about American innocents abroad. Sometimes they, too, overplay the part. But all this is a very old genre of diplomatic theater; John Adams and the Comte de Vergennes set the archetypes more than two hundred years ago. It has since been performed on many stages with a wonderful variety of scripts and sets. No doubt it will be played often in future, as well.

Beyond problems of tone and execution, the real differences in the multilateral strategies of the Bush Administration and other friendly states are not about "unilateralism" versus "multilateralism." They have to do instead with five contrasting ways of conceiving and operationalizing multilateral action.

First, the Bush Administration prefers an inductive method that draws ideas from many sources and adapts them to specific conditions. Alternative deductive strategies develop abstract principles and develop them into generic, universal solutions. For instance, the painstakingly crafted individual agreement drawn up for handling war crimes in Sierra Leone, blending local and foreign judicial traditions, is preferable to the one-size-fits-all approach taken in designing the International Criminal Court. The former is designed to solve real problems and get real results on a case-by-case basis. The latter, unfortunately, aims at more rarefied ambitions.

Second, the administration prefers international institutions that judge performance and stress accountability rather than those that maintain a detached neutrality in order to preserve a friendly consensus. Too often, for example, the institutions charged with preventing the proliferation of weapons of mass destruction seemed to choose courtesy over candor. The history of IAEA inspections in Iraq—going back to the 1970s—is exemplary

Third, the administration prefers multilateral strategies that rely on the sovereign accountability of states instead of strategies that limit sovereignty in order to link states together in a common enterprise—but which thereby dissipate responsibility. As the European Union attempts to develop a "common foreign and security policy", it can hardly avoid a lowest common denominator approach, for example.

Fourth, the administration takes a view of international law that emphasizes democratic accountability, plainly linking the authority of international officials to constitutional sources of political authority that are essentially national in character. Other nations contemplate and encourage much broader delegation of sovereign powers, where only the initial delegation need occur through a democratic process, so that international officials can have greater freedom of action. Hence the U.S. government believes that the International Criminal Court, as a permanent yet essentially stateless entity, might grow ever more distant from the democratic sources of legitimacy that are an essential source for its claimed right to administer global justice.

Fifth, the administration prefers functional institutions that produce concrete results instead of symbolic measures that might rally more support for an ideal, but at the cost of not doing much to further its attainment. Sometimes, as in the case of the Kyoto Accords, well-meaning but dysfunctional efforts can be worse than useless if they complicate attempts to develop a more effective and sustainable solution. We cannot afford such indulgences in a world that in some ways is more threatening today than it was during the Cold War.

Redefining National Security As A Function Of Time

IMAGINE THREATS as having a cadence—rhythm plus speed. In the past, threats tended to emerge slowly, often visibly, as weapons were forged, armies were conscripted, units were trained and deployed, and enemy forces were massed in position to move. The greatest threats, too, came from large states that could raise and equip the mass armies of the industrial age. Precisely because of their size and elaborate structures, these large states had much to lose in a war. Doctrines of deterrence were developed to confront such states, first in the pre-nuclear and then in the nuclear age.

In today's world, threats can emerge more quickly, without having to accumulate a mass of men and metal. Nor do the greatest threats necessarily come from large states that have much to lose.[7] It is thus hard to quarrel with the essential premise of the Bush Administration's open willingness to consider pre-emption, which is that the strategic military doctrines developed for the Cold War must be adapted to the circumstances of the 21st century.

Note, too, the interaction of the new cadence of threat with the new geography of national security. The fault-lines are now inside societies, where even small political factions might have access to weapons of unprecedented destructiveness. The line between internal and international security becomes blurry in parallel with the acceleration of the cadence of threat.

The basic critique of the Bush strategy of pre-emption is that it is better to wait until threats are so acute and universally apparent that a consensus can form in favor of forceful action against them. The flaw in this argument is that there is today a kind of inverse continuum of threat and vulnerability. As a potential enemy's WMD capability becomes more threatening, it becomes less vulnerable to military disruption. Such programs are most vulnerable when they are immature, but that is when the threats they pose are so ambiguous that it is harder to rally allies to act against them. Yet at the point where such threats are so evident that coalitions readily arise, it may be too late—weapons will have already been used, programs will be difficult or impossible to destroy, or outsiders will be themselves deterred by their fear of retaliation.[8]

There is a long tradition in American history of keeping dangerous threats at bay, if necessary by pre-emptive action. Defending Andrew Jackson's pre-emptive invasion of Spanish Florida, his occupation of Pensacola and his execution of individuals inciting Seminole and Creek raids against the United States, Secretary of State John Quincy Adams informed the Spanish government that

> by all the laws of neutrality and of war, as well as of prudence and humanity, [Jackson] was warranted in anticipating his enemy by the amicable and, that being refused, by the forcible occupation.... There will need no citations from printed treatises on international law to prove the correctness of this principle. It is engraved in adamant on the common sense of mankind. No writer upon the laws of nations ever pretended to contradict it. None, of any reputation or authority, ever omitted to assert it.[9]

More recently the United States was prepared to deal with acute dangers by taking pre-emptive or preventive military action against states with which we were technically at peace—Cuba and the Soviet Union in 1962 and North Korea in 1994—if diplomacy had failed to remove or contain specified dangers. The great debates in the Kennedy Administration were over whether or how to even give diplomacy a chance. But there is abundant evidence of President Kennedy's underlying resolve to remove the Soviet missiles from Cuba, one way or another, to prevent an even more dangerous nuclear crisis from arising the next month over Berlin. In 1993 and 1994 the Clinton Administration readied military options against North Korea with evident purpose, but President Clinton never had to make the final decision on what to do if his diplomatic gambit failed. Actually, Iraq is a less clear case of pre-emptive strategy than either of these historical examples, since the United States and the most concerned allies have been in a state of hostilities against Iraq for years. The United States and others have long been conducting constant, if mostly low-intensity, combat operations ever since Iraq conclusively broke the terms of the 1991 military truce that suspended, but never concluded, the Gulf War.

The Bush Administration's strategy is a more explicit adaptation to the new conditions of international life. That explicitness has drawn fire from some who argue that while it might be proper to hold these views, it is not wise to call attention to them. They do not reject the administration's thinking, and can themselves list many examples of real and potential cases of pre-emption. They rather accuse the administration of tactlessness.

The new strategy is somewhat provocative, but it is deliberately so. It must be provocative if it is to foster the painful worldwide debate that must occur in order to condition the international community to think hard about these new dangers, and about how the cadence of security threats has changed. Although the debate is still raging, the older habits of thought are already changing, subtly, around the world. It remains to be seen how the argument over what is and is not tactful will ultimately be settled.

In any event, the United States has not arrogated to itself some vague right to pursue and pummel anyone it dislikes, as many critics contend. The strategy document lists five criteria that must be met for a state or an organization to cast itself outside of ordinary international protection. Condoleezza Rice observed in her Wriston lecture:

> [T]his approach must be treated with great caution. The number of cases in which it might be justified will always be small. It does not give a green light—to the United States or any other nation—to act first without exhausting other means, including diplomacy. Pre-emptive action does not come at the beginning of a long chain of effort. The threat must be very grave. And the risks of waiting must far outweigh the risks of action.

Moreover, the administration's focus is not just on security threats, but also on opportunities. The terrorists have no vision of the future that can assure Muslim parents that their children will lead a better life; the United States does. There are only a few states that could start a new wave of dangerous WMD proliferation, and deflecting them now may nudge history in the right direction. Failure to do so, however, may condemn millions to needless suffering and the American people to years of living in fear.

Grand Strategy and Small Realities

THE BUSH Administration has helped spur worldwide debate not only about the purposes of American power, but about the objectives of the international system as a whole. The United States is not challenging the necessity of international institutions for common action, but it is pressing other nations to decide what they want, to reconsider how to get it, and to re-evaluate old habits in light of new realities. This is an uncomfortable and necessarily disruptive process, and the U.S. government is bound to add its share of distracted stumbles to its progress. But the international agenda is already changing in positive ways.

Grand strategy usually disappoints when it is carried into action. Measured against this or that phrase, this or any administration will come up short. But it is possible to offer general direction, to set different chains of action and reaction into motion. Critics of the Bush Administration's emerging ideas must either accept the new definitions of national security presented or articulate coherent alternatives, working through the implications of present—not past—realities.

There are always too many problems. But the great powers are working together more than anyone would have considered possible half a century ago. New approaches to economic and humanitarian cooperation offer great promise. The advances in human

---

Patience, Flexibility, Intelligence

For progress there is no cure. Any attempt to find automatically safe channels for the present explosive variety of progress must lead to frustration. The only safety possible is relative, and it lies in an intelligent exercise of day-to-day judgment.

The problems created by the combination of the presently possible forms of nuclear warfare and the rather unusually unstable international situation are formidable and not to be solved easily. Those of the next decades are likely to be similarly vexing, 'only more so.' The U.S.-USSR tension is bad, but when other nations begin to make felt their full offensive potential weight, things will not become simpler.

Present awful possibilities of nuclear warfare may give way to others even more awful...[and] we should not deceive ourselves: once such possibilities become actual, they will be exploited. It will, therefore, be necessary to develop suitable new political forms and procedures. All experience shows that even smaller technological changes than those now in the cards profoundly transform political and social relationships. Experience also shows that these transformations are not *a priori* predictable and that most contemporary 'first guesses' concerning them are wrong. For all these reasons, one should take neither present difficulties nor presently proposed reforms too seriously.

The one solid fact is that the difficulties are due to an evolution that, while useful and constructive, is also dangerous. Can we produce the required adjustments with the necessary speed? The most hopeful answer is that the human species has been subjected to similar tests before and seems to have a congenital ability to come through, after varying amounts of trouble. To ask in advance for a complete recipe would be unreasonable. We can specify only the human qualities required: patience, flexibility, intelligence.

—John von Neumann
"Can We Survive Technology?"
Fortune (June 1955)

---

liberty and material well-being just since 1990 have been staggering enough to encourage ambitious thoughts about what might be possible. With new understandings about the real problems the United States and its friends face together, and a flexible, pragmatic approach to achieving them, the prospects have never been better for the project of building a commonwealth of freedom.

Philip Zelikow is the White Burkett Miller Professor of History and Director of the Miller Center of Public Affairs at the University of Virginia. He contributed unofficially to the preparation of the *National Security Strategy of the United States*, but the views expressed in this article are solely his own.

Notes

1. This document represents a more comprehensive articulation of themes that President Bush had already begun to introduce, notably in his speeches to a joint session of Congress (September 20, 2001), on the State of the Union (January 29, 2002) and to graduating cadets at West Point (June 1, 2002). Less well known, but displaying a related set of ideas that antedate the 9/11 attacks, were the President's remarks to the World Bank on July 17, 2001 and, subsequent to those attacks, the address to the Inter-American Development Bank on March 14, 2002 (followed eight days later by the Millennium Challenge Account initiative in Monterrey). Condoleezza Rice also added an important summary of the strategy in the Wriston Lecture she delivered in New York City on October 1, 2002.

2. See Richard Koebner and Helmut Dan Schmidt, *Imperialism: The Story and Significance of a Political Word*, 1840-1960 (Cambridge: Cambridge University Press, 1964).

3. Murray, *Morality and Modern War* (New York: Council on Religion and International Affairs, 1959).

4. From an address in Ottawa delivered after the Quebec strategy conference of 1943. The language was FDR's own, added in his own hand at the end of the draft his speechwriters had prepared. Samuel I. Rosenman, *Working with Roosevelt* (London: Rupert Hart-Davis, 1952), p.356.

5. I have elaborated. these ideas in "American Engagement in Asia", in Robert D. Blackwill & Paul Dibb, eds., *America's Asian Alliances* (Cambridge, MA: MIT Press, 2000); and "The New Concert of Europe", *Survival* (Summer 1992).

6. Quotations are drawn from Bob Woodward, *Bush At War* (New York; Simon & Schuster, 2002), pp. 44, 341-2.

7. An essential argument in Charles Krauthammer, "The Unipolar Moment Revisited", *The National Interest* (Winter 2002/03).

8. The current cases of North Korea and Iraq offer interesting illustrations. For a general consideration of preventive/preemptive strategies, see my essay on "Offensive Military Options", in Robert D. Blackwill & Albert Carnesale, eds., *New Nuclear Nations: Consequences for US. Policy* (New York: Council on Foreign Relations, 1993).

9. Adams to Erving, November 28, 1818, in Worthington Chauncey Ford, ed., *Writings of John Quincy Adams* (New íYork: Macmillan, 1916), p. 483. John Lewis Gaddis called my attention to the significance of Adams' influence in the Goldman Lectures he delivered in New York City in September 2002.

Nuclear Nightmares

Experts on terrorism and proliferation agree on one thing:
Sooner or later, an attack will happen here. When and how is what robs them of sleep.

By Bill Keller

The panic that would result from contaminating the Magic Kingdom with a modest amount of cesium would probably shut the place down for good and constitute a staggering strike at Americans' sense of innocence.

Not If But When Everybody who spends much time thinking about nuclear terrorism can give you a scenario, something diabolical and, theoretically, doable. Michael A. Levi, a researcher at the Federation of American Scientists, imagines a homemade nuclear explosive device detonated inside a truck passing through one of the tunnels into Manhattan. The blast would crater portions of the New York skyline, barbecue thousands of people instantly, condemn thousands more to a horrible death from radiation sickness and—by virtue of being underground—would vaporize many tons of concrete and dirt and river water into an enduring cloud of lethal fall-

out. Vladimir Shikalov, a Russian nuclear physicist who helped clean up after the 1986 Chernobyl accident, envisioned for me an attack involving highly radioactive cesium-137 loaded into some kind of homemade spraying device, and a target that sounded particularly unsettling when proposed across a Moscow kitchen table—Disneyland. In this case, the human toll would be much less ghastly, but the panic that would result from contaminating the Magic Kingdom with a modest amount of cesium—Shikalov held up his teacup to illustrate how much—would probably shut the place down for good and constitute a staggering strike at Americans' sense of innocence. Shikalov, a nuclear enthusiast who thinks most people are ridiculously squeamish about radiation, added that personally he would still be happy to visit Disneyland after the terrorists struck, although he would pack his own food and drink and destroy his clothing afterward.

Another Russian, Dmitry Borisov, a former official of his country's atomic energy ministry, conjured a suicidal pilot. (Suicidal pilots, for obvious reasons, figure frequently in these fantasies.) In Borisov's scenario, the hijacker dive-bombs an Aeroflot jetliner into the Kurchatov

Institute, an atomic research center in a gentrifying neighborhood of Moscow, which I had just visited the day before our conversation. The facility contains 26 nuclear reactors of various sizes and a huge accumulation of radioactive material. The effect would probably be measured more in property values than in body bags, but some people say the same about Chernobyl. Maybe it is a way to tame a fearsome subject by Hollywoodizing it, or maybe it is a way to drive home the dreadful stakes in the arid-sounding business of nonproliferation, but in several weeks of talking to specialists here and in Russia about the threats an amateur evildoer might pose to the homeland, I found an unnerving abundance of such morbid creativity. I heard a physicist wonder whether a suicide bomber with a pacemaker would constitute an effective radiation weapon. (I'm a little ashamed to say I checked that one, and the answer is no, since pacemakers powered by plutonium have not been implanted for the past 20 years.) I have had people theorize about whether hijackers who took over a nuclear research laboratory could improvise an actual nuclear explosion on the spot. (Expert opinions differ, but it's very unlikely.) I've been instructed how to disperse plutonium into the ventilation system of an office building.

The realistic threats settle into two broad categories. The less likely but far more devastating is an actual nuclear explosion, a great hole blown in the heart of New York or Washington, followed by a toxic fog of radiation. This could be produced by a black-market nuclear warhead procured from an existing arsenal. Russia is the favorite hypothetical source, although Pakistan, which has a program built on shady middlemen and covert operations, should not be overlooked. Or the explosive could be a homemade device, lower in yield than a factory nuke but still creating great carnage.

The second category is a radiological attack, contaminating a public place with radioactive material by packing it with conventional explosives in a "dirty bomb" by dispersing it into the air or water or by sabotaging a nuclear facility. By comparison with the task of creating nuclear fission, some of these schemes would be almost childishly simple, although the consequences would be less horrifying: a panicky evacuation, a gradual increase in cancer rates, a staggeringly expensive cleanup, possibly the need to demolish whole neighborhoods. Al Qaeda has claimed to have access to dirty bombs, which is unverified but entirely plausible, given that the makings are easily gettable.

Nothing is really new about these perils. The means to inflict nuclear harm on America have been available to rogues for a long time. Serious studies of the threat of nuclear terror date back to the 1970's. American programs to keep Russian nuclear ingredients from falling into murderous hands—one of the subjects high on the agenda in President Bush's meetings in Moscow this weekend—were hatched soon after the Soviet Union disintegrated a decade ago. When terrorists get around to trying their first nuclear assault, as you can be sure they will, there will be plenty of people entitled to say I told you so.

All Sept. 11 did was turn a theoretical possibility into a felt danger. All it did was supply a credible cast of characters who hate us so much they would thrill to the prospect of actually doing it—and, most important in rethinking the probabilities, would be happy to die in the effort. All it did was give our nightmares legs.

Tom Ridge cupped his hands prayerfully and pressed his fingertips to his lips. "Nuclear," he said simply.

And of the many nightmares animated by the attacks, this is the one with pride of place in our experience and literature—and, we know from his own lips, in Osama bin Laden's aspirations. In February, Tom Ridge, the Bush administration's homeland security chief, visited The Times for a conversation, and at the end someone asked, given all the things he had to worry about—hijacked airliners, anthrax in the mail, smallpox, germs in crop-dusters—what did he worry about most? He cupped his hands prayerfully and pressed his fingertips to his lips. "Nuclear," he said simply.

My assignment here was to stare at that fear and inventory the possibilities. How afraid should we be, and what of, exactly? I'll tell you at the outset, this was not one of those exercises in which weighing the fears and assigning them probabilities laid them to rest. I'm not evacuating Manhattan, but neither am I sleeping quite as soundly. As I was writing this early one Saturday in April, the floor began to rumble and my desk lamp wobbled precariously. Although I grew up on the San Andreas Fault, the fact that New York was experiencing an earthquake was only my second thought.

The best reason for thinking it won't happen is that it hasn't happened yet, and that is terrible logic. The problem is not so much that we are not doing enough to prevent a terrorist from turning our atomic knowledge against us (although we are not). The problem is that there may be no such thing as "enough."

25,000 Warheads, and It Only Takes One My few actual encounters with the Russian nuclear arsenal are all associated with Thomas Cochran. Cochran, a physicist with a Tennessee lilt and a sense of showmanship, is the director of nuclear issues for the Natural Resources Defense Council, which promotes environmental protection and arms control. In 1989, when glasnost was in flower, Cochran persuaded the Soviet Union to open some of its most secret nuclear venues to a roadshow of American scientists and congressmen and invited along a couple of reporters. We visited a Soviet missile cruiser bobbing in the Black Sea and drank vodka with physicists and engineers in the secret city where the Soviets first produced plutonium for weapons.

Not long ago Cochran took me cruising through the Russian nuclear stockpile again, this time digitally. The days of glasnost theatrics are past, and this is now the only way an outsider can get close to the places where Russians store and deploy their nuclear weapons. On his office computer in Washington, Cochran has installed a detailed United States military map of Russia and superimposed upon it high-resolution satellite photographs. We spent part of a morning mouse-clicking from missile-launch site to submarine base, zooming in like voyeurs and contemplating the possibility that a terrorist could figure out how to steal a nuclear warhead from one of these places.

"Here are the bunkers," Cochran said, enlarging an area the size of a football stadium holding a half-dozen elongated igloos. We were hovering over a site called Zhukovka, in western Russia. We were pleased to see it did not look ripe for a hijacking.

"You see the bunkers are fenced, and then the whole thing is fenced again," Cochran said. "Just outside you can see barracks and a rifle range for the guards. These would be troops of the 12th Main Directorate. Somebody's not going to walk off the street and get a Russian weapon out of this particular storage area."

In the popular culture, nuclear terror begins with the theft of a nuclear weapon. Why build one when so many are lying around for the taking? And stealing tends to make better drama than engineering. Thus the stolen nuke has been a staple in the literature at least since 1961, when Ian Fleming published "Thunderball," in which the malevolent Spectre (the Special Executive for Counterintelligence, Terrorism, Revenge and Extortion, a strictly mercenary and more technologically sophisticated precursor to al Qaeda) pilfers a pair of atom bombs from a crashed NATO aircraft. In the movie version of Tom Clancy's thriller "The Sum of All Fears," due in theaters this week, neo-Nazis get their hands on a mislaid Israeli nuke, and viewers will get to see Baltimore blasted to oblivion.

Eight countries are known to have nuclear weapons—the United States, Russia, China, Great Britain, France, India, Pakistan and Israel. David Albright, a nuclear-weapons expert and president of the Institute for Science and International Security, points out that Pakistan's program in particular was built almost entirely through black markets and industrial espionage, aimed at circumventing Western export controls. Defeating the discipline of nuclear nonproliferation is ingrained in the culture. Disaffected individuals in Pakistan (which, remember, was intimate with the Taliban) would have no trouble finding the illicit channels or the rationalization for diverting materials, expertise—even, conceivably, a warhead.

But the mall of horrors is Russia, because it currently maintains something like 15,000 of the world's (very roughly) 25,000 nuclear warheads, ranging in destructive power from

about 500 kilotons, which could kill a million people, down to the one-kiloton land mines that would be enough to make much of Manhattan uninhabitable. Russia is a country with sloppy accounting, a disgruntled military, an audacious black market and indigenous terrorists.

It's easier to take the fuel and build an entire weapon from scratch than it is to make one of these things go off.

There is anecdotal reason to worry. Gen. Igor Valynkin, commander of the 12th Main Directorate of the Russian Ministry of Defense, the Russian military sector in charge of all nuclear weapons outside the Navy, said recently that twice in the past year terrorist groups were caught casing Russian weapons-storage facilities. But it's hard to know how seriously to take this. When I made the rounds of nuclear experts in Russia earlier this year, many were skeptical of these near-miss anecdotes, saying the security forces tend to exaggerate such incidents to dramatize their own prowess (the culprits are always caught) and enhance their budgets. On the whole, Russian and American military experts sound not very alarmed about the vulnerability of Russia's nuclear warheads. They say Russia takes these weapons quite seriously, accounts for them rigorously and guards them carefully. There is no confirmed case of a warhead being lost. Strategic warheads, including the 4,000 or so that President Bush and President Vladimir Putin have agreed to retire from service, tend to be stored in hard-to-reach places, fenced and heavily guarded, and their whereabouts are not advertised. The people who guard them are better paid and more closely vetted than most Russian soldiers.

Eugene E. Habiger, the four-star general who was in charge of Amer-

ican strategic weapons until 1998 and then ran nuclear antiterror programs for the Energy Department, visited several Russian weapons facilities in 1996 and 1997. He may be the only American who has actually entered a Russian bunker and inspected a warhead *in situ*. Habiger said he found the overall level of security comparable to American sites, although the Russians depend more on people than on technology to protect their nukes.

The image of armed terrorist commandos storming a nuclear bunker is cinematic, but it's far more plausible to think of an inside job. No observer of the unraveling Russian military has much trouble imagining that a group of military officers, disenchanted by the humiliation of serving a spent superpower, embittered by the wretched conditions in which they spend much of their military lives or merely greedy, might find a way to divert a warhead to a terrorist for the right price. (The Chechen warlord Shamil Basayev, infamous for such ruthless exploits as taking an entire hospital hostage, once hinted that he had an opportunity to buy a nuclear warhead from the stockpile.) The anecdotal evidence of desperation in the military is plentiful and disquieting. Every year the Russian press provides stories like that of the 19-year-old sailor who went on a rampage aboard an Akula-class nuclear submarine, killing eight people and threatening to blow up the boat and its nuclear reactor; or the five soldiers at Russia's nuclear-weapons test site who killed a guard, took a hostage and tried to hijack an aircraft, or the officers who reportedly stole five assault helicopters, with their weapons pods, and tried to sell them to North Korea.

The Clinton administration found the danger of disgruntled nuclear caretakers worrisome enough that it considered building better housing for some officers in the nuclear rocket corps. Congress, noting that the United States does not build housing for its own officers, rejected the idea out of hand.

If a terrorist did get his hands on a nuclear warhead, he would still face the problem of setting it off. American warheads are rigged with multiple PAL's ("permissive action links")—codes and self-disabling devices designed to frustrate an unauthorized person from triggering the explosion. General Habiger says that when he examined Russian strategic weapons he found the level of protection comparable to our own. "You'd have to literally break the weapon apart to get into the gut," he told me. "I would submit that a more likely scenario is that there'd be an attempt to get hold of a warhead and not explode the warhead but extract the plutonium or highly enriched uranium." In other words, it's easier to take the fuel and build an entire weapon from scratch than it is to make one of these things go off.

Then again, Habiger is not an expert in physics or weapons design. Then again, the Russians would seem to have no obvious reason for misleading him about something that important. Then again, how many times have computer hackers hacked their way into encrypted computers we were assured were impregnable? Then again, how many computer hackers does al Qaeda have? This subject drives you in circles.

The most troublesome gap in the generally reassuring assessment of Russian weapons security is those tactical nuclear warheads—smaller, short-range weapons like torpedoes, depth charges, artillery shells, mines. Although their smaller size and greater number makes them ideal candidates for theft, they have gotten far less attention simply because, unlike all of our long-range weapons, they happen not to be the subject of any formal treaty. The first President Bush reached an informal understanding with President Gorbachev and then with President Yeltsin that both sides would gather and destroy thousands of tactical nukes. But the agreement included no inventories of the stockpiles, no outside monitoring, no verification of any kind. It was one of those trust-me deals that, in the

hindsight of Sept. 11, amount to an enormous black hole in our security.

Did I say earlier there are about 15,000 Russian warheads? That number includes, alongside the scrupulously counted strategic warheads in bombers, missiles and submarines, the commonly used estimate of 8,000 tactical warheads. But that figure is at best an educated guess. Other educated guesses of the tactical nukes in Russia go as low as 4,000 and as high as 30,000. We just don't know. We don't even know if the Russians know, since they are famous for doing things off the books. "They'll tell you they've never lost a weapon," said Kenneth Luongo, director of a private antiproliferation group called the Russian-American Nuclear Security Advisory Council. "The fact is, they don't know. And when you're talking about warhead counting, you don't want to miss even one."

And where are they? Some are stored in reinforced concrete bunkers like the one at Zhukovka. Others are deployed. (When the submarine Kursk sank with its 118 crewmen in August 2000, the Americans' immediate fear was for its nuclear armaments. The standard load out for a submarine of that class includes a couple of nuclear torpedoes and possibly some nuclear depth charges.) Still others are supposed to be in the process of being dismantled under terms of various formal and informal arms-control agreements. Some are in transit. In short, we don't really know.

The other worrying thing about tactical nukes is that their anti-use devices are believed to be less sophisticated, because the weapons were designed to be employed in the battlefield. Some of the older systems are thought to have no permissive action links at all, so that setting one off would be about as complicated as hot-wiring a car.

Efforts to learn more about the state of tactical stockpiles have been frustrated by reluctance on both sides to let visitors in. Viktor Mikhailov, who ran the Russian

Ministry of Atomic Energy until 1998 with a famous scorn for America's nonproliferation concerns, still insists that the United States programs to protect Russian nuclear weapons and material mask a secret agenda of intelligence-gathering. Americans, in turn, sometimes balk at reciprocal access, on the grounds that we are the ones paying the bills for all these safety upgrades, said the former Senator Sam Nunn, co-author of the main American program for securing Russian nukes, called Nunn-Lugar.

People in the field talk of a nuclear 'conex' bomb, using the name of those shack-size steel containers—2,000 of which enter America every hour on trains, trucks and ships. Fewer than 2 percent are cracked open for inspection.

"We have to decide if we want the Russians to be transparent—I'd call it cradle-to-grave transparency with nuclear material and inventories and so forth," Nunn told me. "Then we have to open up more ourselves. This is a big psychological breakthrough we're talking about here, both for them and for us."

The Garage Bomb One of the more interesting facts about the atom bomb dropped on Hiroshima is that it had never been tested. All of those spectral images of nuclear coronas brightening the desert of New Mexico— those were to perfect the more complicated plutonium device that was dropped on Nagasaki. "Little Boy," the Hiroshima bomb, was a rudimentary gunlike device that shot one projectile of highly enriched uranium into another, creating a critical mass that exploded. The mechanics were so simple that few doubted it would work, so the first experiment was in the sky over Japan.

The closest thing to a consensus I heard among those who study nuclear terror was this: building a nuclear bomb is easier than you think, probably easier than stealing one. In the rejuvenated effort to prevent a terrorist from striking a nuclear blow, this is where most of the attention and money are focused.

A nuclear explosion of any kind "is not a sort of high-probability thing," said a White House official who follows the subject closely. "But getting your hands on enough fissile material to build an improvised nuclear device, to my mind, is the least improbable of them all, and particularly if that material is highly enriched uranium in metallic form. Then I'm really worried. That's the one."

To build a nuclear explosive you need material capable of explosive nuclear fission, you need expertise, you need some equipment, and you need a way to deliver it.

Delivering it to the target is, by most reckoning, the simplest part. People in the field generally scoff at the mythologized suitcase bomb; instead they talk of a "conex bomb," using the name of those shack-size steel containers that bring most cargo into the United States. Two thousand containers enter America every hour, on trucks and trains and especially on ships sailing into more than 300 American ports. Fewer than 2 percent are cracked open for inspection, and the great majority never pass through an X-ray machine. Containers delivered to upriver ports like St. Louis or Chicago pass many miles of potential targets before they even reach customs.

"How do you protect against that?" mused Habiger, the former chief of our nuclear arsenal. "You can't. That's scary. That's very, very scary. You set one of those off in Philadelphia, in New York City, San Francisco, Los Angeles, and you're going to kill tens of thousands of

people, if not more." Habiger's view is "It's not a matter of *if*; it's a matter of *when*"—which may explain why he now lives in San Antonio.

The Homeland Security office has installed a plan to refocus inspections, making sure the 2 percent of containers that get inspected are those without a clear, verified itinerary. Detectors will be put into place at ports and other checkpoints. This is good, but it hardly represents an ironclad defense. The detection devices are a long way from being reliable. (Inconveniently, the most feared bomb component, uranium, is one of the hardest radioactive substances to detect because it does not emit a lot of radiation prior to fission.) The best way to stop nuclear terror, therefore, is to keep the weapons out of terrorist hands in the first place.

Fabricating a nuclear weapon is not something a lone madman—even a lone genius—is likely to pull off in his hobby room.

The basic know-how of atom-bomb-building is half a century old, and adequate recipes have cropped up in physics term papers and high school science projects. The simplest design entails taking a lump of highly enriched uranium, about the size of a cantaloupe, and firing it down a big gun barrel into a second lump. Theodore Taylor, the nuclear physicist who designed both the smallest and the largest American nuclear-fission warheads before becoming a remorseful opponent of all things nuclear, told me he recently looked up "atomic bomb" in the World Book Encyclopedia in the upstate New York nursing home where he now lives, and he found enough basic information to get a careful reader started. "It's accessible all over the place," he said. "I don't

mean just the basic principles. The sizes, specifications, things that work."

Most of the people who talk about the ease of assembling a nuclear weapon, of course, have never actually built one. The most authoritative assessment I found was a paper, "Can Terrorists Build Nuclear Weapons?" written in 1986 by five experienced nuke-makers from the Los Alamos weapons laboratory. I was relieved to learn that fabricating a nuclear weapon is not something a lone madman—even a lone genius—is likely to pull off in his hobby room. The paper explained that it would require a team with knowledge of "the physical, chemical and metallurgical properties of the various materials to be used, as well as characteristics affecting their fabrication; neutronic properties; radiation effects, both nuclear and biological; technology concerning high explosives and/or chemical propellants; some hydrodynamics; electrical circuitry; and others." Many of these skills are more difficult to acquire than, say, the ability to aim a jumbo jet.

The schemers would also need specialized equipment to form the uranium, which is usually in powdered form, into metal, to cast it and machine it to fit the device. That effort would entail months of preparation, increasing the risk of detection, and it would require elaborate safeguards to prevent a mishap that, as the paper dryly put it, would "bring the operation to a close."

Still, the experts concluded, the answer to the question posed in the title, while qualified, was "Yes, they can."

David Albright, who worked as a United Nations weapons inspector in Iraq, says Saddam Hussein's unsuccessful crash program to build a nuclear weapon in 1990 illustrates how a single bad decision can mean a huge setback. Iraq had extracted highly enriched uranium from research-reactor fuel and had, maybe, barely enough for a bomb. But the manager in charge of casting the metal was so afraid the stuff would spill or get contaminated that he decided to melt it in tiny batches. As a result, so much of the uranium was wasted that he ended up with too little for a bomb.

"You need good managers and organizational people to put the elements together," Albright said. "If you do a straight-line extrapolation, terrorists will all get nuclear weapons. But they make mistakes."

On the other hand, many experts underestimate the prospect of a do-it-yourself bomb because they are thinking too professionally. All of our experience with these weapons is that the people who make them (states, in other words) want them to be safe, reliable, predictable and efficient. Weapons for the American arsenal are designed to survive a trip around the globe in a missile, to be accident-proof, to produce a precisely specified blast.

But there are many corners you can cut if you are content with a big, ugly, inefficient device that would make a spectacular impression. If your bomb doesn't need to fit in a suitcase (and why should it?) or to endure the stress of a missile launch; if you don't care whether the explosive power realizes its full potential; if you're willing to accept some risk that the thing might go off at the wrong time or might not go off at all, then the job of building it is immeasurably simplified.

"As you get smarter, you realize you can get by with less," Albright said. "You can do it in facilities that look like barns, garages, with simple machine tools. You can do it with 10 to 15 people, not all Ph.D.'s, but some engineers, technicians. Our judgment is that a gun-type device is well within the capability of a terrorist organization."

All the technological challenges are greatly simplified if terrorists are in league with a country—a place with an infrastructure. A state is much better suited to hire expertise (like dispirited scientists from decommissioned nuclear installations in the old Soviet Union) or to send its own scientists for M.I.T. degrees.

Thus Tom Cochran said his greatest fear is what you might call a bespoke nuke—terrorists stealing a quantity of weapons-grade uranium and taking it to Iraq or Iran or Libya, letting the scientists and engineers there fashion it into an elementary weapon and then taking it away for a delivery that would have no return address.

That leaves one big obstacle to the terrorist nuke-maker: the fissile material itself.

To be reasonably sure of a nuclear explosion, allowing for some material being lost in the manufacturing process, you need roughly 50 kilograms—110 pounds—of highly enriched uranium. (For a weapon, more than 90 percent of the material should consist of the very unstable uranium-235 isotope.) Tom Cochran, the master of visual aids, has 15 pounds of depleted uranium that he keeps in a Coke can; an eight-pack would be plenty to build a bomb.

Only 41 percent of Russia's weapon-usable material has been secured… So the barn door is still pretty seriously ajar. We don't know whether any horses have gotten out.

The world is awash in the stuff. Frank von Hippel, a Princeton physicist and arms-control advocate, has calculated that between 1,300 and 2,100 metric tons of weapons-grade uranium exists—at the low end, enough for 26,000 rough-hewed bombs. The largest stockpile is in Russia, which Senator Joseph Biden calls "the candy store of candy stores."

Until a decade ago, Russian officials say, no one worried much about the safety of this material. Vik-

tor Mikhailov, who ran the atomic energy ministry and now presides over an affiliated research institute, concedes there were glaring lapses.

"The safety of nuclear materials was always on our minds, but the focus was on intruders," he said. "The system had never taken account of the possibility that these carefully screened people in the nuclear sphere could themselves represent a danger. The system was not designed to prevent a danger from within."

Then came the collapse of the Soviet Union and, in the early 90's, a few frightening cases of nuclear materials popping up on the black market.

If you add up all the reported attempts to sell highly enriched uranium or plutonium, even including those that have the scent of security-agency hype and those where the material was of uncertain quality, the total amount of material still falls short of what a bomb-maker would need to construct a single explosive.

But Yuri G. Volodin, the chief of safeguards at Gosatomnadzor, the Russian nuclear regulatory agency, told me his inspectors still discover one or two instances of attempted theft a year, along with dozens of violations of the regulations for storing and securing nuclear material. And as he readily concedes: "These are the detected cases. We can't talk about the cases we don't know." Alexander Pikayev, a former aide to the Defense Committee of the Russian Duma, said: "The vast majority of installations now have fences. But you know Russians. If you walk along the perimeter, you can see a hole in the fence, because the employees want to come and go freely."

The bulk of American investment in nuclear safety goes to lock the stuff up at the source. That is clearly the right priority. Other programs are devoted to blending down the highly enriched uranium to a diluted product unsuitable for weapons but good as reactor fuel. The Nuclear Threat Initiative, financed by Ted Turner and led by Nunn, is studying ways to double the rate of this diluting process.

Still, after 10 years of American subsidies, only 41 percent of Russia's weapon-usable material has been secured, according to the United States Department of Energy. Russian officials said they can't even be sure how much exists, in part because the managers of nuclear facilities, like everyone else in the Soviet industrial complex, learned to cook their books. So the barn door is still pretty seriously ajar. We don't know whether any horses have gotten out.

And it is not the only barn. William C. Potter, director of the Center for Nonproliferation Studies at the Monterey Institute of International Studies and an expert in nuclear security in the former Soviet states, said the American focus on Russia has neglected other locations that could be tempting targets for a terrorist seeking bomb-making material. There is, for example, a bomb's worth of weapons-grade uranium at a site in Belarus, a country with an erratic president and an anti-American orientation. There is enough weapons-grade uranium for a bomb or two in Kharkiv, in Ukraine. Outside of Belgrade, in a research reactor at Vinca, sits sufficient material for a bomb—and there it sat while NATO was bombarding the area.

"We need to avoid the notion that because the most material is in Russia, that's where we should direct all of our effort," Potter said. "It's like assuming the bank robber will target Fort Knox because that's where the most gold is. The bank robber goes where the gold is most accessible."

Weapons of Mass Disruption The first and, so far, only consummated act of nuclear terrorism took place in Moscow in 1995, and it was scarcely memorable. Chechen rebels obtained a canister of cesium, possibly from a hospital they had commandeered a few months before. They hid it in a Moscow park famed for its weekend flea market and called the press. No one was hurt. Authorities

treated the incident discreetly, and a surge of panic quickly passed.

The story came up in virtually every conversation I had in Russia about nuclear terror, usually to illustrate that even without splitting atoms and making mushroom clouds a terrorist could use radioactivity—and the fear of it—as a potent weapon.

The idea that you could make a fantastic weapon out of radioactive material without actually producing a nuclear bang has been around since the infancy of nuclear weaponry. During World War II, American scientists in the Manhattan Project worried that the Germans would rain radioactive material on our troops storming the beaches on D-Day. Robert S. Norris, the biographer of the Manhattan Project director, Gen. Leslie R. Groves, told me that the United States took this threat seriously enough to outfit some of the D-Day soldiers with Geiger counters.

No country today includes radiological weapons in its armories. But radiation's limitations as a military tool—its tendency to drift afield with unplanned consequences, its long-term rather than short-term lethality—would not necessarily count against it in the mind of a terrorist. If your aim is to instill fear, radiation is anthrax-plus. And unlike the fabrication of a nuclear explosive, this is terror within the means of a soloist.

If your aim is to instill fear, radiation is anthrax-plus. And unlike the fabrication of a nuclear explosive, this is terror within the means of a soloist.

That is why, if you polled the universe of people paid to worry about weapons of mass destruction (W.M.D., in the jargon), you would find a general agreement that this is

probably the first thing we'll see. "If there is a W.M.D. attack in the next year, it's likely to be a radiological attack," said Rose Gottemoeller, who handled Russian nuclear safety in the Clinton administration and now follows the subject for the Carnegie Endowment. The radioactive heart of a dirty bomb could be spent fuel from a nuclear reactor or isotopes separated out in the process of refining nuclear fuel. These materials are many times more abundant and much, much less protected than the high-grade stuff suitable for bombs. Since Sept.11, Russian officials have begun lobbying hard to expand the program of American aid to include protection of these lower-grade materials, and the Bush administration has earmarked a few million dollars to study the problem. But the fact is that radioactive material suitable for terrorist attacks is so widely available that there is little hope of controlling it all.

The guts of a dirty bomb could be cobalt-60, which is readily available in hospitals for use in radiation therapy and in food processing to kill the bacteria in fruits and vegetables. It could be cesium-137, commonly used in medical gauges and radiotherapy machines. It could be americium, an isotope that behaves a lot like plutonium and is used in smoke detectors and in oil prospecting. It could be plutonium, which exists in many research laboratories in America. If you trust the security of those American labs, pause and reflect that the investigation into the great anthrax scare seems to be focused on disaffected American scientists.

Back in 1974, Theodore Taylor and Mason Willrich, in a book on the dangers of nuclear theft, examined things a terrorist might do if he got his hands on 100 grams of plutonium—a thimble-size amount. They calculated that a killer who dissolved it, made an aerosol and introduced it into the ventilation system of an office building could deliver a lethal dose to the entire floor area of a large skyscraper. But plutonium dispersed outdoors in the open air, they estimated, would be far less effective. It would blow away in a gentle wind.

The Federation of American Scientists recently mapped out for a Congressional hearing the consequences of various homemade dirty bombs detonated in New York or Washington. For example, a bomb made with a single footlong pencil of cobalt from a food irradiation plant and just 10 pounds of TNT and detonated at Union Square in a light wind would send a plume of radiation drifting across three states. Much of Manhattan would be as contaminated as the permanently closed area around the Chernobyl nuclear plant. Anyone living in Manhattan would have at least a 1-in-100 chance of dying from cancer caused by the radiation. An area reaching deep into the Hudson Valley would, under current Environmental Protection Agency standards, have to be decontaminated or destroyed.

Frank von Hippel, the Princeton physicist, has reviewed the data, and he pointed out that this is a bit less alarming than it sounds. "Your probability of dying of cancer in your lifetime is already about 20 percent," he said. "This would increase it to 20.1 percent. Would you abandon a city for that? I doubt it."

Indeed, some large portion of our fear of radiation is irrational. And yet the fact that it's all in your mind is little consolation if it's also in the minds of a large, panicky population. If the actual effect of a radiation bomb is that people clog the bridges out of town, swarm the hospitals and refuse to return to live and work in a contaminated place, then the impact is a good deal more than psychological. To this day, there is bitter debate about the actual health toll from the Chernobyl nuclear accident. There are researchers who claim that the people who evacuated are actually in worse health over all from the trauma of relocation, than those who stayed put and marinated in the residual radiation. But the fact is, large swaths of developed land around the Chernobyl site still lie abandoned,

much of it bulldozed down to the subsoil. The Hart Senate Office Building was closed for three months by what was, in hindsight, our society's inclination to err on the side of alarm.

There are measures the government can take to diminish the dangers of a radiological weapon, and many of them are getting more serious consideration. The Bush administration has taken a lively new interest in radiation-detection devices that might catch dirty-bomb materials in transit. A White House official told me the administration's judgment is that protecting the raw materials of radiological terror is worth doing, but not at the expense of more catastrophic threats.

"It's all over," he said. "It's not a winning proposition to say you can just lock all that up. And then, a bomb is pretty darn easy to make. You don't have to be a rocket scientist to figure about fertilizer and diesel fuel." A big fertilizer bomb of the type Timothy McVeigh used to kill 168 people in Oklahoma City, spiced with a dose of cobalt or cesium, would not tax the skills of a determined terrorist.

"It's likely to happen, I think, in our lifetime," the official said. "And it'll be like Oklahoma City plus the Hart Office Building. Which is real bad, but it ain't the World Trade Center."

The Peril of Power Plants Every eight years or so the security guards at each of the country's 103 nuclear power stations and at national weapons labs can expect to be attacked by federal agents armed with laser-tag rifles. These mock terror exercises are played according to elaborate rules, called the "design basis threat," that in the view of skeptics favor the defense. The attack teams can include no more than three commandos. The largest vehicle they are permitted is an S.U.V. They are allowed to have an accomplice inside the plant, but only one. They are not allowed to improvise. (The mock assailants at one Department of Energy

lab were ruled out of order because they commandeered a wheelbarrow to cart off a load of dummy plutonium.) The mock attacks are actually announced in advance. Even playing by these rules, the attackers manage with some regularity to penetrate to the heart of a nuclear plant and damage the core. Representative Edward J. Markey, a Massachusetts Democrat and something of a scourge of the nuclear power industry, has recently identified a number of shortcomings in the safeguards, including, apparently, lax standards for clearing workers hired at power plants.

One of the most glaring lapses, which nuclear regulators concede and have promised to fix, is that the design basis threat does not contemplate the possibility of a hijacker commandeering an airplane and diving it into a reactor. In fact, the protections currently in place don't consider the possibility that the terrorist might be willing, even eager, to die in the act. The government assumes the culprits would be caught while trying to get away.

A nuclear power plant is essentially a great inferno of decaying radioactive material, kept under control by coolant. Turning this device into a terrorist weapon would require cutting off the coolant so the atomic furnace rages out of control and, equally important, getting the radioactive matter to disperse by an explosion or fire. (At Three Mile Island, the coolant was cut off and the reactor core melted down, generating vast quantities of radiation. But the thick walls of the containment building kept the contaminant from being released, so no one died.)

One way to accomplish both goals might be to fly a large jetliner into the fortified building that holds the reactor. Some experts say a jet engine would stand a good chance of bursting the containment vessel, and the sheer force of the crash might disable the cooling system—rupturing the pipes and cutting off electricity that pumps the water through the core. Before nearby residents had begun to evacuate, you could have a meltdown that would spew a volcano of radioactive isotopes into the air, causing fatal radiation sickness for those exposed to high doses and raising lifetime cancer rates for miles around.

This sort of attack is not as easy, by a long shot, as hitting the World Trade Center. The reactor is a small, low-lying target, often nestled near the conspicuous cooling towers, which could be destroyed without great harm. The reactor is encased in reinforced concrete several feet thick, probably enough, the industry contends, to withstand a crash. The pilot would have to be quite a marksman, and somewhat lucky. A high wind would disperse the fumes before they did great damage.

Invading a plant to produce a meltdown, even given the record of those mock attacks, would be more complicated, because law enforcement from many miles around would be on the place quickly, and because breaching the containment vessel is harder from within. Either invaders or a kamikaze attacker could instead target the more poorly protected cooling ponds, where used plutonium sits, encased in great rods of zirconium alloy. This kind of sabotage would take longer to generate radiation and would be far less lethal.

Discussion of this kind of potential radiological terrorism is colored by passionate disagreements over nuclear power itself. Thus the nuclear industry and its rather tame regulators sometimes sound dismissive about the vulnerability of the plants (although less so since Sept.11), while those who regard nuclear power as inherently evil tend to overstate the risks. It is hard to sort fact from fear-mongering.

Nuclear regulators and the industry grumpily concede that Sept. 11 requires a new estimate of their defenses, and under prodding from Congress they are redrafting the so-called design basis threat, the one plants are required to defend against. A few members of Congress have proposed installing ground-to-air missiles at nuclear plants, which most experts think is a recipe for a disastrous mishap.

"Probably the only way to protect against someone flying an aircraft into a nuclear power plant," said Steve Fetter of the University of Maryland, "is to keep hijackers out of cockpits."

Being Afraid For those who were absorbed by the subject of nuclear terror before it became fashionable, the months since the terror attacks have been, paradoxically, a time of vindication. President Bush, whose first budget cut $100 million from the programs to protect Russian weapons and material (never a popular program among conservative Republicans), has become a convert. The administration has made nuclear terror a priority, and it is getting plenty of goading to keep it one. You can argue with their priorities and their budgets, but it's hard to accuse anyone of indifference. And resistance—from scientists who don't want security measures to impede their access to nuclear research materials, from generals and counterintelligence officials uneasy about having their bunkers inspected, from nuclear regulators who worry about the cost of nuclear power, from conservatives who don't want to subsidize the Russians to do much of anything—has become harder to sustain. Intelligence gathering on nuclear material has been abysmal, but it is now being upgraded; it is a hot topic at meetings between American and foreign intelligence services, and we can expect more numerous and more sophisticated sting operations aimed at disrupting the black market for nuclear materials. Putin, too, has taken notice. Just before leaving to meet Bush in Crawford, Tex., in November, he summoned the head of the atomic energy ministry to the Kremlin on a Saturday to discuss nuclear security. The subject is now on the regular agenda when Bush and Putin talk.

These efforts can reduce the danger but they cannot neutralize the fear, particularly after we have been so vividly reminded of the hostility some of the world feels for us, and of our vulnerability.

Fear is personal. My own—in part, because it's the one I grew up with, the one that made me shiver through the Cuban missile crisis and "On the Beach"—is the horrible magic of nuclear fission. A dirty bomb or an assault on a nuclear power station, ghastly as that would be, feels to me within the range of what we have survived. As the White House official I spoke with said, it's basically Oklahoma City plus the Hart Office Building. A nuclear explosion is in a different realm of fears and would test the country in ways we can scarcely imagine.

A mushroom cloud of irradiated debris would blossom more than two miles into the air. Then highly lethal fallout would begin drifting back to earth, riding the winds into the Bronx or Queens or New Jersey.

As I neared the end of this assignment, I asked Matthew McKinzie, a staff scientist at the Natural Resources Defense Council, to run a computer model of a one-kiloton nuclear explosion in Times Square, half a block from my office, on a nice spring workday. By the standards of serious nuclear weaponry, one kiloton is a junk bomb, hardly worthy of respect, a fifteenth the power of the bomb over Hiroshima.

A couple of days later he e-mailed me the results, which I combined with estimates of office workers and tourist traffic in the area. The blast and searing heat would gut buildings for a block in every direction, incinerating pedestrians and crushing people at their desks. Let's say 20,000 dead in a matter of seconds. Beyond this, to a distance of more than a quarter mile, anyone directly exposed to the fireball would die a gruesome death from radiation sickness within a day—anyone, that is, who survived the third-degree burns. This larger circle would be populated by about a quarter million people on a workday. Half a mile from the explosion, up at Rockefeller Center and down at Macy's, unshielded onlookers would expect a slower death from radiation. A mushroom cloud of irradiated debris would blossom more than two miles into the air, and then, 40 minutes later, highly lethal fallout would begin drifting back to earth, showering injured survivors and dooming rescue workers. The poison would ride for 5 or 10 miles on the prevailing winds, deep into the Bronx or Queens or New Jersey.

A terrorist who pulls off even such a small-bore nuclear explosion will take us to a whole different territory of dread from Sept. 11. It is the event that preoccupies those who think about this for a living, a category I seem to have joined.

"I think they're going to try," said the physicist David Albright. "I'm an optimist at heart. I think we can catch them in time. If one goes off, I think we will survive. But we won't be the same. It will affect us in a fundamental way. And not for the better."

Bill Keller is a Times columnist and a senior writer for the magazine.

From the _New York Times_ Magazine, May 26, 2002, pp. 22, 24-29, 51, 54-55, 57. © 2002 by Bill Keller. Distributed by The New York Times Special Features. Reprinted by permission.

Lifting the Veil
Understanding the Roots of Islamic Militancy

Henry Munson

In the wake of the attacks of September 11, 2001, many intellectuals have argued that Muslim extremists like Osama bin Laden despise the United States primarily because of its foreign policy. Conversely, US President George Bush's administration and its supporters have insisted that extremists loathe the United States simply because they are religious fanatics who "hate our freedoms." These conflicting views of the roots of militant Islamic hostility toward the United States lead to very different policy prescriptions. If US policies have caused much of this hostility, it would make sense to change those policies, if possible, to dilute the rage that fuels Islamic militancy. If, on the other hand, the hostility is the result of religious fanaticism, then the use of brute force to suppress fanaticism would appear to be a sensible course of action.

Groundings for Animosity

Public opinion polls taken in the Islamic world in recent years provide considerable insight into the roots of Muslim hostility toward the United States, indicating that for the most part, this hostility has less to do with cultural or religious differences than with US policies in the Arab world. In February and March 2003, Zogby International conducted a survey on behalf of Professor Shibley Telhami of the University of Maryland involving 2,620 men and women in Egypt, Jordan, Lebanon, Morocco, and Saudi Arabia. Most of those surveyed had "unfavorable attitudes" toward the United States and said that their hostility to the United States was based primarily on US policy rather than on their values. This was true of 67 percent of the Saudis surveyed. In Egypt, however, only 46 percent said their hostility resulted from US policy, while 43 percent attributed their attitudes to their values as Arabs. This is surprising given that the prevailing religious values in Saudi Arabia are more conservative than in Egypt. Be that as it may, a plurality of people in all the countries surveyed said that their hostility toward the United States was primarily based on their opposition to US policy.

The issue that arouses the most hostility in the Middle East toward the United States is the Israeli-Palestinian conflict and what Muslims perceive as US responsibility for the suffering of the Palestinians. A similar Zogby International survey from the summer of 2001 found that more than 80 percent of the respondents in Egypt, Kuwait, Lebanon, and Saudi Arabia ranked the Palestinian issue as one of the three issues of greatest importance to them. A survey of Muslim "opinion leaders" released by the Pew Research Center for the People and the Press in December 2001 also found that the US position on the Israeli-Palestinian conflict was the main source of hostility toward the United States.

It is true that Muslim hostility toward Israel is often expressed in terms of anti-Semitic stereotypes and conspiracy theories—think, for example, of the belief widely-held in the Islamic world that Jews were responsible for the terrorists attacks of September 11, 2001. Muslim governments and educators need to further eliminate anti-Semitic bias in the Islamic world. However, it would be a serious mistake to dismiss Muslim and Arab hostility toward Israel as simply a matter of anti-Semitism. In the context of Jewish history, Israel represents liberation. In the context of Palestinian history, it represents subjugation. There will always be a gap between how the West and how the Muslim societies perceive Israel. There will also always be some Muslims (like Osama bin Laden) who will refuse to accept any solution to the Israeli-Palestinian conflict other than the destruction of the state of Israel. That said, if the United States is serious about winning the so-called "war on terror," then resolution of the Israeli-Palestinian conflict should be among its top priorities in the Middle East.

Eradicating, or at least curbing, Palestinian terrorism entails reducing the humiliation, despair, and rage that drive many Palestinians to support militant Islamic groups like Hamas and Islamic Jihad. When soldiers at an Israeli checkpoint prevented Ahmad Qurei (Abu al Ala), one of the principal negotiators of the Oslo accords and president of the Palestinian Authority's parliament, from traveling from Gaza to his home on the West Bank, he declared, "Soon, I too will join Hamas." Qurei's words reflected his outrage at the subjugation of his people and the humiliation that Palestinians experience every day at the checkpoints that surround their homes. Defeating groups like Hamas requires diluting the rage that fuels them. Relying on force alone tends to increase rather than weaken their appeal. This is demonstrated by some of the unintended consequences of the US-led invasion and occupation of Iraq in the spring of 2003.

On June 3, 2003, the Pew Research Center for the People and the Press released a report entitled *Views of a*

Changing World June 2003. This study was primarily based on a survey of nearly 16,000 people in 21 countries (including the Palestinian Authority) from April 28 to May 15, 2003, shortly after the fall of Saddam Hussein's regime. The survey results were supplemented by data from earlier polls, especially a survey of 38,000 people in 44 countries in 2002. The study found a marked increase in Muslim hostility toward the United States from 2002 to 2003. In the summer of 2002, 61 percent of Indonesians held a favorable view of the United States. By May of 2003, only 15 percent did. During the same period of time, the decline in Turkey was from 30 percent to 15 percent, and in Jordan it was from 25 percent to one percent.

Indeed, the Bush administration's war on terror has been a major reason for the increased hostility toward the United States. The Pew Center's 2003 survey found that few Muslims support this war. Only 23 percent of Indonesians did so in May of 2003, down from 31 percent in the summer of 2002. In Turkey, support dropped from 30 percent to 22 percent. In Pakistan, support dropped from 30 percent to 16 percent, and in Jordan from 13 percent to two percent. These decreases reflect overwhelming Muslim opposition to the war in Iraq, which most Muslims saw as yet another act of imperial subjugation of Muslims by the West.

The 2003 Zogby International poll found that most Arabs believe that the United States attacked Iraq to gain control of Iraqi oil and to help Israel. Over three-fourths of all those surveyed felt that oil was a major reason for the war. More than three-fourths of the Saudis and Jordanians said that helping Israel was a major reason, as did 72 percent of the Moroccans and over 50 percent of the Egyptians and Lebanese. Most Arabs clearly do not believe that the United States overthrew Saddam Hussein out of humanitarian motives. Even in Iraq itself, where there was considerable support for the war, most people attribute the war to the US desire to gain control of Iraqi oil and help Israel.

Not only has the Bush administration failed to win much Muslim support for its war on terrorism, its conduct of the war has generated a dangerous backlash. Most Muslims see the US fight against terror as a war against the Islamic world. The 2003 Pew survey found that over 70 percent of Indonesians, Pakistanis, and Turks were either somewhat or very worried about a potential US threat to their countries, as were over half of Jordanians and Kuwaitis.

This sense of a US threat is linked to the 2003 Pew report's finding of widespread support for Osama bin Laden. The survey of April and May 2003 found that over half those surveyed in Indonesia, Jordan, and the Palestinian Authority, and almost half those surveyed in Morocco and Pakistan, listed bin Laden as one of the three world figures in whom they had the most confidence "to do the right thing." For most US citizens, this admiration for the man responsible for the attacks of September 11, 2001, is incomprehensible. But no matter how outrageous this widespread belief may be, it is vitally important to understand its origins. If one does not understand why people think the way they do, one cannot induce them to

think differently. Similarly, if one does not understand why people act as they do, one cannot hope to induce them to act differently.

The Appeal of Osama bin Laden

Osama bin Laden first engaged in violence because of the occupation of a Muslim country by an "infidel" superpower. He did not fight the Russians in Afghanistan because he hated their values or their freedoms, but because they had occupied a Muslim land. He participated in and supported the Afghan resistance to the Soviet occupation from 1979 to 1989, which ended with the withdrawal of the Russians. Bin Laden saw this war as legitimate resistance to foreign occupation. At the same time, he saw it as a *jihad*, or holy war, on behalf of Muslims oppressed by infidels.

When Saddam Hussein invaded Kuwait in August 1990, bin Laden offered to lead an army to defend Saudi Arabia. The Saudis rejected this offer and instead allowed the United States to establish bases in their kingdom, leading to bin Laden's active opposition to the United States. One can only speculate what bin Laden would have done for the rest of his life if the United States had not stationed hundreds of thousands of US troops in Saudi Arabia in 1990. Conceivably, bin Laden's hostility toward the United States might have remained passive and verbal instead of active and violent. All we can say with certainty is that the presence of US troops in Saudi Arabia did trigger bin Laden's holy war against the United States. It was no accident that the bombing of two US embassies in Africa on August 7, 1998, marked the eighth anniversary of the introduction of US forces into Saudi Arabia as part of Operation Desert Storm.

Part of bin Laden's opposition to the presence of US military presence in Saudi Arabia resulted from the fact that US troops were infidels on or near holy Islamic ground. Non-Muslims are not allowed to enter Mecca and Medina, the two holiest places in Islam, and they are allowed to live in Saudi Arabia only as temporary residents. Bin Laden is a reactionary Wahhabi Muslim who undoubtedly does hate all non-Muslims. But that hatred was not in itself enough to trigger his *jihad* against the United States.

Indeed, bin Laden's opposition to the presence of US troops in Saudi Arabia had a nationalistic and anti-imperialist tone. In 1996, he declared that Saudi Arabia had become an American colony. There is nothing specifically religious or fundamentalist about this assertion. In his book *Chronique d'une Guerre d'Orient*, Gilles Kepel describes a wealthy whiskey-drinking Saudi who left part of his fortune to bin Laden because he alone "was defending the honor of the country, reduced in his eyes to a simple American protectorate."

In 1996, bin Laden issued his first major manifesto, entitled a "Declaration of Jihad against the Americans Occupying the Land of the Two Holy Places." The very title focuses on the presence of US troops in Saudi Arabia, which bin Laden calls an "occupation." But this manifesto also refers to other examples of what bin Laden sees as the oppression of Muslims by infidels. "It is no secret that the people of Islam have suffered from the oppression, injustice, and aggression of the alliance of Jews and Christians and their collaborators to the point that the blood of the Muslims became the cheapest and their wealth was loot in the hands of the enemies," he writes. "Their blood was spilled in Palestine and Iraq."

Bin Laden has referred to the suffering of the Palestinians and the Iraqis (especially with respect to the deaths caused by sanctions) in all of his public statements since at least the mid-1990s. His 1996 "Declaration of Jihad" is no exception. Nonetheless, it primarily focuses on the idea that the Saudi regime has "lost all legitimacy" because it "has permitted the enemies of the Islamic community, the Crusader American forces, to occupy our land for many years." In this 1996 text, bin Laden even contends that the members of the Saudi royal family are apostates because they helped infidels fight the Muslim Iraqis in the Persian Gulf War of 1991.

A number of neo-conservatives have advocated the overthrow of the Saudi regime because of its support for terrorism. It is true that the Saudis have funded militant Islamic movements. It is also true that Saudi textbooks and teachers often encourage hatred of infidels and allow the extremist views of bin Laden to thrive. It is also probably true that members of the Saudi royal family have financially supported terrorist groups. The fact remains, however, that bin Laden and his followers in Al Qaeda have themselves repeatedly called for the overthrow of the Saudi regime, saying that it has turned Saudi Arabia into "an American colony."

If the United States were to send troops to Saudi Arabia once again, this time to overthrow the Saudi regime itself, the main beneficiaries would be bin Laden and those who think like him. On January 27, 2002, a *New York Times* article referenced a Saudi intelligence survey conducted in October 2001 that showed that 95 percent of educated Saudis between the ages of 25 and 41 supported bin Laden. If the United States were to overthrow the Saudi regime, such people would lead a guerrilla war that US forces would inevitably find themselves fighting. This war would attract recruits from all over the Islamic world outraged by the desecration of "the land of the two holy places." Given that US forces are already fighting protracted guerrilla wars in Iraq and Afghanistan, starting a third one in Saudi Arabia would not be the most effective way of eradicating terror in the Middle East.

Those who would advocate the overthrow of the Saudi regime by US troops seem to forget why bin Laden began his holy war against the United States in the first place. They also seem to forget that no one is more committed to the overthrow of the Saudi regime than bin Laden himself. Saudi Arabia is in dire need of reform, but yet another US occupation of a Muslim country is not the way to make it happen.

In December 1998, Palestinian journalist Jamal Abd al Latif Isma'il asked bin Laden, "Who is Osama bin Laden, and what does he want?" After providing a brief history of his life, bin Laden responded to the second part of the question, "We demand that our land be liberated from the enemies, that our land be liberated from the Americans. God almighty, may He be praised, gave all living beings a natural desire to reject external intruders. Take chickens, for example. If an armed soldier enters a chicken's home wanting to attack it, it fights him even though it is just a chicken." For bin Laden and millions of other Muslims, the Afghans, the Chechens, the Iraqis, the Kashmiris, and the Palestinians are all just "chickens" defending their homes against the attacks of foreign soldiers.

In his videotaped message of October 7, 2001, after the attacks of September 11, 2001, bin Laden declared, "What America is tasting now is nothing compared to what we have been tasting for decades. For over 80 years our *umma* has been tasting this humiliation and this degradation. Its sons are killed, its blood is shed, its holy places are violated, and it is ruled by other than that which God has revealed. Yet no one hears. No one responds."

Bin Laden's defiance of the United States and his criticism of Muslim governments who ignore what most Muslims see as the oppression of the Palestinians, Iraqis, Chechens, and others, have made him a hero of Muslims who do not agree with his goal of a strictly Islamic state and society. Even young Arab girls in tight jeans praise bin Laden as an anti-imperialist hero. A young Iraqi woman and her Palestinian friends told Gilles Kepel in the fall of 2001, "He stood up to defend us. He is the only one."

Looking ahead

Feelings of impotence, humiliation, and rage currently pervade the Islamic world, especially the Muslim Middle East. The invasion and occupation of Iraq has exacerbated Muslim concerns about the United States. In this context, bin Laden is seen as a heroic Osama Maccabeus descending from his mountain cave to fight the infidel oppressors to whom the worldly rulers of the Islamic world bow and scrape.

The violent actions of Osama bin Laden and those who share his views are not simply caused by "hatred of Western freedoms." They result, in part at least, from US policies that have enraged the Muslim world. Certainly, Islamic zealots like bin Laden do despise many aspects of Western culture. They do hate "infidels" in general, and Jews in particular. Muslims do need to seriously examine the existence and perpetuation of such hatred in their societies and cultures. But invading and occupying their countries simply exacerbates the sense of impotence, humiliation, and rage that induce them to support people like bin Laden. Defeating terror entails diluting the rage that fuels it.

Henry Munson is Chair of the Department of Anthropology at the University of Maine.

The Great War on Militant Islam

You may have heard it called "The War on Terror." Let's stop kidding ourselves.

BY ANDREW C. McCARTHY

Allahu akbar! IT HAS BECOME A DRUMBEAT, a soundtrack of atrocity. *Allahu akbar! God is great!* Not a celebration of divine bounty and mercy. Instead, their antithesis, the relentless coda of terror. Nor just any kind of terror, but a terror sprung from a very particular and poisonous fount—militant Islam, the central challenge of our age.

It has brutalized us for a dozen years. And now, once again, militant Islam has symbolically, and barbarously, targeted individuals. First Daniel Pearl, now Nicholas Berg—Jewish Americans slain, two years apart, while reaching out peacefully to the Muslim world.

The wild-eyed chant—*Allahu akbar!*—should have seared into the American consciousness in 1994, when the first Islamic militants to crave destruction of New York City's Twin Towers were sentenced. At high noon on February 26, 1993, they had detonated a powerful urea nitrate bomb as well over 50,000 innocents packed the World Trade Center complex. Miraculously, only six (including a pregnant woman) were killed. The bombing, however, served as militant Islam's declaration of war against America, and the jihadists' intention had been for the towers to collapse, one into the next, killing thousands upon thousands. As a district judge meted out 240-year jail terms to the bombers, one, Palestinian-born Mohammed Salameh, rabidly pounded the table, shouting *Allahu akbar, Allahu akbar!*

Militant Islam, the same centrifugal force that spurs suicide bombers to sacrifice all just to kill a relative few, had plainly galvanized Salameh, not merely in the execution of a heinous crime but in his chiding disdain for the power that presumed to impose a crushing penalty. Unbowed, he saw himself as a successful jihad warrior, his

sentence a minor setback in militant Islam's inevitable march to hegemony.

It had been no different in spring 1993, for a group of mostly Sudanese jihadists huddled in a dank New York City garage, constructing the bombs they hoped would destroy the United Nations complex and the Lincoln and Holland Tunnels. *Allahu akbar*, their commander exhorted, concurrently annealing his charges and dehumanizing the countless thousands they were bent on slaughtering—and would have, had not an informer infiltrated their circle, recording their exertions and frustrating their designs.

So it was throughout relentless cannonades of the 1990s. *Allahu akbar*, the leitmotif stringing together a 1995 plot to blow U.S. airliners out of the sky over the Pacific, during a test-run for which a Japanese tourist was callously killed and the rest of the crowded flight nearly taken down; the 1996 bombing of the Khobar Towers in Dhahran, Saudi Arabia, killing 19 U.S. Air Force servicemen; the 1998 obliteration of American embassies in Kenya and Tanzania, killing over 250; the failed attempt on the eve of the Millennium to destroy Los Angeles International Airport; and the 2000 suicide strike on the USS Cole as it docked in Aden, Yemen, killing 17 American sailors. All the handiwork of zealots, jihad warriors, some suicidal and all galvanized by militant Islam.

THE APOGEE OF THIS ONSLAUGHT was the cataclysm that became September 11, 2001, in which over 3,000 Americans were slaughtered by 19 suicide hijackers who plunged aircraft into the Twin Towers and the Pentagon, and whose attempt to destroy yet another symbolic target was foiled by heroic passengers who forced a crash landing in Pennsylvania. For a short time, this seemed an epoch-shattering event: changing everything, forcing the U.S. and the West, finally, to come to grips with the potent force that was actively seeking their demise. That, however, was largely an illusion.

Concededly, there have been some meaningful tactical adjustments. Military power, at least for now, has supplanted the judicial system as the point of America's responsive spear. This is as it must be. For all the hullabaloo over perfect conviction records, the Justice Department's terror trials from 1993 through mid-2001 managed to halt fewer than three dozen jihadists—during an interval when the international militant ranks swelled into the tens of thousands, and more. Yet, in less than three years since 9/11, even as the militants' atrocities have continued worldwide and even as our military fights on in Afghanistan and Iraq, resolve has waned and focus on the true enemy—to the limited extent it was ever there—has evaporated.

Why? Because, though the stormy skies now thunder "*Allahu akbar!*" into a second decade, the West, led by the United States, remains mulishly frozen in the politesse of the pre-9/11 world. We allow ourselves to see only the symptom, terror. Our willful blinders eschew even recognition, let alone treatment, of the disease. But we have met the enemy and it, surely, is militant Islam. Declining to confront that fact is not merely self-defeating. In the long haul, it may prove suicidal. Nevertheless, the blind-eye persists, driven by a doctrinaire reticence so desperate to be loved, so anxious to be perceived tolerant, it resists any inspection bearing the faintest whiff of judgment. Thus do we weave willful ignorance into policy.

Our war, we insist, is not with Islam. The official version of history is written thus: Nineteen suicide terrorists hijacked a great religion, a religion of peace; therefore, Islamic doctrine itself requires none of our attention. Nearly three years later, the aftershocks of 9/11 yet quake in the mounting body-counts of Djerba, Bali, Casablanca, Istanbul, Baghdad, and now Madrid. Still, we avert our eyes—and our ears. With literally thousands dead and the airwaves braying promises of greater carnage to come, we remain deaf to the chant—*Allahu akbar, Allahu akbar!*—even as it overwhelms the glottal agony of Nicholas Berg, decapitated on film in early May. By militant Islam.

NICK BERG, OF COURSE, was only the most recent victim beheaded for no better reasons than that he was an American and a Jew. While questions about his ordeal abound, his savage murder parallels eerily with that of *Wall Street Journal* reporter Daniel Pearl two years earlier. In small compass, the failure to learn the lessons of Pearl has led to the travesty of Berg, just as, writ large, nescience about the real war—the one with militant Islam—may lead to disarray and defeat in the so-called "War on Terror."

Danny Pearl was on to something as his research took him to Islamabad, Pakistan, in January 2002. He was probing what was then the most notorious near-miss in militant Islam's defiant answer to America's post-9/11 campaign: the attempt by a Briton named Richard C. Reid to bring down an American Airlines flight as it sailed over the Atlantic. With characteristically myopic focus on terror *tactics*, and assiduous avoidance of terror, causes the U.S. government and the Western media instantly turned Reid into "the Shoe Bomber," rather than "the jihadist."

Danny Pearl, however, had far more insight. He was, first and foremost, a journalist driven by curiosity about truth and culture, not hype. As the *Jerusalem Post* reported in a post-mortem profile, though Jewish he was not particularly observant, and, far from predisposed to hostility, he befriended Muslims and walked easily—if too unwarily—in the Islamic world. His cautious reporting over the years bespoke a sophisticated awareness of the tension between hardliners and moderates, and of that tension's salience for the future of Islam. Implicitly, he saw that to examine the existence and rationale of Islamic militancy was essential to understanding the threat facing not only the West but the millions of Muslim moderates struggling to construct an Islam that assimilates peacefully and tolerantly in the modern world. To conduct such an examination was plainly not to declare oneself an enemy of Islam itself—a libel that is a commonplace among the enemy, for obvious reasons, and the faint of heart, who cleave to the "War on Terror" rhetoric because, for them, the fear of being slandered as bigots trumps the national security imperative of concretely identifying that which we must defeat.

Our war, we insist, is not with Islam. The official version of history is written thus: Nineteen suicide terrorists hijacked a great religion, a religion of peace; therefore, Islamic doctrine itself requires none of our attention.

Taking a lit match to explosive-laden footwear was self-evidently "terrorism." Pearl was far more interested in *why* Richard Reid would do such a thing. It led him to Islamabad, in search of Sheikh Mubarak Shah Gilani, a Pakistani cleric who led an organization called al-Fuqra, one of dozens of such groups proselytizing Islamic militancy in the U.S., Europe, and elsewhere. It was under Gilani's tutelage that Reid had become radicalized—had become inspired to incinerate a fuselage full of men, women, and children who meant him no harm, whom he did not even know.

THIS IS a phenomenon unique to Islam in the modern world. Other religions familiar to the West—branches of

Christianity and Judaism in particular—regard themselves, like Islam regards itself, as a final, divinely revealed truth. Yes, they too proselytize, and they have their occasional religiously motivated murderers. But those are aberrational and instantly condemned by the rest of the faithful. On the Planet Earth today, only Islam sports an unbridled faction that systematically inculcates hatred, systematically dehumanizes non-adherents, and systematically kills massively and indiscriminately. Moreover, that faction, militant Islam, is plainly far more robust and extensive than the scant lunatic fringe the U.S. delusionally comforts itself to limn; and its killings, far from condemnation, provoke tepid admiration if not outright adulation in a further, considerable cross-section of the Muslim world. Militant Islam murders repeatedly, unabashedly, and globally. Why is this so, and what does it mean for us?

For the most part, maddeningly, we don't seem to want to know what Danny Pearl appears to have died trying to understand.

For the most part, maddeningly, we don't seem to want to know what Danny Pearl appears to have died trying to understand. The ostensible reason for not scrutinizing Islam is our traditional deference to religious freedom. In the U.S., belief is sacrosanct, and religious exercise is regarded as beyond the watchful eye of regulation so long as it does not transgress civil laws of general application. Insofar as it does, moreover, we are quick to stress—instinctively, defensively and, indeed, apologetically—that our sole interest lies in the legal transgression itself, not in the animating belief system. We wall off church from state, a doctrine much of Europe has transmogrified into such official hostility to the ecclesiastical that religion's dwindling vestiges have nearly been driven from the public square, and thus from the public consciousness. Any urge to examine doctrine is thus dismissed, either as *ultra vires* or quaint.

Even if this were a sound approach—and it is anything but—Islam should be viewed as exceptional. Offhandedly, we categorize it as a "religion" and place it in the blinders bucket with other faiths. Islam, however, is not just a religion. It aspires to be much more: a self-contained, comprehensive religious, political, and social system. As a result, when we put our blinders on, we are effectively averting our glance from many matters that are not, strictly speaking, issues of mere religious doctrine: the legal status of women; the rights of non-Muslims; the manner for assessing credibility in court proceedings (so dependent on whether witnesses happen to be Muslims or non-Muslims); the propriety of charging interest in financial transactions; purportedly charitable giving that in fact subsidizes much—including terror—that has little to do with what we think of as "charity"; the source of authority to make law; the legitimacy of rule by consent of the governed (as opposed to dictate of the divine—as interpreted, naturally, by influential Muslims); and the duty to take up arms. Furthermore, unlike other transcendent creeds for which the earthly realm is a mere preparation phase for life in the next, Islam is extensively fixed on the events of this world in all their comparative minutia.

From this hospitable soil sprouts militant Islam which, as the estimable scholar Daniel Pipes has observed, is not so much a religion as a "radical utopian movement." Like Islam, the militant strain has a universalist vision. It goes, however, far beyond this ambitious premise. All the world, it holds, must either adopt Islam or submit to the authority of the Muslim state—not in the afterlife but, urgently, in the here and now. And of course, there is jihad.

JIHAD IS NOT, AS MANY WISH TO BELIEVE and as our government obstinately suggests, a modern corruption. It is, instead, a central tenet of Islam, an enduring pillar of the system in its seventh century origins. The Prophet Mohammed was himself a warrior, and Islam's early history is one of virtually constant military engagements and conquests—to the point of lasting domination in what is now entrenched as the Islamic world (Mesopotamia, the Middle East, much of Eastern Europe and Africa, etc.), and of episodic and unended conflict with Christendom that did not see Muslim armies finally turned away from the gates of Western Europe until the late 15th century. Indeed, armed conflict is at the core of the militant Islamic Weltanschauung, which divides the world into two spheres: *Dar al Islam*, the domain of the Muslims, and *Dar al Harb*, the realm of war.

It should therefore come as no surprise that *jihad* is often rendered as *holy war*, although it more literally means to *strive* or *struggle* (*jihad fi sabil Allah*—striving for the sake of God). Understandably and admirably, moderate Muslims, seeking an Islam whose tenets are more congenial to modern sensibilities, assiduously bowdlerize this obligation: re-interpreting jihad as a striving to become a better person, the struggle to rid neighborhoods of menaces like drug-trafficking, and the like. Tradition, though, stubbornly complicates their effort: jihad, classically, is a military concept. Thus, it is readily portrayed and taught as a duty to do violence.

When Danny Pearl sought out the afflatus of militancy in the person of a charismatic Islamic cleric, he was thus treading the epicenter of this age's fundamental, and most irresponsibly ignored, quake. It is this: When influential imams exhort Muslims to violent jihad, they are—far from hijacking a peaceful religion—speaking with history on their side. And it is a history that will not easily fade away because it is relentlessly invigorated. For years, countless millions of dollars, mostly from the Saudi Wahabist strongholds of the oil-rich Gulf, have sluiced through mosques, madrassas, and Islamic social centers worldwide, promoting brutality and inculcating jihad as a forcible duty in children, who thus learn to hate before they know how to reason. What they are taught to hate are infidels—Jews, Christians, and Americans. What the militants have wrought is not a fringe. It is an ascendant, intimidating, carefully nurtured cultural force.

It is a force, all the more, because of another little understood feature of this largely unexamined creed. Though militants serially commit what must objectively be seen as atrocities, they are not roundly condemned and marginalized in the Muslim world because there is no dynamic consensus that they are wrong. Terrorist acts are not sensibly labeled as heterodoxy. Heresy, as the preeminent Bernard Lewis has written, is a Christian idea that has "little or no relevance to the history of Islam, which has no synods, churches, or councils to define orthodoxy, and therefore none to define and condemn departures from orthodoxy." As a result, an influential cleric, such as Sheik Omar Abdel Rahman—the inspiration behind the 1993 World Trade Center bombing who holds a doctorate from al Azhar University in Egypt, the world's premier center of Islamic learning—is free to pronounce, virtually without challenge, that "[t]here is no such thing as commerce, industry, and science in jihad.... 'Jihad'...means jihad with the sword, with the cannon, with the grenades, and with the missile. This is jihad." Many Muslims—no doubt most—disagree, but few indeed will condemn with the authority and gravitas so urgently needed if the militants are to be isolated and marginalized.

Though militants serially commit what must objectively be seen as atrocities, they are not roundly condemned and marginalized in the Muslim world

---

Danny Pearl never got a chance to meet with Sheik Ali Gilani, to draw out and perhaps challenge his views. He was lured to Karachi on the false promise of an interview by a militant named Omar Ahmed Saeed Sheikh. Sheikh, like Richard Reid and like so many others, was educated and radicalized in the slumbering West—a Briton born of Pakistani immigrants in London. He traveled to the badlands of Afghanistan and Pakistan, trained on the militants' proving grounds, became further enmeshed in the early 1990s by the Bosnian Muslims' jihad against the Serbs, and relocated again to Pakistan. There, he fulfilled a widely followed model, becoming a leader in Jaish-e-Mohammed (the "Army of Mohammed"), among countless militant groups that orbit together—plotting ad hoc terror conspiracies against perceived "enemies of Islam"—within the global network known as al Qaeda. Sheikh, in fact, personally became a trusted confederate of the network's leader, Osama bin Laden, and thus enjoyed access to its highest echelons.

Sheikh was an experienced kidnapper, having been jailed for five years by the Indian government—a major target of militant Islam's Pakistani gangs—for detaining American and British tourists. India, however, falling into an all too familiar appeasement pattern, made the historic error of negotiating with Islamic terror. In 1999, Sheikh and other militants were traded by India in exchange for the freedom of 155 hostages hijacked to Kanadahar in an India Airlines plane. The message to the militants, as always, was clear: terrorism works. Plus, Sheikh was free to terrorize again.

Pearl's efforts to escape during the ensuing week were thwarted—he was physically abused and chained to a car engine.

And so he did. Sheikh, working in conjunction with at least two other Pakistani groups, Lashkar-e-Jhangvi and Harkat-ul-Mujahideen-Al Alami, devised a compartmented plan in which one cell was responsible for Danny Pearl's kidnapping—which was accomplished when the victim was snatched near a Karachi restaurant on January 22, 2002, having been lured there on the false promise of an interview with the cleric, Gilani—and another cell was responsible for his detention at a compound a short distance away. Pearl's efforts to escape during the ensuing week were thwarted—he was physically abused and chained to a car engine.

FINALLY CAME ANOTHER standard and monstrous agitprop: the capital show trial. With such a high-value target in militant clutches—a Jewish American journalist whose captivity had spawned intense international media coverage—the owner of the compound, Saud Memon, is said to have arrived on January 31 or February 1 with three Arabic-speaking men, one of whom is believed to have been Khalid Sheikh Mohammed, al Qaeda's top operations commander, generally credited with masterminding the 9/11 attacks. The journalist was forced to admit his "crimes" with the camera rolling: "I am a Jew," he acknowledged, adding that his parents, too, were Jews.

For maximum propaganda purposes, Pearl was further made to criticize American policies, particularly the detention in Guantanamo Bay, Cuba, of terrorist combatants seized from the battlefield in Afghanistan after the 9/11 attacks. As the militants have learned, the complacent Western press is guaranteed to follow the script—to report any subsequent atrocity, in morally equivalent terms, as "retaliation" for the cited American action, as opposed to what it manifestly is. Thus would Pearl's grisly murder be cast as turnabout for the mere imprisonment of Muslim warriors, rather than as further evidence of militant Islam's inhumanity and its blood-soaked designs on world hegemony.

Sentence, then, was not so much pronounced as executed: Daniel Pearl's throat was slashed (by K.S. Mohammed, according to knowledgeable statements reportedly made to Pakistani authorities) and he was savagely decapitated, his severed head held aloft as a trophy by his killers—Allahu akbar! The victim having served all his captors' execrable purposes, he was dismembered and interred in a shallow grave outside Karachi.

More than two years later, Danny Pearl's unspeakable murder has not been solved. At least 16 militants were implicated, but only four—Sheikh and three other kidnappers—were tried and convicted, with Sheikh sentenced to death and the others to life imprisonment. Appeals, however, have been interminably delayed, with the Pakistani government seemingly in no hurry either to have sentences executed or to bring the actual murderers to justice. Adding to this tawdry state of affairs, the renowned French philosopher Bernard-Henri Levy, after spending a year investigating the case, published a book last year, *Who Killed Daniel Pearl?*, theorizing that the Pakistani Intelligence Service—long suspected of ties to militant Islam—was complicit in the homicide (and that Sheikh was in fact one of its covert agents). Whatever the viability of such a claim, it serves only to obscure the incontestable truth glaring defiantly before us: the root evil is militant Islam.

As if to underscore that reality, it is now on shocking display again in the inhuman hacking of Nicholas Berg, a 26-year-old American Jew whose decapitated body was found on May 8 strewn on the outskirts of Baghdad. Far less is known about his captivity than about Danny Pearl's, but the parallels are striking. During a videotaped execution, Berg was made to identify himself. Islamic militants then read a scathing indictment of President Bush and American foreign policy—purporting to justify the barbarism they were about to visit on Berg by the American abuse of Iraqi prisoners at the Abu Ghraib prison. Presciently, they were confident that the mainstream American media would report the atrocity as retaliation rather than as innate evil. Berg was then beheaded, with his minute of chilling, tortured screams answered by his assassins' hypnotic antiphony: *Allahu akbar, Allahu akbar, Allahu akbar!*

The war is not on "terror." The war is on militant Islam. Like the totalitarian ideologies of the 20th century that strove for cruel hegemony, militant Islam must be identified, acknowledged, and defeated utterly. Until we come to grips with, and act on, that truth, we sentence ourselves to hearing that refrain over and over again.

Andrew C. McCarthy is a former chief assistant U.S. attorney who led the 1995 terrorism prosecution of Sheik Omar Abdel Rahman et al. in connection with the first World Trade Center bombing.

CHANGING COURSE ON CHINA

Elizabeth Economy

"Relations between China and the United States are perhaps the best they have been since 1989...What accounts for this seemingly dramatic transformation?"

In the immediate aftermath of his election in November 2000, President George W. Bush proclaimed China a strategic competitor and asserted that US policy in Asia should be reoriented toward American allies in the region, including Taiwan. Consensus withinthe administration on how to implement this new policy, however, remained elusive.

Indeed, for much of the early tenture of the Bush White House, the administration seemed divided on how best to approach China. The US trade representative and members of the State Department preached the virtues of engagement with the mainland while the Pentagon formulated its own policy to enhance US relations with Taiwan. Some within the administration went so far as to place the mantle of the former Soviet Union on China, calling the People's Republic the next great threat to US security. Within months after Bush took office, an ugly altercation over a US spy plane, a sizeable arms sale to Taiwan, and aggressive talk of American missile defense increased tensions markedly.

Nearly three years later, relations between China and the United States are perhaps the best they have been since 1989. President Bush appears to be following in the footsteps of his predecessors in recognizing both the importance of China to US foreign policy interests and the benefits of a more proactive approach to the mainland. Apparent divisions within the Bush administration over how to approach China have resolved themselves, at least temporarily, in favor of a more engagement-oriented policy.

What accounts for this seemingly dramatic transformation in US policy toward China? Above all else, the evolution in policy reflects a new set of strategic realities that have confronted America since September 11, 2001. The events of September 11 caused the Bush adminstration to reorient its international priorities, both diminishing the centrality of China in US foreign policy as a potential long-term threat and offering a new and important opportunity for the two countries to cooperate.

Soon after September 11, when the Bush administration began its campaign to enforce Iraqi compliance with United Nations weapons inspections, US officials actively courted the Chinese leadership. The administration first sought China's support for stepped-up inspections of Iraq's weapons arsenal, then for a UN resolution authorizing military action against Iraq. In the process, the administration turned its agenda with China on its head. More recently, the nuclear crisis on the Korean peninsula has reinforced the importance of China to US strategic interests while opening a new avenue for cooperation. For the United States, China has become an essential partner in meeting its new geostrategic challenges.

China's leaders, in turn, have recognized an opportunity to use these new strategic realities to meet their own fundamental objectives: stability in the Sino-American relationship to ensure continued economic modernization, and the maintenance of China's own domestic security. As Chinese President Hu Jintao has argued, the importance of the United States to China's economic development requires a flexible and accommodating posture that keeps US–China relations on an even keel.

This does not mean that significant policy differences between the two countries have disappeared, or that this partnership will continue indefinitely at its current level of mutual accommodation. The Sino-American relationship remains fragile and continues to require high-level intervention to ensure its stability. Critical differences still mark the manner in which the two countries approach international relations generally, as well as specific issues such as Taiwan, missile defense, and human rights. In addition, President Bush does not make China policy in a vacuum. Differences within the administration over relations with China remain. An active congressional lobby on China and Taiwan, while quiescent for most of 2002 and 2003, may again become energized in an effort to redefine US-China policy.

THE FIRST SIX MONTHS

During the presidential campaign, candidate Bush set the tone for a distinctly new China policy. He promised to refocus US attention in the region away from the mainland and toward Japan, South Korea, and Taiwan. The Bush team stressed that these East Asian powers were democratic and capitalist and thus natural allies of the United

States. Candidate Bush also referred to China as a strategic competitor, emphasizing its human rights abuses and role in the proliferation of missile technology.

For the United States, China has become an essential partner in meeting its new geostrategic challenges.

While this rhetoric received significant media attention, subtle hints also signaled a degree of continuity with the previous administration's policies. Candidate Bush argued that the United States had to remain deeply engaged on the trade front with China, noting that the development of an entrepreneurial class and the advent of the Internet were cornerstones in a process of long-term political liberalization in China and that American farmers would benefit from China's entry into the World Trade Organization. Bush even reiterated the classic engagement line: "I think if we make China an enemy, they'll wind up being an enemy."

The first testing ground for the new administration and its approach to China came less than three months after Bush assumed the office. In late March 2001, a US Navy EP-3 surveillance aircraft flying near China's Hainan Island collided with a Chinese fighter jet. The Chinese pilot was lost at sea and the EP-3's flight crew was forced to make an emergency landing on Hainan. It was not until April 4, five days after the accident, that diplomatic exchanges occurred.

This was a tense and difficult time, with many members of Congress calling for tough action and some policy analysts predicting military conflict between China and the United States. Once China opened the door to US diplomats, however, President Bush moved quickly to effect a resolution. On April 5, Secretary of State Colin Powell issues a statement of "regret" that progressed to "sorry" and "very sorry" by April 7. On April 11, the US flight crew departed Hainan to return to the United States. In the aftermath of the incident, the consensus in the United States was that President Bush had handled a difficult situation well by preventing the conflict from escalating. Still, some conservative policy analysts critized the administration for appeasing Chinese misbehavior. They argued too that the strategy President Bush adopted represented a victory for Secretary of State Powell and National Security Adviser Condoleezza Rice over Vice President Dick Chaney and Secretary of Defense Donald Rumsfeld.

Still, the EP-3 incident, as well as the initial difficulty in reaching and negotiating with senior Chinese military and party officials, provided the rationale that China hawks within the administration needed to take a series of significant steps. The Defense Department broke off all informal and regular military-to-military contacts, noting that such contacts would have to be approved on a case-by-case basis. The administration announced an arms sale package to Taiwan worth as much as $4 billion that included up to eight diesel submarines and four guided-missile destryoers. It cited as justification China's continued missile buildup opposite Taiwan and the mainland's refusal to renounce the right to use force to reunify with Taiwan. For the first time, the Defense Department allowed Taiwanese military officials to participate in courses at its Hawaii-based think tank, the Asia-Pacific Center for Security Studies.

Meanwhile, discussions of the Bush administration's plans for missile defense, which the Chinese regarded as extremely threatening, filled the news media. And when asked whether he would use military force to defend Taiwan in case of a Chinese attack, the president said, "Whatever it takes to help Taiwan defend itself." This assertion caused great consternation in mainland China, where it was perceived by some as dropping the American commitment to "strategic ambiguity" (whereby the United States had refused to discuss various scenarios in which conflict between the mainland and Taiwan might emerge in order to preserve the full range of options for US action).

Yet even as the security relationship clearly deteriorated, Bush did not permit the EP-3 incident to become entangled with other areas of the Sino-American relationship, such as China's impending entry into the World Trade Organization. Morever, the manner in which the EP-3 incident itself was handled demonstrated that, whatever the administration's rhetoric, the White House was committed to keeping its relationship with China on track and to preventing tensions in the security realm from spilling over into other arenas. In fact, during the summer of 2001 Secretary of State Powell pursued a noticeably more engagement-oriented approach than others within the administration, visiting the People's Republic and elucidating a new foundation for US dialogue with China dubbed the "Three C's": candid, cooperative, and constructive. Thus, for much of the early part of the Bush administration's tenure, China policy proceeded along two distinct tracks: one directed by the Department of Defense and vice president's office and a second navigated primarily by the State Department.

AFTER SEPTEMBER 11

The devasting terrorist attcks on the World Trade Center in New York and on the Pentagon in Virginia caused the United States to reorient radically its foreign policy priorities. The central focus shifted from promotion of free trade, democracy, and stability to a global war against terrorism. No longer was the Bush administration preoccupied with defining the next Soviet-like menace; it needed to identify terrorist cells throughout the world.

For Sino-American relations, this transformation is US foreign policy had two important implications. First, China policy became simply one of many issues rather than a top preoccupation for the administration, Congress, and the media. In a world in which America's physical integrity had been so violently breached, concern over the potential economic and security threat posed by China greatly diminished. Second, September 11 provided a clear opportunity for China to establish common interest with the United States on the latter's number one priority: combating terrorism.

China initially hedged its bets, calling for the United States to support China in battling terrorists in Xinjiang, Tibet, and Taiwan. Just one week after the attacks, the government spokesman Zhu Bangzao remarked that "the United States has asked us to help it fight terrorism. Equally, we have reasons for asking the United States to lend us its support and understanding in our struggle against terrorism and separatism. There can be no double standards. We are not suggesting any horse trading; but China and the United States has a common interest in opposing the Taiwanese independence movement which constitutes the main threat to stability in the [Taiwan] Strait."

China soon amended its request to ask for assistance only in fighting the terrorist threat in Xinjiang Autonomous Region, perhaps realizing that equating Al Qaeda with separatist movements in Taiwan and Tibet would be poorly received by many in the United States. Still, China reiterated its concerns that America establish concrete proof of Osama Bin Laden's guilt, that any military strike it might carry out accord with UN rules, and that any action taken be in the long-term interest of world peace and development. Since September 11, the Bush administration has generally given China high marks for its help in tracking down terrorist financing, cooperating on law enforcement, and providing humanitarian aid to Afghanistan.

But, even as China and the United States began to forge new bonds on the issue of global terrorism, Taiwan remained a sticking point. In March 2002, for example, the US Taiwan Business Council hosted in the United States Taiwanese Defense Minister Tang Yiau-ming and other military officials—including US Deputy Secretary of Defense Paul Wolfowitz—thereby violating an undeclared prohibition against meetings between senior Taiwanese and senior US officials. During the meeting, moreover, Wolfowitz repeated President Bush's earlier controversial statement that the United States would do "whatever it takes" to defend Taiwan from military strikes by China. Wolfowitz's affirmation that "China is not an enemy" was lost in the firestorm that his other rhetoric provoked.

At the same time, the Defense Department's Nuclear Posture Review, a summary of strategic planning for America's nuclear weapons over the next decade, was leaked to the media. The document noted that a conflict between China and Taiwan could lead to the use of nuclear weapons by the United States. This added to the growing tensions in Sino-American relations. Congress, too, always an active player on Taiwan issues, became energized by the Bush administration's more proactive Tai-

wan policy. On April 9, 2002—the twenty-third anniversary of the Taiwan Relations Act—the House of Representatives established the Taiwan Caucus. One of the forum's primary goals was to promote US military ties with Taiwan. (The 1979 act guarantees continued trade and cultural relations with the island and provides assurances for Taiwan's security.)

Beijing was quick to respond to the Defense Department's increased attention to Taiwan. In reply to Wolfowitz's remarks, Foreign Ministry spokeswoman Zhang Qiyue stated, "The comments of the senior US defense official seriouly violated the clear-cut promises laid out in the three joint communiques [that have framed Sino-American relations] and moreover rudely interfered in China's internal political affairs." In addition, Beijing denied the destroyer the USS *Curtis Wilbur* a port call in Hong Kong for early April, and threatened to cancel the impending visit of then Vice President Hu Juntao later in the month. Thus, only weeks before Hu's visit to the United States, the Pentagon's priority on enhancing ties with Taiwan at the expense of relations with the mainland threatened yet again to derail the Sino-American relationship.

SETTING A NEW AGENDA

While the war against terrorism opened the door to the creation of a new foundation for Sino-American relations, the most significant factor in the evolving relationship was the Bush administration's almost singular focus on Iraq. In working to secure backing for a US-led attack on Iraq, the Bush administration arrived at an entirely new agenda with China. Beginning in April 2002 with the visit of then Vice President Hu Jintao, the White House moved quietly and effectively to set the stage for a newfound unity in US-China relations.

The first sign of this emerged during Hu's visit. Both before and after the visit, the White House downplayed the contentious issues of human rights, Taiwan, and weapons proliferation. Although these concerns came up during a session between Vice President Chaney and Vice President Hu, the thrust of the meetings was positive. One high-ranking but unidentified White House official said: "Mr. Hu is bright, amenable, and very pleasant ...she is a pleasure to do business with." Another senior administration official offered even higher praise: "He can come across as warm and even flexible, yet gives nothing away. I can see the day when Mr. Bush feels he can pick up the phone and call Mr. Hu. I don't think he has ever quite felt that way with [President] Jiang Zemin. They are of different eras."

Just one month after Hu's visit, Pentagon hawks, who had defined China as the next Soviet Union, were flying to the People's Republic to discuss conditions for restoring military-to-military contacts. Senior administration officials no longer identified China as a strategic competitor but rather as Secretary of State Powell had—as a participant in a "candid, cooperative, and constructive" dialogue. Deputy Secretary of Defense Wolfowitz began to speak of the need to engage China.

In August 2002 the Bush administration labeled the East Turkestan Islamic Movement—a small group committed to the independence of China's Xinjiang region—a terrorist organization. Beijing, which had long claimed connections between Al Qaeda and ethnic Uighur separatists in Xinjiang, welcomed the move. Perhaps most telling, the Bush administration also indicated in August that it would grant President Jiang Zemin his long-sought invitation to President Bush's ranch in Crawford, Texas, an invitation previously extended only to President Bush's closest perceived allies, such as British Prime Minister Tony Blair and Russian President Vladimir Putin. Moreover, the administration virtually ignored the release of a long-anticipated report by the congressional US-China Security Review Commission in fall 2002, which widely condemned China for its indirect sponsorship of rogue states through the sale of missle technology and called for a range of new sanctions. The report was an embarrassment for an administration intent on making a new friend in China.

China responded positively to th US steps to improve relations. In late August 2002, the Chinese government issued a new set of regulations governing the export of missile technology and reacted only moderately to strong separist rhetoric by Taiwan's leader, Chen Shui-bian, and a visit to the Pentagon by a senior Taiwanese defense official. In October Beijing also announced regulations to control material and technology for biological weapons and took steps to improve the situation of a few political dissidents, including AIDS activist Wan Yanhai, who was permitted to establish his own non-governmental organization to combat AIDS.

On the issue of greatest importance to the United States—reining in any Iraqi program to develop weapons of mass destruction—China supported the United States in its initial effort to press for more aggressive sanctions. It voted in favor of UN Security Council Resolution 1441, which warned Iraq that it would face "serious consequences" if it did not comply with UN weapons inspections. Prime minister Zhu Rongji insisted that "Iraq must cooperate unconditionally with the United Nations" on weapons inspections "At the same time," he continued, "we must respect the sovereignty and territorial integrity of Iraq. If arms inspections do not take place, if there is not clear proof and if there is no authorization from the Security Council, there cannot be a military attack on Iraq."

As the United States pressed for an additional resolution authorizing the automatic use of force in the face of Iraqi noncompliance, China stood with France and Russia in opposition to the proposal. But it did so much less forcefully and publicly than its Security Council partners; there was nothing to be gained from unnecessary antagonizing the United States, particularly in the midst of such a positive overall bilateral relationship. In return, China

escaped the far more vocal criticism the United States directed at France and Russia.

Iraq engendered a new US approach toward China, but it has been the crisis on the Korean peninsula that has served to reinforce the importance of a US-China partnership. In October 2002, North Korea admitted that it had not forsaken its nuclear program, triggering a crisis in US-North Korean relations and setting off shock waves throughout the rest of East Asia. The United States and China had long professed "peace and stability on the Korean peninsula" as a common policy goal. Yet rarely had the policy been put to such a test, and the two countries soon found themselves articulating signficiantly different approaches to the resolution of the crisis.

China, like Japan and South Korea, favored a strategy that engaged North Korea, offering economic incentives for compliance. The United States argued for a much tougher policy, possibly involving sanctions and not ruling out the use of force. China also supported North Korea's desire to resolve the issue through bilateral US-North Korea negotiations; the United States insisted on multilateral talks. As North Korea pushed the envelope, withdrawing from the nuclear Non-Proliferation Treaty, the United States sought a resolution within the UN security Council condemning North Korea for its actions. But China resisted resorting to the United Nations, believing that doing so would only isolate North Korea further and put China in the awkward position of having to ally itself openly with either North Korea or the United States.

Despite their different policy prescriptions, the United States and China continued to work together behind the scenes at the behest of the United States. Some Chinese scholars within elite policy circles also began to suggest that China needed to reevaluate its position, arguing that a nuclear North Korea posed a threat to China. In February 2003, China shut down an oil supply line to North Korea for three days—ostensibly due to "technical problems" but more likely in an effort to signal to North Korea that its belligerent approach was costing it China's support. Soon thereafter, China brokered an agreement between the United States adn North Korea to hold three-way talks in which China would be the third participant. While the negotiations produced no tangible results, the United States and China had moved their relationship to a new level of partnership in the process and agreed to continue to work together to resolve the crisis.

MARGINALIZED: HUMAN RIGHTS, TAIWAN, TRADE

After 1989, the issues that defined the Sino-American relationship were human rights, Taiwan and trade. These concerns no longer dominate the bilateral agenda. Securing stability on the Korean peninsula, jointly fighting terrorism, and meeting US objectives in Iraq, as well as China's desire for continued stability to ensure economic growth, are now the dominant factors in the relationship.

This transformation in the bilateral agenda has been marked by substantive changes in policy. For the first time since 1989, the United States in April 2003 did not pursue a resolution at the UN Human Rights Commission in Geneva condemning China for its human rights practices. The United States has been noticeably quiet about human rights violations that otherwise would have given rise to comment from the White House, including China's decision to prevent labor activists in Liaoning province from meeting with US diplomats and foreign reporters, and the continued imprisonment without councel of long-time American resident Yang Jianli, a prominent democracy activist who illegally entered China to observe mass labor protests in northeast China. Whereas previously the Bush administration might have exploited Beijing's initial mismanagement of the SARS crisis as an opportunity to criticize the regime, at least publicly, the Bush team instead congratulated the Chinese leadership on its handling of what was feared might become a global epidemic.

With regard to Taiwan, the Bush administration continues to act as the island's primary interlocutor in international forums over the objections of the People's Republic: supporting Taiwan's accession to the World Health Organization (blocked by China) in the midst of the SARS crisis, for instance, and backing Taiwan as it wrangles with Beijing over its formal name within the World Trade Organization. The United States continues to promote military relations with Taiwan through exchanges and the sale of advanced weaponry.

Yet the administration has taken steps to reassure Beijing of its commitment to a one China policy, moving beyond previous US administration to state definitively that America does not support Taiwan independence. Beijing, in turn, has begun to develop policy proposals that might persuade the Bush administration to reduce its commitment to Taiwan. In October 2002, for example, President Jiang proposed withdrawing some of the several hundred missiles targeted at Taiwan in exchange for US cutbacks on arms sales. Although the administration has yet to reply, Bush has reportedly asked his staff to develop a formal response. At the same time, within some quarters of China's elite, growing economic integration and personal links between the mainland and Taiwan have begun to foster a new sense of confidence concerning the eventuality of reunification.

In meetings between US and Chinese officials, Taiwan remains an important matter, but there is little evidence of the rancor that marked bilateral exchanges on the issue prior to September 11. In June 2003 at the Group of Eight summit in Evian, France, Hu Jintao—who replaced Jiang as president this March—stated that he appreciated President Bush's declaration not to support Taiwanese independence.

Unlike human rights and Taiwan, trade issues typically have had an ameliorative effective on Sino-American relations. China's accession to the World Trade Organization in December 2001 was celebrated both in China and the United States as signaling signficant future trade and economic opportunities. America is China's largest export market, with overall exports from China to the United States last year totaling more than $100 billion.

Still, many in the American business community and within the US trade representative's office have begun to express concern with a growing trade deficit that, according to US Trade Representative Robert Zoellick, will exceed $100 billion in 2003. In several critical areas such as agriculture, telecommunications, and finance, Zoellick and various China watchers have noted an increasing number of bureaucratic impediments to foreign access to China's markets.

In addition, while successive US administrations had felt comfortable with Prime Minister Zhu Rongji and his committment to effective implementation of China's World Trade Organization obligations, his retirement in April 2003, as well as the elimination of the Ministry of Foreign Trade and Economic Cooperation, has left some uncertainly in the trade relationship. In May 2003, the United States sanctioned a major Chinese conglomerate, NORINCO, for transferring dual-use technology to an Iranian company known to produce missiles. (The Chinese government's muted response lent weight to the US claim.) Thus, although trade remains one of the pillars of the Sino-American relationship, it is possible that friction will increase over time, particularly if China's WTO implementation proves problematic. Indeed, the United States is the third-largest source of foreign direct investment (FDI) in China after Hong Kong and the Virgin Islands, through which FDI from a number of countries flows. (In 2002, total US FDI contracted totaled more than $7 billion.)

WHAT NEXT?

Even as China and the United States continue to strengthen their bilateral relationship by working closely on issues of global security and downplaying traditional rifs, their recent accommodation may well prove ephemeral. Fundamental divisions persist. The United States coninued to desire, and work toward, evolution in China's political system. Many in China still perceive America as a signficant obstacle to China's growing status as a regional, if not global power.

The uncertainty generated by the recent leadership transition in China and the potential for a new policy agenda in Beijing could also open the door to significant change in China's approach to the United States. Former President Jiang Zemin continues to exercise power beind the scenes, often in conflict with Hu Jintao. While unlikely, it is not impossible that policy toward the United States could fall victim to elite power politics as one side or the other attempts to play an anti-US/nationalism card in hopes of currying popular support. President Hu's new emphassi on slowing down the pace of economic reform to redress the vast social inequities that emerged over the course of Jiang's tenure might also contribute to a slowdown in the implementation of China's trade commitments and a consequent downturn in Sino-American relations.

Important opportunities remain, however, for strengthening the foundation of the relationship and helping it endure beyond the current, potentially transitory accommodation that has resulted from new geostrategic realities. Cooperative ventures in areas where common interest exists naturally, such as public health and the environment, should be fostered. Even more critically, the administration should seek opportunities to advance Sino-American relations in the area most difficult to negotiate: security, human rights, and Taiwan.

In the security realm, the Department of Defense should move quickly to reestablish military-to-military relations. Although the EP-3 incident reinforced the importance of maintaining open channels of communication and fostering military exchange, the Pentagon has yet to follow through on President Bush's directive to restart these exchanges. American military officers must still obtain approval for every meeting or exchange with their Chinese counterparts. This is inefficient and does little to help develop the open dialogue and personal relationships that can prove effective for crisis management as well as longer-term understanding of each country's security priorities and appraches. Military contracts are particularly critical as the United States and China continue to work closely on a highly sensitivie security issue such as North Korea.

In working to secure backing of a US-led attack on Iraq, the Bush administration arrived at an entirely new agenda with China.

With regard to human rights, the new accommodation in Sino-American relations offers the Bush administration the opportunity to call directly on President Hu to embrace a more aggressive program of rights protection. The United States is already cooperating, both directly and indirectly, with the Chinese government on law enforcement and the development of the rule of law. Recent high-profile cooperation in dismantling a major drug-trafficking network in Asia and North America, for example, may mark a breakthrough for future cooperation on law enforcement. Various US and NGO efforts focus on the training of Chinese judges and lawyers, cooperation on intellectual property enforcement, and support for more open media in the hopes of assisting enforcement. The State Department's Bureau of Human Rights, De-

mocracy and Labor in particular has mounted a variety of ambitious initiatives designed to advance the cause of human rights, rule of law, development of NGOS, and freedom of expression.

The administration should also renew the dialogue on human rights at the highest level. While the long-stalled dialogue on human rights between the State Department and the Chinese Foreign Ministry resumed formally in December 2002, it has failed to produce follow-up visits or any tangible results, ostensibly because of personnel shuffling within China. The United States should make it clear to Beijing that dialogue and action on human rights are as important a part of the bilateral relationship as discussions on restoring direct military-to-military relations or providing additional assistance to combat terrorism.

Finally, the administration appears to be successfully navigating the often treacherous waters of cross-strait politics. As the White House presses forward with various security initiatives in Asia, including the restationing of troops in South Korea and Japan and the development of missile defense, administration officials should take the opportunity to consult and even cooperate with the People's Republic. If the United States pursues a missile defense architecture that permits targeting of Chinese missiles, for example, early consultation with China to establish mutually acceptable limits on the number of Chinese missiles deployed in response would be extremely useful in avoiding a potential regional arms race and a downward spiral in US-China relations. In the spirit of the new Sino-American relationship, China as indicated that, while it is concerned about the development of theater missile defense, it is willing to conduct "constructive dialogue" on the issue. The only aspect of a potential US missile defense system that China considers non-negotiable is the inclusion of Taiwan behind the shield.

For the foreseeable future, a strong and stable bilateral relationship serves both Chinese and American interests. China considers US support central to its own continued economic development and security. The United States considers China's support necessary for addressing challenges in North Korea and Iraq, and to combat terrorism. As long as these strategic realities continue to dominate each country's domestic and foreign policy agendas, positive relations between China and the United States are likely to remain a high priority for the leaders of both countries. By developing long-term strategies on key contentious issues, the leaders could improve chances that this stability will continue.

ELIZABETH ECONOMY is a senior fellow and director of Asia studies at the Council on Foreign Relations.

From *Current History*, September 2003, pp. 243-249. © 2003 by Current History, Inc. Reprinted by permission.

THE KOREA CRISIS

North Korea is not crazy, near collapse, nor about to start a war. But it is dangerous, not to mention dangerously misunderstood. Defusing the threat that North Korea poses to its neighbors and the world will require less bluster, more patience, and a willingness on the part of the United States to probe and understand the true sources of the North's conduct.

By Victor D. Cha and David C. Kang

"North Korea Belongs in the Axis of Evil'"

No. The only link between North Korea and Iran and Iraq, the other two members of the "axis of evil" identified by President George W. Bush in his 2002 State of the Union speech, is financial. North Korea has sold missile technology to Iran, as it has to a number of countries, including U.S. allies Pakistan and Egypt. Unlike the original Axis powers, Japan, Germany, and Italy, which were joined formally by the Tripartite Pact of 1940, North Korea, Iran, and Iraq do not coordinate or work together beyond the sale of goods to one another. Furthermore, North Korea does not share any religious, ideological, or strategic goals with Iran and Iraq. North Korea's concerns focus solely on the peninsula and do not extend to the Middle East. Although it does nasty things like sell drugs and make counterfeit money, North Korea has not engaged in terrorism in the last 16 years, and there has never been any link, nor any suggested, between North Korea and al Qaeda.

Iran, Iraq, and North Korea do share some common traits, the main one being an adversarial relationship with the United States. They are also authoritarian, have allegedly supported or sponsored terrorism, and have programs to develop weapons of mass destruction. However, using those latter criteria, several other countries could fit in the axis. Why not U.S. allies Pakistan or Saudi Arabia, for example?

"Kim Jong Il Is Crazy, Unpredictable, and Undeterrable"

Wrong. Kim Jong Il is as rational and calculating as he is brutal. Dictators generally want to survive, and Kim is no exception. He has not launched a war, because he has good reason to think he would face fatal opposition from the United States and South Korea. In fact, like his father Kim Il Sung, Kim has clearly shown he is deterrrable: North Korea has not started a war in five decades.

Dictators do not survive without sophisticated political skills. Kim has maintained power despite intelligence assessments that his leadership would not survive the death of his father in July 1994. And he has endured despite famine, floods, economic collapse, nuclear crises, the loss of two major patrons in Russia and China, and U.S. pressure. There has been no palace or military coup, no extensive social unrest, no obvious chaos in the military, and no wholesale purge of various officials. Moreover, Kim's decision to proceed with North Korea's tentative and measured economic reforms is further proof that, however morally repugnant he may be, he is also quite capable of assessing costs and benefits.

But his rationality does not make him any less dangerous. Under Kim's rule, North Korea has engaged in a coercive bargaining strategy designed to ratchet up a crisis with the United States. Provocations such as test-firing missiles, shadowing spy planes, and walking away from treaties can grab attention and even force the United States and its allies to provide inducements persuading North Korea back from the brink. A risky approach, perhaps—but rational, too. If you have little to negotiate with, it makes sense to leverage the status quo for maximum bargaining advantage.

"North Korea Poses a Direct Nuclear Threat to the United States"

Calm **down.** What sparked the current crisis over North Korea's nuclear intentions were revelations last October that Pyongyang has pursued a secret program to produce highly enriched uranium that could be used to make nuclear bombs. That effort violated the Agreed Framework negotiated between North Korea and the Clinton administration in 1994, under which the North had agreed to freeze its nuclear program and accept international inspections in return for fuel oil shipments and, eventually, two "proliferation-resistant" nuclear reactors. The October revelations prompted a stiff U.S. response that included a cutoff of fuel oil deliveries. North Korea, meanwhile, has kicked out international inspectors, withdrawn from the Non-Proliferation Treaty, and begun to restart its nuclear reactors—prompting fears that the country will soon have much more than the one or two bombs' worth of nuclear material typically cited by U.S. intelligence analysts.

But lost amid all the alarm and bluster is the reality that the logic of deterrence will prevail even if North Korea develops and deploys a nuclear force. North Korea pursues nuclear weapons not for leverage but for the same reason that other highly vulnerable nations arm themselves: to deter an adversary, in this case a superpower that is armed with nuclear weapons. But even if the North develops nuclear weapons, the threat of a devastating U.S. response will prevent it from ever using them—after all, unlike shadowy terrorist cells, nations cannot hide from a retaliatory strike.

What about the North's missile threat? Some analysts claim that North Korea already possesses a long-range nuclear missile capability. That's false. The longest-range missile currently deployed by North Korea is the No Dong missile, which can carry a 1,500-pound payload approximately 800 miles. However, North Korea has reportedly tested the No Dong only once. The untested Taepo Dong 2 can potentially carry a several hundred–pound payload between 6,000 and 9,000 miles—far enough to reach the West Coast of the United States. But without adequate testing, such a nuclear missile would be highly unreliable.

The fact is, North Korea could blow up terrorist bombs in downtown Seoul every week if it had the desire to do so. It could smuggle a nuclear device into Japan, given the extensive network of Koreans in that country with ties to the North. For that matter, why should North Korea develop an expensive ballistic missile to shoot at the United States when smuggling a nuclear weapon in a shipping container would be so much easier? The primary value of the North's missiles is as a military deterrent, not as an offensive weapon.

The only nuclear threat to the United States from North Korea is indirect, in the potential transfer of such capabilities to third parties. Pyongyang has shown no aversion to selling weapons to anyone with the hard currency or barter to pay for them. North Korean nuclear weapons or fissile material hidden in tens of thousands of underground caves would likely go undetected even by the most intrusive inspections. But a transfer of nuclear material would be a risky proposition for a regime that values survival above all else. Given the preemptive mind-set of a post–September 11 United States, the North would have to be confident that any transfer would escape U.S. detection and therefore the threat of a massive U.S. retaliation.

"North Korea Does Not Honor International Agreements"

Mostly **true.** Heralded for a half century as an outlaw state, North Korea has maintained some of its international commitments. It is a member of the Conference on Disarmament, Biological Weapons Convention, and Geneva Protocol. After the attacks of September 11, 2001, the country signed on to two U.N. antiterrorism protocols. During the negotiations for the Agreed Framework, the United States required the North to improve relations with South Korea. Pyongyang eventually responded by agreeing to a summit (just prior to Kim Il Sung's death). As of this writing, the North has also honored its 1999 ballistic missile-testing moratorium for four years.

But the North also has a history of engaging in strategic deception"—signing agreements to convey reliability but purposefully cheating on them to its own advantage. The history of inter-Korean relations, for example, is littered with pacts that Pyongyang has not honored, including the 1992 denuclearization declaration in which North Korea agreed to forgo developing nuclear and nuclear-reprocessing facilities. The United States may have been slow to implement the Agreed Framework, but the North is blatantly breaking the framework's spirit, if not letter, with its covert uranium enrichment program. But perhaps the best evidence of strategic deception occurred in June 1950: On the eve of the Korean War, North Korea put forth a major peace initiative to the South.

"North Korea's Political and Economic Collapse Is Imminent"

Don't **bet on it.** Observers have predicted an imminent North Korean collapse since the fall of the Berlin Wall in 1989. The country's economic situation is desperate, but signs of political collapse are absent. The best indicator of regime stability is that social control, however vicious, remains solid. Although the flow of refugees from the North is increasing, there is no widespread internal migration, and few observable signs of protest.

Some evidence suggests that North Korea is serious about normal political and economic relations with South Korea and the rest of the world. By December 2002, North Korea had cleared land mines from sections of the demilitarized zone (DMZ) separating North and South Korea. Tracks are being connected on the Kyonggui Railway, which would run from South Korea through the western corridor of the DMZ into North Korea. Pyongyang has also begun work on a four-lane highway on the eastern corridor as well. In July 2002, the central government formally abandoned the centrally planned economy and allowed prices and wages to be set by the market. The government has also created three special economic zones to exploit tourism and investment and amended its laws on foreign ownership, land leases, and taxes and tariffs. Although these reforms have been halting and only marginally successful, they are also becoming increasingly hard to reverse.

Though the regime appears resilient, there are two sources of potential fissures. First, the decidedly mixed results of several recent initiatives by Kim—among them, his decisions to lift price controls and to acknowledge North Korea's kidnapping of Japanese nationals in the 1970s—have exposed "Dear Leader" to potential disgruntlement in the top ranks. Second, the process of reform could create cracks in the regime's foundation. As Montesquieu observed, revolutions don't occur when the people's conditions are at rock bottom but when reform creates a spiral of expectations that spurs people to action against the old stultifying system.

"China Has the Most Influence on North Korea"

Yes, **but good luck** getting the Chinese to use it. If a state's influence on North Korea is merely a function of the North's material dependence on it, then China holds the trump cards. Seventy to 90 percent of North Korea's annual energy supplies, roughly 30 percent of its total outside assistance, and 38 percent of its imports reportedly come from China. Beijing played a quiet but critical role in inter-Korean dialogue leading up to the June 2000 summit. It also influenced Kim Jong Il and his decision to tentatively reform the North Korean economy by hosting Kim in Shanghai in 2002 and backing the creation of special economic zones.

Notwithstanding this close history, Chinese Foreign Ministry officials deny any influence on North Korea, complaining that "North Korea doesn't listen to us, it doesn't listen to anyone." But China's protestations largely reflect its unwillingness to put real pressure on its neighbor. China's traditional stake in North Korea has rested in keeping the regime afloat as a geostrategic buffer against U.S. influence on China's border. It also has no desire to provoke a regime collapse that would send millions of North Korean refugees flooding across the border.

A different set of Chinese interests may now come to the fore. Beijing opposes nuclear weapons on the Korean peninsula and delivered a dressing-down to North Korean embassy officials in Beijing in January about the country's cheating on the Agreed Framework. For China, nothing good comes from a nuclear North Korea. Such an outcome could prompt Japan to move from merely developing missile defense capabilities to acquiring ballistic missiles or nuclear weapons. And Taiwan might also cross the nuclear threshold if the country's leaders see North Korea successfully guaranteeing its security this way. While the rest of Asia provides China's economic lifeblood, Beijing continues to throw good money, food, and fuel down a rat hole in North Korea with little prospect of major reform.

Chinese policy may change under a new cadre of leaders such as President Hu Jintao who are less wedded to the Cold War relationship with Pyongyang, which used to be characterized as close as "lips and teeth." Or more likely, Beijing's hesitation to intervene may be tactical, as it waits for the United States to do the heavy lifting with North Korea (despite Bush's rhetoric to the contrary) and then swoops in to help close the deal and maximize its influence.

"The DMZ Is the Scariest Place in the World"

Yes, **if looks could kill.** When former U.S. President Bill Clinton called the border between the two Koreas the world's scariest place, he was referring to the massive forward deployment of North Korean forces around the DMZ and the shaky foundations of the 50-year-old armistice—not peace treaty—that still keeps the peace between the two former combatants. Since the end of the Korean War in 1953, there have been more than 1,400 incidents across the DMZ, resulting in the deaths of 899 North Koreans, 394 South Koreans, and 90 U.S. soldiers. Tensions have been so high that in 1976 the United States mobilized bombers and an aircraft carrier battle group to trim one tree in the DMZ. The deployments and operational battle plans on both sides suggest that if a major outbreak of violence were to start, a rapid escalation of hostilities would likely ensue.

In practice, however, no such outbreak has occurred. North Korea has faced both a determined South Korean military, and more important, U.S. military deployments that at their height comprised 100,000 troops and nuclear-tipped Lance missiles and even today include 37,000 troops, nuclear-capable airbases, and naval facilities that guarantee U.S. involvement in any Korean conflict.

The balance of power has held because any war would have disastrous consequences for both sides. Seoul and Pyongyang are less than 150 miles apart—closer than New York is to Washington, D.C. Seoul is 30 miles from the DMZ and easily within reach of North Korea's artillery tubes. Former Commander of U.S Forces Korea Gen. Gary Luck estimated that a war on the Korean peninsula would cost $1 trillion in economic damage and result in 1 million casualties, including 52,000 U.S. military casualties. As one war gamer described, the death toll on the North Korean side would be akin to a "holocaust," and Kim Jong Il and his 1,000 closest generals would surely face death or imprisonment. As a result, both sides have moved cautiously and avoided major military mobilizations that could spiral out of control.

Ironically enough, as for the DMZ itself, although bristling with barbed wire and sown with land mines, it has also become a remarkable nature preserve stretching across the peninsula that is home to wild birds and a trove of other rare species.

"The Clinton Administration's Policies Toward North Korea Failed"

No. The North's breach of the Agreed Framework may make Clinton's policies look ineffective, but consider the counterfactual proposition. If Clinton had not succeeded in freezing North Korea's main nuclear facilities at Yongbyon for nine years, North Korea would today have enough plutonium for at least 30 nuclear weapons rather than one or two bombs' worth.

Clinton's engagement with North Korea also provided a useful test of North Korean intentions and expectations. Previously, the United States had little sense of the North's interest in swapping its proliferation threat for external assistance. True, the debate between hawks and doves over this question still has a "he said, she said" quality to it: Hawks see North Korea's violations of the Agreed Framework as evidence of the North's lack of interest in such a deal; doves see those same violations as a reaction to the U.S. failure to fulfill the framework and still believe Pyongyang will give up nukes in return for outside support of economic reform. But now there is a baseline or "data" for a debate that previously took place at a theological and ideological level. Before Clinton, there was also no way to use leverage on a country with which the United States had next to no contact for five decades. Since 1994, the North has gained food, fuel, economic assistance, and diplomatic relations not just with South Korea but also Japan, the European Union, Australia, Canada, and others. Ironically, Clinton's carrots have become Bush's sticks, enabling the latter to pursue a harder-line policy by threatening to withhold what was once previously promised.

"The Bush Administration Caused the Current Crisis"

No. Bush's "axis of evil" speech and his professed loathing of Kim may have exacerbated the current crisis, but they certainly did not cause it.

First, North Korea started its covert uranium enrichment program for nuclear weapons long before Bush took office. As far back as 1997, Pakistani nuclear scientists were shuttling to Pyongyang, providing technology for uranium enrichment in return for North Korean missile systems.

Second, prior to the October 2002 revelations and despite Bush's occasional negative statements on North Korea, the United States had offered a string of consistent assurances at lower levels that it would pursue some form of engagement. These assurances included the creation of a package of new incentives and the expressed willingness to meet "any time, any place, and without preconditions." In addition, the Bush administration abandoned several initiatives cited as attempts to derail North Korean engagement—revision of the Agreed Framework and a push for conventional force reductions—after they proved to be nonstarters with U.S. allies. Compared with the Clinton administration's effusive advances to North Korea, Bush's aggressive posturing was portrayed by some media as a dramatic shift, but the U.S. predisposition for engagement remained. The North Koreans' response? They refused to engage in direct bilateral dialogue with the United States, accusing Washington of high-handedness.

Third, there is no denying the harder turn in both U.S. statements and policy after October 2002. North Korea's perception of the preemptive language in the Bush administration's new national security strategy and nuclear posture review could only have heightened North Koreans' worst fears. But Bush's unconditional refusal to talk with North Korea didn't create the crisis. The administration believes North Korea stands so far outside the nonproliferation regime that negotiating its return would be tantamount to blackmail. Should Pyongyang first make compliance gestures, however, then the United States would be willing to discuss incentives including security assurances, energy, and economic assistance. Sounds like a negotiating position to us.

"The United States Should Pull Its Troops Out of an Ungrateful South Korea"

Not yet. Massive demonstrations, Molotov cocktails hurled into U.S. bases, and American soldiers stabbed on the streets of Seoul have stoked anger in Congress and on the op-ed pages of major newspapers about South Korea. As North Korea appears on the nuclear brink, Americans are puzzled by the groundswell of anti-Americanism. They cringe at a younger generation of Koreans who tell

CBS television's investigative program *60 Minutes* that Bush is more threatening than Kim, and they worry about reports that South Korea's new president, Roh Moo-hyun, was avowedly anti-American in his younger days.

Most Koreans have complicated feelings about the United States. Some of them are anti-American, to be sure, but many are grateful. South Korea has historically been one of the strongest allies of the United States. Yet it would be naive to dismiss the concerns of South Koreans about U.S. policy and the continued presence of U.S. forces as merely emotional. Imagine, for example, how Washingtonians might feel about the concrete economic impact of thousands of foreign soldiers monopolizing prime real estate downtown in the nation's capital, as U.S. forces do in Seoul.

But hasty withdrawal of U.S. forces is hardly the answer to such trans-Pacific anxiety, particularly as the U.S.–South Korean affiance enters uncharted territory. The North Koreans would claim victory, and the United States would lose influence in one of the most dynamic economic regions in the world—an outcome it neither wants nor can afford. In the long term, such a withdrawal would also pave the way for Chinese regional domi-nance. Some South Koreans might welcome a larger role for China—a romantic and uninformed notion at best. Betting on China, after all, did not make South Korea the 12th largest economy and one of the most vibrant liberal democracies in the world.

The alternatives to the alliance are not appealing to either South Koreans or Americans. Seoul would have to boost its relatively low level of defense spending (which, at roughly 3 percent of gross domestic product, is less than that of Israel and Saudi Arabia, for example). Washington would run the risk of jeopardizing its military presence across East Asia, as a U.S. withdrawal from the peninsula raised questions about the raison d'être for keeping its troops in Japan. A revision in the U.S. military presence in Korea is likely within the next five years, but withdrawal of that presence and abrogation of its alliance are not.

Victor D. Cha is associate professor of government and D.S. Song-Korea foundation chair at Georgetown University's Edmund Walsh School of Foreign Service. David C. Kang is associate professor of government and adjunct associate professor at the Tuck School of Business at Dartmouth College. They are coauthors of Nuclear North Korea: A Debate On Engagement Strategies *(New York: Columbia University Press, October 2003).*

Want to Know More?

On North Korean history, the classic remains Chong-sik Lee and Robert Scalapino's *Communism in Korea* (Berkeley: University of California Press, 1972). The best recent history is Don Obderdorfer's *The Two Koreas* (New York: Basic Books, 2001). On North Korea's leadership, one of the few good works is Dae-Sook Suh's *Kim Il-Sung: The North Korean Leader* (New York: Columbia University Press, 1988). The inauguration speech of South Korea's new president, Roh Moo-hyun, can be found on the Web site of the *Korean Information Service*. The Korean Central News Agency Web site offers the North's perspective, served up zany and fresh.

On U.S.–North Korean relations, see Samuel Kim's, ed., *North Korean Foreign Relations in the Post–Cold War Era* (New York: Oxford University Press, 1998), Leon Sigal's *Disarming Strangers: Nuclear Diplomacy With North Korea* (Princeton: Princeton University Press, 1998), and Michael May's, ed., **"Verifying the Agreed Framework"** (Livermore: Center for Global Security Research, April 2001). For a recent roundtable featuring the views of 28 top Korea experts, see the February 2003 report of the Task Force of U.S. Korea Policy, **"Turning Point in Korea: New Dangers and New Opportunities for the United States,"** available on the Web site of the Center for International Policy. A deja vu–inducing snapshot of U.S.–North Korean relations circa 1979 can be found in journalist Gareth Porter's **"Time to Talk With North Korea"** (FOREIGN POLICY, Spring 1979).

On the history of nuclear and missile proliferation on the Korean peninsula, see James Clay Moltz and Alexandre Mansourov's, eds., *The North Korean Nuclear Program* (New York: Routledge, 1999), Peter Hayes's **Pacific Powderkeg** (Lexington: Lexington Books, 1991), and David Albright and Kevin O'Neill's, eds., **Solving the North Korean Nuclear Puzzle** (Washington: Institute for Science and International Security, 2000). David Shambaugh argues that China does not ultimately want to end North Korea's nuclear program in **"China and the Korean Peninsula: Playing for the Long Term"** (The Washington Quarterly, Spring 2003).

On North Korea's economy, see Marcus Noland's *Avoiding the Apocalypse: The Future of the Two Koreas* (Washington: Institute for International Economics., 2000), Nicholas Eberstadt's *The End of North Korea* (Washington: American Enterprise Institute Press, 1999), and Hy-Sang Lee's **North Korea: A Strange Socialist Fortress** (Westport: Praeger, 2001).

For links to relevant Web sites, access to the *FP* Archive, and a comprehensive index of related FOREIGN POLICY articles, go to *www.foreignpolicy.com*.

UNIT 6
Cooperation

Unit Selections

Key Points to Consider

- Itemize the products you own that were manufactured in another country.

- What recent contacts have you had with people from other countries? How was it possible for you to have these contacts?

- Do you use the Internet to access people or Web sites in other countries?

- How do you use the World Wide Web to learn about other countries and cultures?

- Identify nongovernmental organizations in your community that are involved in international cooperation (e.g., Rotary International).

- What are the prospects for international governance? How do trends in this direction enhance or threaten American values and constitutional rights?

- What new strategies for cooperation can be developed to fight terrorism, international narcotics trafficking and other criminal threats?

- How can conflict and rivalry be transformed into meaningful cooperation?

 **Links: www.dushkin.com/online/**
These sites are annotated in the World Wide Web pages.

Carnegie Endowment for International Peace
http://www.ceip.org
Commission on Global Governance
http://www.sovereignty.net/p/gov/gganalysis.htm
OECD/FDI Statistics
http://www.oecd.org/statistics/
U.S. Institute of Peace
http://www.usip.org

An individual can write a letter to another, and assuming it is properly addressed, the sender can be relatively certain that the letter will be delivered to just about any location in the world. This is true even though the sender pays for postage only in the country of origin and not in the country where it is delivered. A similar pattern of international cooperation is true when a traveler boards an airplane and never gives a thought to the issue of potential language and technical barriers, even though the flight's destination is halfway around the world.

Many of the most basic activities of our lives are the direct result of multinational cooperation. International organizational structures, for example, have been created to monitor threats to public health and to scientifically evaluate changing weather conditions. The flow of mail, the safety of airlines, and the monitoring of changing atmospheric conditions are just some of the examples of individual governments recognizing that their self-interest directly benefits from cooperation (in some cases by giving up some of their sovereignty through the creation of international governmental organizations, or IGOs).

Transnational activities are not limited to the governmental level. There are now tens of thousands of international nongovernmental organizations (INGOs). The activities of INGOs range from staging the Olympic Games to organizing scientific meetings to actively discouraging the hunting of seals. The number of INGOs along with their influence has grown tremendously in the past 50 years.

During the same period in which the growth in importance of IGOs and INGOs has taken place, there also has been a parallel expansion of corporate activity across international borders. Most U.S. consumers are as familiar with Japanese or German brand-name products as they are with items made in their own country. The multinational corporation (MNC) is an important non-state actor. The value of goods and services produced by the biggest MNCs is far greater than the gross domestic product (GDP) of many countries. The international structures that make it possible to buy a Swedish automobile in Sacramento or a Korean television in Argentina have been developed over many years. They are the result of governments negotiating treaties that create IGOs to implement the agreements (e.g., the World Trade Organization). As a result, corporations engaged in international trade and manufacturing have created complex transnational networks of sales, distribution, and service that employ millions of people.

To some observers these trends indicate that the era of the nation-state as the dominant player in international politics is passing. Other experts have observed these same trends and have concluded that the state system has a monopoly of power and that the diverse variety of transnational organizations depends on the state system and, in significant ways, perpetuates it.

In many of the articles that appear elsewhere in this book, the authors have concluded their analysis by calling for greater international cooperation to solve the world's most pressing problems. The articles in this section provide examples of successful cooperation. In the midst of a lot of bad news, it is easy to overlook the fact that we are surrounded by international cooperation and that basic day-to-day activities in our lives often directly benefit from it.

STRATEGIES FOR WORLD PEACE:

THE VIEW OF THE UN SECRETARY-GENERAL

By Kofi A. Annan

Today, in Afghanistan, a girl will be born. Her mother will hold her and feed her, comfort her and care for her, just as any mother would anywhere in the world. In these most basic acts of human nature, humanity knows no divisions.

But to be born a girl in today's Afghanistan is to begin life centuries away from the prosperity that one small part of humanity has achieved. It is to live under conditions that many of us would consider inhuman. Truly, it is as if it were a tale of two planets.

I speak of a girl in Afghanistan, but I might equally well have mentioned a baby boy or girl in Sierra Leone. No one today is unaware of this divide between the world's rich and poor. No one today can claim ignorance of the cost that this divide imposes on the poor and dispossessed who are no less deserving of human dignity, fundamental freedoms, security, food, and education than any of us. The cost, however, is not borne by them alone. Ultimately, it is borne by all of us—North and South, rich and poor, men and women of all races and religions.

Today's real borders are not between nations, but between powerful and powerless, free and fettered, privileged and humiliated. Today, no walls can separate humanitarian or human-rights crises in one part of the world from national-security crises in another.

Scientists tell us that the world of nature is so small and interdependent that a butterfly flapping its wings in the Amazon rain forest can generate a violent storm on the other side of the earth. This principle is known as the "Butterfly Effect." Today, we realize, perhaps more than ever, that the world of human activity also has its own "Butterfly Effect"—for better or for worse.

The Universal Bond of Tragedy

We have entered the third millennium through a gate of fire. If today, after the horror of September 11, we see better and we see further, we will realize that humanity is indivisible. New threats make no distinction among races, nations, or regions. A new insecurity has entered every mind, regardless of wealth or status. A deeper awareness of the ties that bind us all—in pain as in prosperity—has gripped young and old.

In the early beginnings of the twenty-first century—a century already violently disabused of any hopes that progress toward global peace and prosperity is inevitable—this new reality can no longer be ignored. It must be confronted.

The twentieth century was perhaps the deadliest in human history, devastated by innumerable conflicts, untold suffering, and unimaginable crimes. Time after time, a group or a nation inflicted extreme violence on others, often driven by irrational hatred and suspicion or by unbounded arrogance and thirst for power and resources. In response to these cataclysms, the leaders of the world came together at mid-century to unite the nations as never before.

A forum was created—the United Nations—where all nations could join forces to affirm the dignity and worth of every person and to secure peace and development for all peoples. Here, states could unite to strengthen the rule of law, recognize and address the needs of the poor, restrain man's brutality and greed, conserve the resources and beauty of nature, sustain the equal rights of men and women, and provide for the safety of future generations.

We thus inherit from the twentieth century the political, scientific, and technological power that—if only we have the will to use them—give us the chance to vanquish poverty, ignorance, and disease.

Redefining the United Nations

In the twenty-first century, I believe the mission of the United Nations will be defined by a new, more profound awareness of the sanctity and dignity of every human life, regardless of race or religion. This will require us to look beyond the framework of states and beneath the surface of nations or communities. We must focus, as never before, on improving the conditions of the individual men and women who give the state or nation its richness and character.

Over the past five years, I have often recalled that the United Nations' Charter begins with the words, "We the peoples." What is not always recognized is that "We the peoples" are made up of individuals whose claims to the most fundamental rights have too often been sacrificed in the supposed interests of the state.

The United Nations' priorities for the future are to promote democracy, prevent conflict, and lessen the burden of global poverty.

A genocide begins with the killing of one man not for what he has done but because of who he is. A campaign of "ethnic cleansing" begins with one neighbor turning on another. Poverty begins when even one child is denied his or her fundamental right to education. What begins with the failure to uphold the dignity of one life all too often ends with a calamity for entire nations.

In this new century, we must start from the understanding that peace belongs not only to states or peoples but to each member of those communities. The sovereignty of states must no longer be used as a shield for gross violations of human rights. Peace must be made real and tangible in the daily existence of every individual in need. Peace must be sought, above all, because it is the condition that enables every member of the human family to live a life of dignity and security.

Focusing on Diversity And Dialogue

From this vision of the role of the United Nations in the next century flow three key priorities for the future: eradicating poverty, preventing conflict, and promoting democracy.

Only in a world that is rid of poverty can all men and women make the most of their abilities. Only where individual rights are respected can differences be channeled politically and resolved peacefully. Only in a democratic environment, based on respect for diversity and dialogue, can individual self-expression and self-government be secured and freedom of association be upheld.

The idea that there is one people in possession of the truth, one answer to the world's ills, or one solution to humanity's needs has done untold harm throughout history—especially in the last century. Today, however, even amid continuing ethnic conflict around the world, there is a growing understanding that human diversity is both the reality that makes dialogue necessary and the basis for that dialogue.

We understand, as never before, that each of us is fully worthy of the respect and dignity essential to our common humanity. We recognize that we are the products of many cultures, traditions, and memories; that mutual respect allows us to study and learn from other cultures; and that we gain strength by combining the foreign with the familiar.

In every great faith and tradition one can find the values of tolerance and mutual understanding. The Qur'an, for example, tells us, "We created you from a single pair of male and female and made you into nations and tribes, that you may know each other." Confucius urged his followers, "When the good way prevails in the State, speak boldly and act boldly. When the State has lost the way, act boldly and speak softly." In the Jewish tradition, the injunction to "love thy neighbor as thyself" is considered to be the very essence of the Torah.

This thought is reflected in the Christian Gospel, which also teaches us to love our enemies and pray for those who wish to persecute us. Hindus are taught, "Truth is one, the sages give it various names." And in the Buddhist tradition, individuals are urged to act with compassion in every facet of life.

Each of us has the right to take pride in our particular faith or heritage. But the notion that what is "ours" is necessarily in conflict with what is "theirs" is both false and dangerous. It has resulted in endless enmity and conflict, leading men to commit the greatest of crimes in the name of a higher power.

It need not be so. People of different religions and cultures live side by side in almost every part of the world, and most of us have overlapping identities that unite us with very different groups. We can love what we are without hating what—and who—we are not. We can thrive in our own tradition, even as we learn from others, and come to respect their teachings.

This will not be possible, however, without freedom of religion, of expression, of assembly, and basic equality under the law. Indeed, the lesson of the past century has been that where the dignity of the individual has been trampled or threatened—where citizens have not enjoyed the basic right to choose their government or the right to change it regularly—conflict has too often followed, with innocent civilians paying the price in lives cut short and communities destroyed.

PEACE HAS NO PARADE: KOFI A. ANNAN ON THE NOBEL PEACE PRIZE

In 1960, the Nobel Peace Prize was awarded for the first time to an African—Albert Luthuli, one of the earliest leaders of the struggle against apartheid in South Africa. In 1961, the Prize was first awarded to a secretary-general of the United Nations—posthumously, because Dag Hammarskjöld had already given his life for peace in Central Africa. For me, as a young African beginning his career in the United Nations a few months later, those two men set a standard that I have sought to follow throughout my working life.

This award belongs not just to me. My own path to service at the United Nations was made possible by the sacrifice and commitment of my family and many friends from all continents—some of whom have passed away—who taught me and guided me. To them, I offer my most profound gratitude.

In a world filled with weapons of war and all too often words of war, the Nobel Committee has become a vital agent for peace. Sadly, a prize for peace is a rarity in this world. Most nations have monuments or memorials to war, bronze salutations to heroic battles, archways of triumph. But peace has no parade, no pantheon of victory.

Only by understanding and addressing the needs of individuals for peace, for dignity, and for security can we at the United Nations hope to live up to fulfill the vision of our founders. This is the broad mission of peace that United Nations staff members carry out every day in every part of the world.

—*Kofi A. Annan*

The obstacles to democracy have less to do with culture or religion than with the desire of those in power to maintain their position at any cost. This is neither a new phenomenon nor one confined to any particular part of the world. People of all cultures value their freedom of choice and feel the need to have a say in decisions affecting their lives.

The United Nations, whose membership comprises almost all the states in the world, is founded on the principle of the equal worth of every human being. It is the nearest thing we have to a representative institution that can address the interests of all states and all peoples. Through this universal, indispensable instrument of human progress, states can serve the interests of their citizens by recognizing common goals and pursuing them in unity. No doubt, that is why the Nobel Committee says that it "wishes, in its centenary year, to proclaim that the only negotiable route to global peace and cooperation goes by way of the United Nations."

Our era of global challenges leaves us no choice but to cooperate at the global level. When states undermine the rule of law and violate the rights of their individual citizens, they become a menace not only to their own people but also to their neighbors and the world. What we need today is better governance—legitimate, democratic governance that allows each individual to flourish and each state to thrive.

A Test of Common Humanity

I began with a reference to the girl born in Afghanistan today. Even though her mother will do all in her power to protect and sustain her, there is a one-in-four risk that she will not live to see her fifth birthday. Whether she does is just one test of our common humanity—of our belief in our individual responsibility for our fellow men and women. But it is the only test that matters.

If we remember this girl, then our larger aims—to fight poverty, prevent conflict, or cure disease—will not seem distant or impossible. Indeed, those aims will seem very near and very achievable—as they should. Beneath the surface of states and nations, ideas and language, lies the fate of individual human beings in need. Answering their needs will be the mission of the United Nations in the century to come.

Kofi A. Annan is secretary-general of the United Nations, #S-8000, New York, New York 10017. Web site www.un.org. This article is drawn from his Nobel Lecture delivered on December 10, 2001, in Oslo, Norway.

Peace in our time

Europe has largely avoided war for nearly six decades, but the European Union no longer gets the credit

IN THE whole of Europe there is probably no more blood-soaked battlefield than Verdun. In 1916 some 800,000 French and German soldiers were killed or wounded, fighting inconclusively over a few square miles of territory near the Franco-German border. The young Charles de Gaulle was wounded three times and captured at Verdun. Louis Delors, a 21-year-old French private, suffered terrible injuries there and was almost killed by a German officer who was finishing off the French wounded with a pistol. In the 1990s his son, Jacques Delors, became the founding father of a monetary union of France, Germany and ten other European countries. Mr Delors's two great collaborators were Helmut Kohl, the German chancellor, whose father had also fought at Verdun, and François Mitterrand, the French president. In 1984 these two leaders, in a historic act of Franco-German reconciliation, walked hand in hand across the battlefield that had been a killing ground for so many young men from both countries.

Many of the European Union's most ardent supporters still see the EU as a crucial bulwark against the return of war to Europe. In pressing the case for monetary union, Mr Kohl argued that adopting the euro was ultimately a question of war and peace in Europe. When efforts to write a European constitution looked like stalling, Elmar Brok, a prominent German member of the EU's constitutional convention (and confidant of Mr Kohl), gave warning that if Europe failed to agree on a constitution, it risked sliding back into the kind of national rivalries that had led to the outbreak of the first world war.

Remember the bad old days?

Such arguments resonate particularly strongly among an older generation of French and German politicians, but also have wider currency. Timothy Garton Ash of St Antony's College, Oxford, one of Britain's most astute observers of European affairs, says in a recent book that the Union is needed "to prevent us falling back into the bad old ways of war and European barbarism which stalked the Balkans into the very last year of the last century." Mr Garton Ash concedes that "we can never prove that a continent-wide collection of independent, fully sovereign European democracies would not behave in the same broadly pacific way without the existence of any European Union. Maybe they would, but would you care to risk it?"

Believers in the pacifying effects of the drive for European unity acknowledge the contributions to peace in post-war Europe made by American troops and by the spread of prosperity and democracy. But they argue that the EU has played the central role, by forcing European leaders to cooperate intensively and continuously, by proving that membership of the Union brings prosperity and by demanding that all EU countries adhere to basic principles of democracy, human rights and the peaceful resolution of disputes.

Seen in this light, EU enlargement is part of the same "peace project" that was initially centred on reconciliation between France and Germany. Countries that apply to join the EU first have to meet a set of basic democratic criteria, and have to put aside old territorial disputes. Eight former

members of the Soviet block were admitted to the EU this year, and Romania and Bulgaria are lined up for entry in 2007. But the idea that enlargement of the European Union will inevitably resolve conflicts and spread freedom and democracy will face even bigger tests in future.

The European Commission in Brussels has already made it clear that all the Balkan countries are potentially eligible for membership, in the hope of encouraging them to make peace and introduce democratic reforms. Croatia, which earlier this year became the first major combatant in the Balkan wars to be formally accepted as a candidate for EU membership, had to step up its co-operation with the International War Crimes Tribunal in The Hague before negotiations became possible.

The EU was rightly castigated for its inability to prevent war in the Balkans in the 1990s, and many Europeans felt humiliated by the need for American military intervention to end the conflicts in Bosnia and Kosovo. But the EU is taking over the peacekeeping mission in Bosnia from NATO later this year in what will be the Union's biggest ever military operation. In the longer term, it is hoped, the prospect of EU membership may help to cement the fragile democracies and the peace settlements now in place across the Balkans.

An ever wider Union

A similarly ambitious logic is being applied to Turkey's aspirations to join the Union. Although many EU politicians and citizens are worried about admitting a large Muslim nation into the Union, the proponents of Turkish membership have the upper hand. Turkey is likely to be invited to start negotiations to join the EU later this year. Once again, the key arguments are about peace and the spread of freedom and democracy.

At a time when relations between the West and the Islamic world are so delicate, most EU leaders seem to feel that refusing to admit a large Islamic country into the Union would be seen as a disastrous confirmation of the "clash of civilisations". European diplomats, for their part, hope that admitting Turkey to the EU will bring confirmation that Islam is not incompatible with western values. They point out with some pride that the prospect of EU membership has already driven forward reforms in Turkey such as increased political and civil rights for the Kurdish minority and the abolition of the death penalty.

For geo-strategic thinkers sitting in foreign ministries in London, Paris and Berlin, the arguments for using the EU to spread peace and democratic stability seem compelling. But ordinary European citizens find them much less convincing. Many fear that rather than exporting stability, the EU will import instability. In western Europe, public debate about EU enlargement has tended to concentrate on fears about competition from low-cost labour and waves of immigration. So far such fears have proved containable, and the admission of the new members from central Europe has not caused too much of a fuss. But the new central European members, though poorer than the European average, are smallish (except for Poland), and all are predominantly Christian.

Turkey, which on current trends will have a larger population than any current EU member by 2020, is a different proposition. Because all EU citizens are free to live and work anywhere in the EU, there could be serious resistance to Turkish membership in France, Germany and the Netherlands, where the rapid growth of Muslim populations in the past 30 years is already a highly sensitive issue.

Even without such worries, the traditional arguments for European integration as a "peace project" have anyway been losing force with the passing years. The current generation of EU leaders still has some memories of the depredations of war in Europe. Gerhard Schröder, the German chancellor, never knew his father, who was killed in the second world war; Jacques Chirac, the French president, lived through the war as a child. But for most younger Europeans, the threat of war in western Europe now seems almost unimaginably remote.

The expansion of the EU beyond the original six also added new countries with different historical experiences. Although some Britons, such as Mr Garton Ash, take the threat of a recurrence of war in Europe seriously, the British have generally approached the Union in a very different spirit from the French and Germans. Whereas statesmen such as Monnet and Schuman considered the 1939-45 war as the final proof that traditional European political structures needed to be radically changed, the British tended to see it as a vindication of their own long-established democracy and as confirmation of their anti-continental prejudices. As Margaret Thatcher, a famously Eurosceptic British prime minister, remarked: "In my lifetime all our problems have come from mainland Europe and all the solutions from the English-speaking nations across the world." The British are wary of dreams for political union in Europe. Unlike their French and German counterparts, British politicians have always wanted the EU to be above all about free trade.

Changing the mix

As the EU has expanded, so Britain has become less isolated in its resistance to the idea that European unity is essential for the maintenance of peace. When in 2003 Sweden held a referendum on whether to join the euro, Goran Persson, the Swedish prime minister, used the peace argument, but watched it fall flat in a neutral country that has been at peace for nearly 200 years. The Swedes voted "no". The new EU members in central Europe also bring a different perspective to the European Union. Whereas the traditional "builders of Europe" were suspicious of nationalism and keen to build up supranational institutions at the expense of the nation-state, many of the central Europeans are still joyfully reasserting their own national identities after decades of Soviet domination.

Nonetheless, all these countries were eager to join the EU. They saw membership as an assertion of their European identity, as well as a ticket to prosperity and some protection against any threat from a resurgent Russia. But they are also much less enthusiastic than western European federalists about an "ever closer union" for Europe as spelled out in the Treaty of Rome. Vaclav Havel, the hero of the Velvet Revolution and former president of the Czech Republic, explains that for countries that have recently thrown off Soviet domination, "the concept of national sovereignty is something inviolable".

For a variety of reasons, then, the most powerful traditional argument for European unity, peace, has been losing force. European federalists are beginning to look for new rationales for European unity.

THE ULTIMATE CROP INSURANCE

A new treaty strives to save 10,000 years of plant breeding

BY JANET RALOFF

In late summer 2002, looters threatened war-engulfed Afghanistan's agricultural heritage. Unknown pillagers dumped stocks of carefully labeled seeds as they ransacked buildings in Ghazni and Jalalabad, where the material had been hidden for safekeeping. All the looters wanted, apparently, were the plastic and glass jars in which the seeds were stored. The scattered seeds weren't the starter for next year's crops but the genetic backup for the agrarian nation's agriculture. The catalogued seeds of various strains of wheat, barley, chickpeas, lentils, almonds, pomegranates, and melons would have been deployed to create new seed supplies if drought, insects, or some other disaster had wiped out the region's production of crop seeds. They also represented the raw material for creating future lines of crops.

Botanists had recorded where each seed type had come from and information about the climate and geography of its place of origin. However, seeds for cold-tolerant wheat didn't look any different from those for a cold-sensitive but disease-tolerant plant. Removed from their labeled jars, these and all the other seeds lost their archival value.

"It's like having a library of books with no titles on them," notes Geoffrey Hawtin, the former director general of the International Plant Genetic Resources Institute in Rome.

Seed repositories, such as those in Afghanistan, sometimes called gene banks, are often casualties of war, says Ruth Raymond of that institute. "Cambodia, Rwanda, Somalia, Iraq—we know of at least 50 or 60 examples of where it's happened," she says. Bank holdings often offer looters the makings for dinner or free seed for a small crop. In most cases, these events haven't turned into long-term disasters because gene banks in other countries house backup stores. That's true for Afghanistan and the other war-struck countries. When hostilities subside, a country can begin rebuilding its gene bank with seeds from half a world away.

Offering such relief to an embattled country is just one of the missions of gene banks, which together form an informal international system. The banks also preserve the genes that have helped plant species thrive in harsh climates, survive diseases, or provide particular flavors or other appealing traits.

Crop diversity constitutes agriculture's "global wealth—the set of genes that have been developed by farmers over 10,000 years," explains Clive Stannard of the United Nations Food and Agriculture Organization in Rome.

Unfortunately, the successes of large agribusiness companies are adding to the need for a strong seed-repository system and worldwide crop diversity.

A small number of popular, high-yielding crop varieties bred by these companies have increasingly edged out landraces, the varieties adapted to localized conditions during millennia of farming. If not archived in a secure, internationally sanctioned gene-banking system, these ancient varieties could disappear, taking with them as-yet-undiscovered genes for important traits.

A decade ago, a United Nations treaty on species conservation had the unintended effect of erecting roadblocks to gene banking and breeding. But on June 29, a new International Treaty on Plant Genetic Resources for Food and Agriculture went into effect. Supporters hope it will break down those obstacles and bolster the gene-banking system, and help preserve the genetic heritage of crops.

TO THE RESCUE In December 1993, the Convention on Biological Diversity entered into force, signed and ratified by most nations. The convention dramatically changed how governments, corporations, farmers, and even tribal chiefs in countries such as Brazil viewed the agricultural and wild plants growing around them.

Gone were the days when bioprospectors from industrial nations could drop into a jungle, pluck any intriguing plant and ship it home for analysis, and then extract a substance for use as a high-value, patentable drug. With the new treaty, companies would have to work with locals and share profits (*SN: 5/29/04, p. 344*).

The treaty established the principle that a nation holds sovereign rights to the genetic resources of any plant within its borders (*SN: 6/20/92, p. 407*). But agricultural policymakers quickly recognized that this seemingly straightforward attempt to protect countries from exploitation was at odds with both the nature of agriculture and the conservation of cultivated resources.

It's not easy to determine who deserves sovereignty over plant genes. Throughout history, people have been trading edible plants, so most farmed crops have murky national origins.

In the wake of the 1993 biodiversity treaty, many nations expressed a new reluctance to continue sharing banked seeds freely, lest they forfeit their legal rights to a plant's genetic resources.

Suketoshi Taba encountered this change in attitude when he tried to head off an apparent crisis affecting seed corn in gene banks throughout Latin America. Taba heads maize genetic resources at the International Maize and Wheat Improvement Center, which is known worldwide by its acronym in Spanish, CIMMYT. It's located near Mexico City.

Budgets for maintaining gene banks had been compromised by a regional economic downturn in the early 1990s. Many seeds were not being sufficiently dried before storage, and refrigeration units in the storage facilities often lost power or broke down. The proportion of banked maize seeds able to germinate at any given time should exceed 85 percent, he says, but had fallen well below 50 percent.

To rescue seeds from banks in Cuba and 12 other Latin American nations, Taba offered to have the imperiled seed stores replanted locally and the next-generation seeds shared between the various nations' banks and CIMMYT. The countries initially balked, arguing that these seed stocks, though imperiled, were their property. In the end, botanists prevailed over politicians, and Taba saved 10,000 sets of seeds at a cost of only slightly more than $1 million.

The 1993 treaty, which essentially nationalized genetic resources, created other potential obstacles to improving crop seeds. It encouraged countries to require a seed company or laboratory to work out an international contract every time it imported seed. The cost of legal review, DNA analysis of each batch and supervision by treaty-enforcement officials would be staggering. "You would just close down plant breeding as a science," Raymond says.

ON THE MENU By the mid-1990s, a groundswell developed among nations for a remedy to the Convention on Biological Diversity. In late 2001, after 7 years of debate, negotiators from about 120 countries finally settled on the outline of the new crop biodiversity treaty.

The new treaty promotes the collection and banking of seeds, the sharing of seeds among countries that pledge to

honor the treaty, and an international survey of genetic resources currently existing in gene banks and farmers' fields around the globe (*SN: 7/17/04, p. 45*). The treaty also promises to improve crops for farmers, especially those living in harsh environments.

During their wrangling, negotiators imposed limits on what plants would be covered by the new treaty. Just 35 food crops and 29 plants for livestock forage made the list. Although reasons for a type of plant not making the new treaty's list vary, politics often proved a major factor, says Patrick Mulvany of ITDG, an international sustainable-development foundation in Rugby, England.

If a company breeds a new plant variety from foreign seeds of one of the 64 plants included, the treaty permits that company to commercialize the crop. However, a share of any profit must be contributed to a treaty fund for programs in developing countries to promote sustainable farming or conservation of genetic diversity in crops.

When the treaty qualified to go into force earlier this year, 40 nations had ratified it. Since then, 15 more nations have joined the group, and more are poised to do so.

The United States remains a major holdout among industrial nations. It was among the negotiators of the latest treaty, but the Senate hasn't taken up the treaty. A leading reservation, U.S. officials say, is concern over preserving commercial rights to a share of the profit for seeds passed between countries. Still to be worked out are the size of those royalties, whether they should be paid when seed is first imported or when a crop is marketed, how long a seed developer must pay royalties, and whether royalties should vary by crop.

Still, says Peter Bretting of the U.S. Agricultural Research Service in Beltsville, Md., "the U.S. government is certainly conceptually supportive of the treaty."

The Convention on Biodiversity continues to govern the commercial exploitation of plants not listed under the new treaty. These include some important food crops, such as soybeans, peanuts, onions, and grapes. In these cases, countries negotiate individual commercial agreements without any contribution to an international program.

BANKERS' BLUES The new treaty's biggest current initiative is creating the Global Crop Diversity Trust, which will be an autonomous agency that coordinates the 1,500-or-so gene banks scattered throughout more than 100 nations. Among the largest are the 11 international facilities affiliated with the Consultative Group on International Agricultural Research (CGIAR). Together, they store some 600,000 genetically distinct seed samples.

All CGIAR members will "sign agreements with the treaty, by which they will put their entire collections under its policy guidance," Stannard says.

The United States has its own system of more than 20 gene banks. They hold some 450,000 samples of roughly 10,000 plant species. Seeds maintained in these facilities at temperatures above freezing might require replanting every 5 to 10 years to stay viable.

The nation's biggest gene bank, located in Fort Collins, Colo., holds duplicates of seeds kept in all the other U.S. facilities and stores most of them at -18°C to -55°C. At these temperatures, metabolic processes slow and the aging of seeds is delayed, so seeds stay viable for 50 to 100 years, explains Bretting.

Unfortunately, Raymond notes, most of the world's seed repositories aren't as well maintained or financed as the U.S. facilities are. Seed maintenance is "not sexy," she laments, so most governments neglect it. "How excited can you get about funding refrigeration?" she asks.

Yet, says Raymond, gene banks are an "essential insurance program to guarantee that genes will remain safe and accessible for future crop improvement." This is becoming increasingly important, she notes, as biological diversity wanes throughout the world.

DIMINISHING DIVERSITY In the 1950s, Chinese farmers grew about 10,000 varieties of wheat. Two decades later, Raymond notes, the number had fallen to 1,000. India experienced a similar loss of rice diversity over the past 3 decades, she says. And throughout the Andes, the cradle of tomatoes, wild-tomato species have become so imperiled that "before too long, the only real examples of tomato diversity will be in a gene bank," she says.

Raymond cites U.N. estimates that one in 12 flowering plants—including crop types such as wild potatoes—will go extinct within 2 decades.

Commercial plant breeding has also been winnowing plant diversity, observes Marilyn Warburton, a molecular geneticist at CIMMYT.

Commercial growers prefer uniformity in crops for the sake of mechanized harvesting, so breeders have focused on developing high-yielding crops that will grow the same size across a field; respond consistently to weather, climate, and nourishment; resist blights; and mature within a tight window of time during harvest season.

Warburton has confirmed great genetic variability in landraces of wheat planted in developing countries before the agricultural movement that came to be known as the green revolution. Her DNA analyses show that in their efforts to achieve the high yields, CIMMYT and other green revolution breeders "reduced diversity in traits sampled throughout the genome."

Warburton notes that wheat yields have lately leveled off in many developed countries, despite intense breeding efforts to raise them. This has fostered speculation, she says, that commercial breeders might have "run out of genetic variation."

Her CIMMYT colleagues, however, have developed what they call synthetic bread wheats, which demonstrate that genetic crop diversity can be restored or even amplified. Over a 15-year period, Abdul Mujeeb-Kazi and his team crossed banked seeds that represented the original wild parents of durum wheat and then crossed their progeny with another wild wheat. This effectively duplicated the natural events that originally gave rise to bread wheat some 10,000 years ago.

WHEAT'S WORTH
— These newly synthesized lines of bread wheat have restored the traditional crop's lost genetic diversity.

CIMMYT scientists have since repeated the process to produce additional synthetic bread-wheat lines.

Warburton's analyses of these lines' DNA now show that genetically, "they are about as diverse as the original landraces of wheat." However, she notes, the new wheats give up to 50 percent higher yields than the original green revolution lines and yields similar to the best commercial yields today. An added benefit: The new wheats are extremely resistant to environmental stresses, such as pests, drought, and salty soils.

PLANT TRIAGE Healthy, well-stocked gene banks make such a success possible. Unfortunately, Raymond observes, those in the United States and the CGIAR system aren't typical. Most banks are ill funded and have outmoded equipment. A few amount to little more than a stash of seeds in a bureaucrat's refrigerator.

The Global Crop Diversity Trust is inventorying gene banks to identify those in need of rescue. Nations, foundations, and individuals are being asked to donate funds to upgrade facilities that are in dire straits, especially banks with important collections. Once these gene banks gain solid footing, they may qualify for annual funds from a new endowment being set up under the trust.

Raymond has been charged with spearheading a campaign to raise an additional $260 million to establish that endowment, which should eventually permit annual disbursements to gene banks totaling some $12 million for boring budget items such as payments for electricity and new refrigerators.

"What we're requesting is chicken feed," she says. However, she adds, even this small endowment could yield big dividends in developing countries. "And it's for such a great cause," she argues ardently. "I mean, we're talking about the [agricultural] security of the world."

Reprinted by permission via the Copyright Clearance Center from *Science News*, September 11, 2004, pp. 170-172. Copyright © 2004 by Science Service, Inc.

Medicine Without Doctors

In Africa, just 2 percent of people with AIDS get the treatment they need.
But drugs are cheap, access to them is improving and a new grass-roots
effort gives reason to hope

By Geoffrey Cowley

The first part of Nozuko Mavuka's story is nothing unusual in sub-Saharan Africa. A young woman comes down with aches and diarrhea, and her strong limbs wither into twigs. As she grows too weak to gather firewood for her family, she makes her way to a provincial hospital, where she is promptly diagnosed with tuberculosis and AIDS. Six weeks of treatment will cure the TB, a medical officer explains, but there is little to be done for her HIV infection. It is destroying her immune system and will soon take her life. Mavuka becomes a pariah as word of her condition gets around the community. Reviled by her parents and ridiculed by her neighbors, she flees with her children to a shack in the weeds beyond the village, where she settles down to die.

In the usual version of this tragedy, the young mother perishes at 35, leaving her kids to beg or steal. But Mavuka's story doesn't end that way. While waiting to die last year, she started visiting a two-room clinic in Mpoza, a scruffy village near her home in South Africa's rural Eastern Cape. Health activists were setting up support groups for HIV-positive villagers, and Medecins sans Frontieres (also known as MSF or Doctors Without Borders) was spearheading a plan to bring lifesaving AIDS drugs to a dozen villages around the impoverished Lusikisiki district. Mavuka could hardly swallow water by the time she got her first dose of anti-HIV medicine in late January. But when I met her at the same clinic in May, I couldn't tell she had ever been sick. The clinic itself felt more like a social club than a medical facility. Patients from the surrounding hills had packed the place for an afternoon meeting, and their spirits and voices were soaring. As they stomped and clapped and sang about hope and survival, Mavuka thumbed through her treatment diary to show me how faithfully she'd taken the medicine and how much it had done for her. Her weight had shot from 104 pounds to 124, and her energy was high. "I feel strong," she said, eyes beaming. "I can fetch water, wash clothes—everything. My sons are glad to see me well again. My parents no longer shun me. I would like to find a job."

It would be rash to call Nozuko Mavuka the new face of AIDS in Africa. The disease killed more than 2 million people on the continent last year, and it could kill 20 million more by the end of the decade. The treatments that have made HIV survivable in wealthier parts of the world still reach fewer than 2 percent of the Africans who need them. Yet mass salvation is no longer a fool's dream. The cost of antiretroviral (ARV) drugs has fallen by 98 percent in the past few years, with the result that a life can be saved for less than a dollar a day. The Bush administration and the Geneva-based Global Fund to Fight AIDS, TB and Malaria are financing large international treatment initiatives, and the World Health Organization is orchestrating a global effort to get 3 million people onto ARVs by the end of 2005—an ambition on the scale of smallpox eradication. What will it take to make this hope a reality? Raising more money and buying more drugs are only first steps. The greater challenge is to mobilize millions of people to seek out testing and treatment, and to build health systems capable of delivering it. Those systems don't exist at the moment, and they won't be built in a year. But as I discovered on a recent journey through southern Africa, there's more than one way to get medicine to people who need it. This crisis may require a whole new approach—a grass-roots effort led not by doctors in high-tech hospitals but by nurses and peasants on bicycles.

Until recently, mainstream health experts despaired at the thought of treating AIDS in Africa. The drugs seemed too costly, the regimens too hard to manage. Unlike meningitis or malaria, which can be cured with a short course of strong medicine, HIV stays with you. A three-drug cocktail can suppress the virus and protect the immune system—but only if you take the medicine on schedule, every day, for life. Used haphazardly, the drugs foster less treatable strains of HIV, which can then spread. Strict adherence is a challenge even in rich countries, the experts reasoned, and it might prove impossible in poor ones. In light of the dangers, prevention seemed a more appropriate strategy.

Caregivers working on the front lines resented the idea that anyone should die for having the wrong address. So they set out to prove that treatment could work in tough settings, and by 2001 they'd succeeded. In a project led by Dr. Paul Farmer of

Harvard, two physicians and a small army of community outreach workers introduced ARVs into 60 villages near the Haitian town of Cange. Around the same time, MSF teamed up with South Africa's Treatment Action Campaign to make the drugs available in an urban slum called Khayelitsha. The upstarts simplified the drug regimens and dialed back on lab tests, and most of the patients were monitored by nurses or outreach workers instead of physicians. But none of this made treatment less effective. The cocktails worked as well in the slums as they did in San Francisco—and the patients were often *more* steadfast than Americans about taking their pills. The obstacle to treatment was not a lack of infrastructure, the activists proclaimed. It was a lack of political will.

The climate has changed since then. Yesterday's unacceptable risk is today's moral imperative, and the world's highest-ranking health authorities are pushing hard to realize it. "We still believe in prevention," the WHO's director-general, Dr. Jong-wook Lee, told me during an interview in Geneva this spring. "But 25 million HIV-positive Africans are facing certain death. If we fail to help them, it can't be because we didn't try." Since Lee took office last year, staffers in the agency's HIV/AIDS department have worked at a furious pace to devise a global treatment strategy and help besieged countries design programs that the Global Fund will pay for. Proposals are rolling in, and the fund is responding favorably. Grants approved so far could finance treatment for 1.6 million people over the next five years.

The trouble is, few of the countries winning those grants are ready to absorb them. Their health systems have withered under austerity plans imposed by foreign creditors. Doctors and nurses have left in droves to take private-sector jobs or work in wealthier countries. And those left behind are overwhelmed and exhausted. While traveling in Zambia, I visited Lusaka's University Teaching Hospital, the 1,600-bed facility at the forefront of the country's two-year-old treatment program. Dr. Peter Mwaba, the hospital's stout, vigorous chief of medicine, detailed the country's strategy for treating 100,000 people (50 times the current number) by the end of next year. Yet his own facility was half abandoned. In 1990 the hospital had 42 nurses for every shift. Today it has 24—and the patients are sicker. "I've been here for 30 years," Violet Nsemiwe, the hospital's grandmotherly head nurse, confided as we walked the dim corridors. "It has never been this bad."

In an ideal world, the clock would stop while countries in this predicament trained tens of thousands of health professionals, quintupled their salaries and dispatched them to underserved areas. But the clock is ticking at a rate of 56,000 deaths a week, so the WHO is embracing a different approach—one rooted in the populism of Cange and Khayelitsha. "AIDS care, as we practice it in the North, is about elite specialists using costly tests to monitor individual patients," says Dr. Charles Gilks, the English physician coordinating the WHO's "3 by 5" treatment initiative. "I've done that and it's great. But it's irrelevant in a place like Uganda, where there is one physician for every 18,000 people and that physician is busy at the moment. If we're going to make a difference in Africa, we've got to simplify the regimens and expand the pool of people who can administer them."

That's precisely the agenda that activists are pursuing in Lusikisiki, the remote South African district where Nozuko Mavuka got her life back. When MSF and the Treatment Action Campaign launched their project there last year, the local hospital was performing the occasional HIV test but had little to offer people who were positive—a population that includes 30 percent of pregnant women. Lusikisiki is the poorest part of the poorest province in South Africa, but the activists used what they found—a struggling hospital and a dozen small day clinics—to start a movement. A small team led by Dr. Hermann Reuter, a veteran of the Khayelitsha project, set up a voluntary testing center at each site, organized support groups for positive people and emboldened them to stand up to stigma. Before long, people like Mavuka were donning HIV-POSITIVE T shirts, singing about the virtue of condoms and quizzing each other on the difference between a nucleoside-analogue reverse transcriptase inhibitor and a non-nucleoside-analogue reverse transcriptase inhibitor.

By the time the first drugs arrived last fall, people in the support groups were poised not to receive treatment but to *claim* it. They shared an almost religious commitment to adherence, and some had become counselors and pharmacy assistants. Twenty-eight-year-old Akona Siziwe was as sick as Mavuka when she joined a support group in Lusikisiki last year. Weary of her husband's incessant criticism (he didn't like the way she limped), she had packed up her 7-year-old son and her HIV-positive toddler and gone home to die with her mom. But her health returned quickly when she started treatment in December, and she went to work as a community organizer. She now runs workshops and counsels patients in three villages. "What's a good CD4 count?" she asks. "The nurses don't have time to explain, but people want to know. When I share information that can help them, they're grateful and happy and full of praise. I can't even sleep because they are knocking on my door! They want testing and treatment tonight!"

The Lusikisiki project has only two nurses and two full-time doctors, but it was treating 255 patients when I visited in May, and people from the villages were flocking to the clinics as the good news spread. Many of them show up expecting a quick test and a jar of pills, but as the program's head nurse, Nozie Ntuli, likes to say, "Giving out pills is the final step in the process." First the patient has to join a support group and get treated for secondary infections such as thrush and TB. A counselor then conducts a home study to make sure the person is ready for a long-term commitment. When the supports are all in place, the counselor takes the patient's case to a community-based selection committee. And everyone shares the joy when a patient succeeds. "I see people transformed every day," Ntuli says. "It is a new dispensation."

This isn't the first time village volunteers have launched a successful health initiative. "Home-based care" is a tradition throughout southern Africa, and a cornerstone of countless successful programs. In rural Malawi, minimally trained community volunteers manage everything from pregnancy to cholera. They work with TB patients to ensure adherence, and they supply vitamins, aspirin and antibiotics to people living with HIV/AIDS. When Malawi's Health Ministry starts distributing antiretrovirals through a national program this fall, the volunteers will help administer those, too.

They'll play an especially important role in Thyolo, a desperately poor district surrounded by tall mountains and jade green tea plantations. Roughly 50,000 of Thyolo's half-million residents are HIV-positive, and 8,000 have reached advanced stages of illness. When I visited Thyolo this spring, MSF was treating several hundred of them at the local district hospital, a converted colonial-era country club run by nurses and clinical officers (non-M.D.s with four years of training). But the hospital was in no position to handle thousands more, even if the government provided the drugs. Its two-person AIDS staff was struggling just to keep up with the MSF program. Many of the untreated patients lived too far away to trek in for routine visits anyway.

Dr. Roger Teck, a fiftyish Belgian physician who runs MSF's Thyolo program, described the predicament during a bumpy jeep ride from the hospital to the outlying village of Kapichi, where 20 volunteers were waiting for us in a freshly painted one-room community center. Some were as young as 20, others as old as 70. After an hour of prayers and introductions and soulful choral chants, the group's leader, 49-year-old Kingsley Mathado, peppered us with facts about the 30 villages in his area (3,000 people living with HIV, 500 in need of treatment) and described the volunteers' program for supporting them. When the government drugs reach Thyolo district hospital, the patients will still have to walk a half day to queue up for an exam and an initial two-week supply. They'll also have to return for their first few refills so that a nurse or doctor can see how they're responding. But the volunteers will take over as soon as patients are stable, refilling prescriptions from a village-based pharmacy and charting adherence and side effects.

Could this strategy work on a grand scale? Lay health workers are already a mainstay of large-scale TB initiatives, and the Malawian government has assigned them a big role in AIDS treatment as well. The country's nascent ARV program uses a regimen simple enough for anyone to administer after a week of intensive training (three generic drugs in one pill—no substitutions). Physicians from Malawi's Ministry of Health are now traveling the country to conduct training courses for lay health workers. The first drugs should arrive in the fall. "We've taken a radical leap to ensure real access," says Dr. William Aldis, the WHO's Malawi representative and one of the plan's many architects. "We're either going to win a Nobel Prize or get shot."

Malawi's challenge is to foster the kind of engagement that has made treatment so effective in places like Cange, Khayelitsha and Lusikisiki. If 25 years of HIV/AIDS has taught us anything, it's that grass-roots involvement is critical. "One set of characteristics runs through nearly all of the success stories," the London-based Panos Institute concludes in a 2003 report on the pandemic: "ownership, participation and a politicized civil society." No one denies the need for trained experts to manage programs and handle medical emergencies. But people from affected communities are often better than experts at raising awareness, shattering stigmas and motivating people to take charge of their health. Reuter, the Lusikisiki project's director, recalls an experiment in which doctors teamed up with activists to extend a hospital-based ARV program into community clinics in the Cape Town slum of Gugulethu, where access would be easier and peer counselors could play a bigger role. The ghetto-based patients achieved 93 percent adherence during the first year. The hospital's program had never topped 63 percent a rate Reuter dismisses as "American-style adherence."

With access to treatment, millions of dying people could soon recover as dramatically as Nozuko Mavuka did in Mpoza—and their salvation could revive farms, schools and economies as well as families. But there are hazards, too, and drug resistance isn't the only one. Successful ARV therapy expands the pool of infected people simply by keeping people alive. Unless the survivors can reduce transmission, the epidemic will grow until the demand for treatment is unmanageable. "We can't focus blindly on treatment," Teck mused as our jeep lurched away from Kapichi. "If we don't reduce the infection rate, we're going to end up in a nightmare situation." The patients and counselors in the clinics I visited weren't singing and stomping only about pills. They were celebrating a shared commitment to ending what is already a nightmare. The rest of the world needs to lock arms with them.

countdown TO ERADICATION

Rotary and its partners move ever closer to a polio-free world

by Anne Stein

In 1985, Rotary International committed to help immunize the world's children against polio, with a target date of a polio-free world by 2005, Rotary's centennial year. Since 1988, when the Global Polio Eradication Initiative was launched, millions of Rotary volunteers have raised funds and assisted in oral polio vaccine delivery, social mobilization, and garnering cooperation from national governments and health ministries—making Rotary the leading private-sector partner in the effort.

The program began with donations from Rotarians of more than US $240 million. By 2005, Rotary's financial contribution will reach an estimated $500 million. Along with its global partners—the World Health Organization (WHO), UNICEF, and the U.S. Centers for Disease Control and Prevention (CDC)—Rotary can take pride in the largest public health initiative in history.

ENDING POLIO Ten polio-endemic nations and a $275 million shortfall stand between failure and success.

According to the most recent data, only 537 cases of polio have been reported worldwide for 2001, although that number will likely increase somewhat as final reports are filed. However, it is sure to be down considerably from 2000, when 2,979 cases were logged. The program partners report an astounding 99 percent decrease in polio cases since 1988, the year WHO resolved to eradicate the disease. But despite the remarkable progress in reaching children and the unprecedented outpouring of time and resources from Rotarians and their partners, there's still much work to do and $1 billion in donor contributions needed to reach the final goal.

So far, $725 million has been pledged or projected, leaving a gap of $275 million. In response, the RI Board of Directors and The Rotary Foundation Trustees have announced a new, $80 million fundraising campaign—Rotary's contribution to closing the gap—that will run through the 2002-03 Rotary year. The campaign, Fulfilling Our Promise: Eradicate Polio, was officially launched in June during the RI Convention in Barcelona by 2001-02 Trustees Chairman Luis V. Giay.

"The very principles of Rotary are on the line," says Past RI Vice President Louis Piconi, chairman of the North American Polio Eradication Fundraising Campaign Committee. "We should never make a promise without fulfilling it; it's our word. To have Rotary be a major factor in eradicating polio will be a very significant milestone for the organization and its history."

"Each of these countries has made tremendous progress, but each has its own unique set of challenges."

About 60 percent of today's Rotarians—including all women members—have joined since the launch of the PolioPlus campaign, observes former RI General Secretary Herbert Pigman, who is now the director of the polio eradication fundraising campaign. "This new campaign gives them an opportunity to play an important part in our great humanitarian cause."

Up to $25 million of the contributions to this campaign can be matched by an equal amount from the Bill & Melinda Gates Foundation. The resulting $50 million can be further augmented by $75 million in assistance from the World Bank. "If we raise this money now, we will be saving all of humankind from this disease for all time, and that's priceless," Giay said in April during a joint news conference with representatives of the CDC, UNICEF, and WHO.

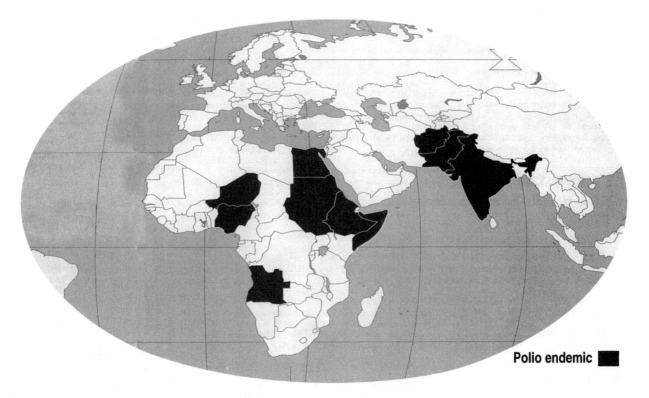

Polio endemic ⬛

Countdown to 2005: The wild poliovirus remains endemic in 10 countries: Afghanistan, Angola, Egypt, Ethiopia, India, Niger, Nigeria, Pakistan, Somalia and Sudan.

What remains to be done? Quite simply, a lot. Intensified immunization activities must take place in the 10 remaining polio-endemic nations (down from 20 nations in 2000 and 127 in 1985) and in several more countries that are considered high risk for outbreaks. Surveillance must also be maintained to ensure polio-free status. "We need to immunize more than 600 million children over the next four years," explains Piconi. "Rotary's campaign will help to buy and get the vaccine to all of those children, no matter where they are."

Areas of greatest focus are the remaining polio-endemic nations: Afghanistan, India, Niger, Nigeria, and Pakistan, which collectively accounted for more than 85 percent of the 2001 polio cases and are considered high-intensity transmission areas, and Angola, Egypt, Ethiopia, Somalia, and Sudan, which are low-intensity transmission areas, accounting for less than 15 percent of new polio cases last year.

The high risk areas have been identified and are receiving special attention during the immunization drives.

"Each of these countries has made tremendous progress, but each has its own unique set of challenges," said Carol Bellamy, executive director of UNICEF, during the April news conference. "Through the battle to rid the

world of polio, we have managed to reach children living in some of the most remote and challenging circumstances imaginable. Over the coming days and months, we must continue this unprecedented effort, using all of our resources to reach the very last child with polio vaccine. We have a unique opportunity to deliver a truly global victory in an uncertain world."

Deepak Kapur, India's National PolioPlus Committee Chairman and a member of the Rotary Club of Delhi South, says: "The job is almost done [in India]. But the most dangerous word in the dictionary is *almost*. We rejoiced when the thousands of cases fell to a few hundred, and today we have cause to celebrate when only a few cases are left. But this new chapter begins with those few cases."

The good news in India is that the geographic distribution of the virus has decreased. In 2001, India reported 268 cases of polio (a slight increase from 2000), about 95 percent of them in the northern states of Bihar and Uttar Pradesh. Kapur says that a range of projects will be undertaken this year, including mobilizing village volunteers and school students to spread awareness of the campaign; continuing the "video van" project and musical street dramas, which are especially useful in explaining PolioPlus to illiterate populations; and updating local media to enlist their editorial support.

According to Kapur, challenges include generally poor immunization rates for routine childhood diseases, overworked public health officials, public indifference because of repeated immunization drives, and outright distrust of immunization efforts by members of certain populations.

"[PolioPlus India] Executive Director Rajiv Tandon and I went to Moradabad on a field visit. It's considered the 'polio capital of the world since more than half the cases have been genetically traced to the virus prevalent in this area," recounts Kapur. "One of the reasons for proliferation is the significant resistance of members of a particular minority community." The two visited a hut in a slum to persuade a mother to let them administer the polio drops to her infant son. Finally, after making the pair promise not to tell her husband, she allowed her baby to be immunized.

"Her husband was convinced that the immunization program was aimed at curtailing the population of their community by making children impotent," Kapur explains. "As we stepped out of the hut, we saw a polio-crippled girl playing in the mud. The neighbors told us that the little girl had contracted polio a few months ago and was that very mother's daughter." Polio workers in India face this kind of resistance because of incomplete information, misinformation, and rumors, says Kapur.

Afghanistan and Pakistan also pose challenges, but progress has been good. Despite the war and refugee crises in the wake of 11 September, Afghanistan reported only 11 cases of polio in 2001 (down from 120 in 2000), clustered in two main areas. Two rounds of National Immunization Days (NIDs) were held in September and November 2001, reaching 35 million children; this year, NIDs began anew in mid-April, coordinated with Pakistan to ensure coverage of cross-border areas.

"The most urgent need is to bring stability to the country [Afghanistan] and start reconstruction work without delay," says Rotarian Abdul Haiy Khan, chairman of Pakistan's National PolioPlus Committee and a member of the Rotary Club of Karachi. "The Afghans are resilient, and during my visits to Afghanistan to monitor their NIDs, I was impressed with their commitment. I think that once they get the opportunity to freely move around the country and have the assurance of obtaining the needed resources, the polio-eradication job will be done in a short time."

In Pakistan, Khan reports, "we have to deal with a large population, 140 million people with about 30 million of them children under age 5, nearly 30 percent living in densely populated or slum areas with poor sanitary conditions." However, the high-risk areas have been identified and are receiving special attention during the immunization drives. The high-transmission season of wild polio-virus runs from May to September.

"This will be the testing period to prove that we are in control of the situation and can hope to come close to the zero polio-case reporting status," says Khan. "The real challenge includes training and retraining health workers, volunteers and monitors, improving the surveillance system, and providing better transport for health workers and volunteers so they can reach the farthest, most isolated areas in the hot summer season."

But ingenuity and creativity bring success. Khan cites one story of an Afghan refugee camp on the outskirts of Karachi, where health workers reported that camp residents refused to let their children be immunized. "We spent some time investigating the matter and were pleasantly surprised to find that the refusal was based on a lack of communication. Our health workers didn't speak [their] language and couldn't convey the purpose of house-to-house visits. And the camp housed a large number of unregistered refugees, so camp elders were reluctant to let government officials (health workers) go door-to-door."

The solution? "We simply located two Afghan girls studying in a local medical college, explained the problem, and they came to the camp to talk to the elders. The next day, the two girls were accompanied by more than 100 fellow students, and in two days, working eight hours a day, they immunized 10,000 Afghan children under the age of 5 whom we had been missing for a couple of years."

The funding gap, along with the continuing threat of conflict worldwide and particularly within the 10 polio-endemic nations, brought a recent warning from experts: While polio has been pushed to the lowest level in history, efforts to complete eradication and immunization must be intensified now before global instability closes the window of opportunity.

"I urge the world to finish the job. Eradicate polio while we still have the opportunity."

– Dr. Gro Harlem Brundtland,
WHO director-general

"When we began the eradication effort in 1988, polio paralyzed more than 1,000 children each day. In 2001, there were far fewer than 1,000 cases the entire year," said Dr. Gro Harlem Brundtland, director-general of WHO, during the joint April news conference. "But we're not finished yet, and the past year has reminded us that we live in a world where security and access to children cannot be guaranteed. So I urge the world to finish the job. Eradicate polio while we still have the opportunity."

Fortunately, in the African nations of Angola, Sierra Leone, and Somalia, warring forces have laid down their arms and allowed health workers to reach the children. In 2000, 16 West African countries coordinated immunization activities, and more than 80 million children were immunized. In 2001, conflict-affected countries in central Africa, including Angola and the Democratic Republic of the Congo, synchronized their NIDs for the first time and 16 million children were vaccinated.

To stop the transmission of the wild poliovirus in Africa by the end of 2002, NIDs will target northern Nigeria along the border with Niger. Major progress has been achieved in Angola, Egypt, Ethiopia, Somalia, and Sudan, and the goal is to stop transmission of the wild poliovirus by mid-2002—given continued political commitment in the five nations and no further deterioration of security.

Seventeen nations are considered at particularly high risk of reestablished wild poliovirus transmission, including Central African Republic, Chad, Congo Republic, Democratic Republic of the Congo, Indonesia, Madagascar, and Mozambique. These countries either are recently endemic or border endemic areas. They also have low routine-immunization coverage and inadequate surveillance capability or both. These nations require continuing, supplementary immunization activities and the assurance that no child has been missed.

"If we don't finish this off," warns Piconi, "this virus can creep back rather quickly. We need to stay focused and eliminate a disease that's vaccine preventable."

Anne Stein is a health and fitness writer in Evanston, Ill., USA, who has written frequently about the polio-eradication effort for THE ROTARIAN.

From *The Rotarian,* July 2002, pp. 22-26. © 2002 by Anne Stein. Reprinted by permission of the author.

The New Containment

An Alliance Against Nuclear Terrorism

Graham Allison & Andrei Kokoshin

DURING THE Cold War, American and Russian policy-makers and citizens thought long and hard about the possibility of nuclear attacks on their respective homelands. But with the fall of the Berlin Wall and the disappearance of the Soviet Union, the threat of nuclear weapons catastrophe faded away from most minds. This is both ironic and potentially tragic, since the threat of a nuclear attack on the United States or Russia is certainly greater today than it was in 1989.

In the aftermath of Osama bin Laden's September 11 assault, which awakened the world to the reality of global terrorism, it is incumbent upon serious national security analysts to think again about the unthinkable. Could a nuclear terrorist attack happen today? Our considered answer is: yes, unquestionably, without any doubt. It is not only a possibility, but in fact the most urgent unaddressed national security threat to both the United States and Russia.[1]

Consider this hypothetical: A crude nuclear weapon constructed from stolen materials explodes in Red Square in Moscow. A 15-kiloton blast would instantaneously destroy the Kremlin, Saint Basil's Cathedral, the ministries of foreign affairs and defense, the Tretyakov Gallery, and tens of thousands of individual lives. In Washington, an equivalent explosion near the White House would completely destroy that building, the Old Executive Office Building and everything within a one-mile radius, including the Departments of State, Treasury, the Federal Reserve and all of their occupants—as well as damaging the Potomac-facing side of the Pentagon.

Psychologically, such a hypothetical is as difficult to internalize as are the plot lines of a writer like Tom Clancy (whose novel *Debt of Honor* ends with terrorists crashing a jumbo jet into the U.S. Capitol on Inauguration Day, and whose *The Sum of All Fears* contemplates the very scenario we discuss—the detonation of a nuclear device in a major American metropolis by terrorists). That these kinds of scenarios are physically possible, however, is an undeniable, brute fact.

After the first nuclear terrorist attack, the Duma, Congress—or what little is left of them—and the press will investigate: Who knew what, when? They will ask what could have been done to prevent the attack. Most officials will no doubt seek cover behind the claim that "no one could have imagined" this happening. But that defense should ring hollow. We have unambiguous strategic warning today that a nuclear terrorist attack could occur at any moment. Responsible leaders should be asking hard questions now. Nothing prevents the governments of Russia, America and other countries from taking effective action immediately—nothing, that is, but a lack of determination.

The argument made here can be summarized in two propositions: first, nuclear terrorism poses a clear and present danger to the United States, Russia and other nations; second, nuclear terrorism is a largely *preventable* disaster. Preventing nuclear terrorism is a large, complex, but ultimately finite challenge that can be met by a bold, determined, but nonetheless finite response. The current mismatch between the seriousness of the threat on the one hand, and the actions governments are now taking to meet it on the other, is unacceptable. Below we assess the threat and outline a solution that begins with a U.S.-Russian led Alliance Against Nuclear Terrorism.

Assessing the Threat

A COMPREHENSIVE threat assessment must consider both the likelihood of an event and the magnitude of its anticipated consequences. As described above, the impact of even a crude nuclear explosion in a city would produce devastation in a class by itself.[2] A half dozen nuclear explosions across the United States or Russia would shift the course of history. The question is: how likely is such an event?

Security studies offer no well-developed methodology for estimating the probabilities of unprecedented events. Contemplating the possibility of a criminal act, Sherlock Holmes investigated three factors: motive, means and opportunity. That framework can be useful for analyzing the question at hand. If no actor simultaneously has motive, means and opportunity, no nuclear terrorist act will occur. Where these three factors are abundant and widespread, the likelihood of a nuclear terrorist attack increases. The questions become: Is anyone *motivated* to instigate a nuclear attack? Could terrorist groups acquire the *means* to attack the United States or Russia with nuclear weapons? Could these groups find or create an *opportunity* to act?

I. Motive

There is no doubt that Osama bin Laden and his associates have serious nuclear ambitions. For almost a decade they have been actively seeking nuclear weapons, and, as President Bush has noted, they would use such weapons against the United States or its allies "in a heartbeat." In 2000, the CIA intercepted a message in which a member of Al-Qaeda boasted of plans for a "Hiroshima" against America. According to the Justice Department indictment for the 1998 bombings of the American embassies in Kenya and Tanzania, "At various times from at least as early as 1993, Osama bin Laden and others, known and unknown, made efforts to obtain the components of nuclear weapons." Additional evidence from a former Al-Qaeda member describes attempts to buy uranium of South African origin, repeated travels to three Central Asian states to try to buy a complete warhead or weapons-usable material, and discussions with Chechen criminal groups in which money and drugs were offered for nuclear weapons.

Bin Laden himself has declared that acquiring nuclear weapons is a religious duty. "If I have indeed acquired [nuclear] weapons," he once said, "then I thank God for enabling me to do so." When forging an alliance of terrorist organizations in 1998, he issued a statement entitled "The Nuclear Bomb of Islam." Characterized by Bernard Lewis as "a magnificent piece of eloquent, at times even poetic Arabic prose," it states: "It is the duty of Muslims to prepare as much force as possible to terrorize the enemies of God." If anything, the ongoing American-led war on global terrorism is heightening our adversary's incentive to obtain and use a nuclear weapon. Al-Qaeda has discovered that it can no longer attack the United States with impunity. Faced with an assertive, determined opponent now doing everything it can to destroy this terrorist network, Al-Qaeda has every incentive to take its best shot.

Russia also faces adversaries whose objectives could be advanced by using nuclear weapons. Chechen terrorist groups, for example, have demonstrated little if any restraint on their willingness to kill civilians and may be tempted to strike a definitive blow to assert independence from Russia. They have already issued, in effect, a radioactive warning by planting a package containing cesium-137 at Izmailovsky Park in Moscow and then tipping off a Russian reporter. Particularly as the remaining Chechen terrorists have been marginalized over the course of the second Chechen war, they could well imagine that by destroying one Russian city and credibly threatening Moscow, they could persuade Russia to halt its campaign against them.

All of Russia's national security documents—its *National Security Concept*, its military doctrine and the recently-updated *Foreign Policy Concept*—have clearly identified international terrorism as the greatest threat to Russia's national security. As President Putin noted in reviewing Russian security priorities with senior members of the Foreign Ministry in January 2001, "I would like to stress the danger of international terrorism and fundamentalism of any, absolutely any stripe." The illegal drug trade and the diffusion of religious extremism throughout Central Asia, relating directly to the rise of the Taliban in Afghanistan, threaten Russia's borders and weaken the Commonwealth of Independent States. The civil war in Tajikistan, tensions in Georgia's Pankisi Gorge, and the conflicts in South Ossetia, Abkhazia and Nagorno-Karabakh—all close to the borders of the Russian Federation—provide feeding grounds for the extremism that fuels terrorism. Additionally, Russia's geographical proximity to South Asia and the Middle East increases concerns over terrorist fallout from those regions. President Putin has consistently identified the dark hue that weapons of mass destruction (WMD) give to the threat of terrorism. In a December 2001 interview in which he named international terrorism the "plague of the 21st century," Putin stated: "We all know exactly how New York and Washington were hit.… Was it ICBMs? What threat are we talking about? We are talking about the use of mass destruction weapons terrorists may obtain."

Separatist militants (in Kashmir, the Balkans and elsewhere) and messianic terrorists (like Aum Shinrikyo, which attacked the Tokyo subway with chemical weapons in 1995) could have similar motives to commit nuclear terrorism. As Palestinians look to uncertain prospects for independent statehood—and never mind whose leadership actually increased that uncertainty in recent years—Israel becomes an ever more attractive target for a nuclear terrorist attack. Since a nuclear detonation in any part of the world would be extremely destabilizing, it threatens American and Russian interests even if few or no Russians or Americans are killed. Policymakers would therefore be foolish to ignore any group with a motive to use a nuclear weapon against any target.

II. Means

To the best of our knowledge, no terrorist group can now detonate a nuclear weapon. But as Secretary of Defense Donald Rumsfeld has stated, "the absence of evidence is not evidence of absence." Are the means beyond terrorists' reach, even that of relatively sophisticated groups like Al-Qaeda?

Over four decades of Cold War competition, the superpowers spent trillions of dollars assembling mass arsenals, stockpiles, nuclear complexes and enterprises that engaged hundreds of thousands of accomplished scientists and engineers. Technical know-how cannot be un-invented. Reducing arsenals that include some 40,000 nuclear weapons and the equivalents of more than 100,000 nuclear weapons in the form of highly enriched uranium (HEU) and plutonium to manageable levels is a gargantuan challenge.

Terrorists could seek to buy an assembled nuclear weapon from insiders or criminals. Nuclear weapons are known to exist in eight states: the United States, Russia, Great Britain, France, China, Israel, India and Pakistan. Security measures, such as "permissive action links" designed to prevent unauthorized use, are most reliable in the United States, Russia, France and the United Kingdom. These safeguards, as well as command-and-control systems, are much less reliable in the two newest nuclear states—India and Pakistan. But even where good systems are in place, maintaining high levels of security requires constant attention from high-level government officials.

Alternatively, terrorists could try to build a weapon. The only component that is especially difficult to obtain is the nuclear fissile material—HEU or plutonium. Although the largest stockpiles of weapons-grade material are predominantly found in the nuclear weapons programs of the United States and Russia, fissile material in sufficient quantities to make a crude nuclear weapon can also be found in many civilian settings around the globe. Some 345 research reactors in 58 states together contain twenty metric tons of HEU, many in quantities sufficient to build a bomb.[3] Other civilian reactors produce enough weapons-grade nuclear material to pose a proliferation threat; several European states, Japan, Russia and India reprocess spent fuel to separate out plutonium for use as new fuel. The United States has actually facilitated the spread of fissile material in the past—over three decades of the Atoms for Peace program, the United States exported 749 kg of plutonium and 26.6 metric tons of HEU to 39 countries.[4]

Terrorist groups could obtain these materials by theft, illicit purchase or voluntary transfer from state control. There is ample evidence that attempts to steal or sell nuclear weapons or weapons-usable material are not hypothetical, but a recurring fact.[5] Just last fall, the chief of the directorate of the Russian Defense Ministry responsible for nuclear weapons reported two recent incidents in which terrorist groups attempted to perform reconnaissance at Russian nuclear storage sites. The past decade has seen repeated incidents in which individuals and groups have successfully stolen weapons material from sites in Russia and sought to export them—but were caught trying to do so. In one highly publicized case, a group of insiders at a Russian nuclear weapons facility in Chelyabinsk plotted to steal 18.5 kg (40.7 lbs.) of HEU, which would have been enough to construct a bomb, but were thwarted by Russian Federal Security Service agents.

In the mid-1990s, material sufficient to allow terrorists to build more than twenty nuclear weapons—more than 1,000 pounds of highly enriched uranium—sat unprotected in Kazakhstan. Iranian and possibly Al-Qaeda operatives with nuclear ambitions were widely reported to be in Kazakhstan. Recognizing the danger, the American government itself purchased the material and removed it to Oak Ridge, Tennessee. In February 2002, the U.S. National Intelligence Council reported to Congress that "undetected smuggling [of weapons-usable nuclear materials from Russia] has occurred, although we do not know the extent of such thefts." Each assertion invariably provokes blanket denials from Russian officials. Russian Atomic Energy Minister Aleksandr Rumyantsev has claimed categori-

cally: "Fissile materials have not disappeared." President Putin has stated that he is "absolutely confident" that terrorists in Afghanistan do not have weapons of mass destruction of Soviet or Russian origin.

For perspective on claims of the inviolable security of nuclear weapons or material, it is worth considering the issue of "lost nukes." Is it possible that the United States or Soviet Union lost assembled nuclear weapons? At least on the American side the evidence is clear. In 1981, the U.S. Department of Defense published a list of 32 accidents involving nuclear weapons, *many of which resulted in lost bombs*.[6] One involved a submarine that sank along with two nuclear torpedoes. In other cases, nuclear bombs were lost from aircraft. Though on the Soviet/Russian side there is no official information, we do know that four Soviet submarines carrying nuclear weapons have sunk since 1968, resulting in an estimated 43 lost nuclear warheads.[7] These accidents suggest the complexity of controlling and accounting for vast nuclear arsenals and stockpiles.

Nuclear materials have also been stolen from stockpiles housed at research reactors. In 1999, Italian police seized a bar of enriched uranium from an organized crime group trying to sell it to an agent posing as a Middle Eastern businessman with presumed ties to terrorists. On investigation, the Italians found that the uranium originated from a U.S.-supplied research reactor in the former Zaire, where it presumably had been stolen or purchased *sub rosa*.

Finally, as President Bush has stressed, terrorists could obtain nuclear weapons or material from states hostile to the United States. In his now-infamous phrase, Bush called hostile regimes developing WMD and their terrorist allies an "axis of evil." He argued that states such as Iraq, Iran and North Korea, if allowed to realize their nuclear ambitions, "could provide these arms to terrorists, giving them the means to match their hatred." The fear that a hostile regime might transfer a nuclear weapon to terrorists has contributed to the Bush Administration's development of a new doctrine of preemption against such regimes, with Iraq as the likeliest test case. It also adds to American concerns about Russian transfer of nuclear technologies to Iran. While Washington and Moscow continue to disagree over whether any safeguarded civilian nuclear cooperation with Iran is justified, both agree on the dangers a nuclear-armed Iran would pose. Russia is more than willing to agree that there should be no transfers of technology that could help Iran make nuclear weapons.

III. Opportunity

Security analysts have long focused on ballistic missiles as the preferred means by which nuclear weapons would be delivered. But today this is actually the least likely vehicle by which a nuclear weapon will be delivered against Russia or the United States. Ballistic weapons are hard to produce, costly and difficult to hide. A nuclear weapon delivered by a missile also leaves an unambiguous return address, inviting devastating retaliation. As Robert Walpole, a National Intelligence Officer, told a Senate subcommittee in March, "Nonmissile delivery means are less costly, easier to acquire, and more reliable and accurate."[8] Despite this assessment, the U.S. government continues to invest much more heavily in developing and deploying missile defenses than in addressing more likely trajectories by which weapons could arrive.

Terrorists would not find it very difficult to sneak a nuclear device or nuclear fissile material into the United States via shipping containers, trucks, ships or aircraft. Recall that the nuclear material required is smaller than a football. Even an assembled device, like a suitcase nuclear weapon, could be shipped in a container, in the hull of a ship or in a trunk carried by an aircraft. After this past September 11, the number of containers that are X-rayed has increased, to about 500 of the 5,000 containers currently arriving daily at the port of New York/New Jersey—approximately 10 percent. But as the chief executive of CSX Lines, one of the foremost container-shipping companies, put it: "If you can smuggle heroin in containers, you may be able to smuggle in a nuclear bomb."

Effectively countering missile attacks will require technological breakthroughs well beyond current systems. Success in countering covert delivery of weapons will require not just technical advances but a conceptual breakthrough. Recent efforts to bolster border security are laudable, but they only begin to scratch the surface. More than 500 million people, 11 million trucks and 2 million rail cars cross into the United States each year, while 7,500 foreign-flag ships make 51,000 calls in U.S. ports. That's not counting the tens of thousands of people, hundreds of aircraft and numerous boats that enter illegally and uncounted. Given this volume and the lengthy land and sea borders of the United States, even a radically renovated and reorganized system cannot aspire to be airtight.

The opportunities for terrorists to smuggle a nuclear weapon into Russia or another state are even greater. Russia's land borders are nearly twice as long as America's, connecting it to more than a dozen other states. In many places, in part because borders between republics were less significant in the time of the Soviet Union, these borders are not closely monitored. Corruption has been a major problem among border patrols. Visa-free travel between Russia and several of its neighbors creates additional opportunities for weapons smugglers and terrorists. The "homeland security" challenge for Russia is truly monumental.

In sum: even a conservative estimate must conclude that dozens of terrorist groups have sufficient motive to use a nuclear weapon, several could potentially obtain nuclear means, and hundreds of opportunities exist for a group with means and motive to make the United States or Russia a victim of nuclear terrorism. The mystery before us is not how a nuclear terrorist attack could possibly occur, but rather why no terrorist group has yet combined motive, means and opportunity to commit a nuclear attack. We have been lucky so far, but who among us trusts luck to protect us in the future?

Chto Delat?[9]

THE GOOD NEWS about nuclear terrorism can be summarized in one line: no highly enriched uranium or plutonium, no

nuclear explosion, no nuclear terrorism. Though the world's stockpiles of nuclear weapons and weapons-usable materials are vast, they are finite. The prerequisites for manufacturing fissile material are many and require the resources of a modern state. Technologies for locking up super-dangerous or valuable items—from gold in Fort Knox to treasures in the Kremlin Armory—are well developed and tested. While challenging, a specific program of actions to keep nuclear materials out of the hands of the most dangerous groups is not beyond reach, *if* leaders give this objective highest priority and hold subordinates accountable for achieving this result.

The starting points for such a program are already in place. In his major foreign policy campaign address at the Ronald Reagan Library, then-presidential candidate George W. Bush called for "Congress to increase substantially our assistance to dismantle as many Russian weapons as possible, as quickly as possible." In his September 2000 address to the United Nations Millennium Summit, Russian President Putin proposed to "find ways to block the spread of nuclear weapons by excluding use of enriched uranium and plutonium in global atomic energy production." The Joint Declaration on the New Strategic Relationship between the United States and Russia, signed by the two presidents at the May 2002 summit, stated that the two partners would combat the "closely linked threats of international terrorism and the proliferation of weapons of mass destruction." Another important result yielded by the summit was the upgrading of the Armitage/Trubnikov-led U.S.-Russia Working Group on Afghanistan to the U.S.-Russia Working Group on Counterterrorism, whose agenda is to thwart nuclear biological and chemical terrorism.

Operationally, however, priority is measured not by words, but by deeds. A decade of Nunn-Lugar Cooperative Threat Reduction Programs has accomplished much in safeguarding nuclear materials. Unfortunately, the job of upgrading security to minimum basic standards is mostly unfinished: according to Department of Energy reports, two-thirds of the nuclear material in Russia remains to be adequately secured.[10] Bureaucratic inertia, bolstered by mistrust and misperception on both sides, leaves these joint programs bogged down on timetables that extend to 2008. Unless implementation improves significantly, they will probably fail to meet even this unacceptably distant target. What is required on both sides is personal, presidential priority measured in commensurate energy, specific orders, funding and accountability. This should be embodied in a new U.S.-Russian led Alliance Against Nuclear Terrorism.

Five Pillars of Wisdom

WHEN IT COMES to the threat of nuclear terrorism, many Americans judge Russia to be part of the problem, not the solution. But if Russia is welcomed and supported as a fully responsible non-proliferation partner, the United States stands to accomplish far more toward minimizing the risk of nuclear terrorism than if it treats Russia as an unreconstructed pariah. As the first step in establishing this alliance, the two presidents should pledge to each other that his government will do everything technically possible to prevent criminals or terrorists from stealing nuclear weapons or weapons-usable material, and to do so on the fastest possible timetable. Each should make clear that he will personally hold accountable the entire chain of command within his own government to assure this result. Understanding that each country bears responsibility for the security of its own nuclear materials, the United States should nonetheless offer Russia any assistance required to make this happen. Each nation—and each leader—should provide the other sufficient transparency to monitor performance.

Archy: "An optimist is a guy that has never had much experience."

—Don Marquis, *Archy and Mehitabel* (1927)

To ensure that this is done on an expedited schedule, both governments should name specific individuals, directly answerable to their respective presidents, to co-chair a group tasked with developing a joint Russian-American strategy within one month. In developing a joint strategy and program of action, the nuclear superpowers would establish a new world-class "international security standard" based on President Putin's Millennium proposal for new technologies that allow production of electricity with low-enriched, non-weapons-usable nuclear fuel.

A second pillar of this alliance would reach out to all other nuclear weapons states—beginning with Pakistan. Each should be invited to join the alliance and offered assistance, if necessary, in assuring that all weapons and weapons-usable material are secured to the new established international standard in a manner sufficiently transparent to reassure all others. Invitations should be diplomatic in tone but nonetheless clear that this is an offer that cannot be refused. China should become an early ally in this effort, one that could help Pakistan understand the advantages of willing compliance.

A third pillar of this alliance calls for global outreach along the lines proposed by Senator Richard Lugar in what has been called the Lugar Doctrine.[11] All states that possess weapons-usable nuclear materials—even those without nuclear weapons capabilities—must enlist in an international effort to guarantee the security of such materials from theft by terrorists or criminals groups. In effect, each would be required to meet the new international security standard and to do so in a transparent fashion. Pakistan is particularly important given its location and relationship with Al-Qaeda, but beyond nuclear weapons states, several dozen additional countries hosting research reactors—such as Serbia, Libya and Ghana—should be persuaded to surrender such material (almost all of it either American or Soviet in origin), or have the material secured to acceptable international standards.

A fourth pillar of this effort should include Russian-American led cooperation in preventing any further spread of nuclear weapons to additional states, focusing sharply on North Korea, Iraq and Iran. The historical record demonstrates that when the United States and Russia have cooperated intensely, nuclear

wannabes have been largely stymied. It was only during periods of competition or distraction, for example in the mid-1990s, that new nuclear weapons states realized their ambitions. India and Pakistan provide two vivid case studies. Recent Russian-American-Chinese cooperation in nudging India and Pakistan back from the nuclear brink suggests a good course of action. The failure and subsequent freeze of North Korean nuclear programs offers complementary lessons about the consequences of competition and distraction. The new alliance should reinvent a robust non-proliferation regime of controls on the sale and export of weapons of mass destruction, nuclear material and missile technologies, recognizing the threat to each of the major states that would be posed by a nuclear-armed Iran, North Korea or Iraq.

Finally, adapting lessons learned in U.S.-Russian cooperation in the campaign against bin Laden and the Taliban, this new alliance should be heavy on intelligence sharing and affirmative counter-proliferation, including disruption and pre-emption to prevent acquisition of materials and know-how by nuclear wannabes. Beyond joint intelligence sharing, joint training for pre-emptive actions against terrorists, criminal groups or rogue states attempting to acquire weapons of mass destruction would provide a fitting enforcement mechanism for alliance commitments.

As FORMER Senator Sam Nunn has noted: "At the dawn of a new century, we find ourselves in a new arms race. Terrorists are racing to get weapons of mass destruction; we ought to be racing to stop them."[12] Preventing nuclear terrorism will require no less imagination, energy and persistence than did avoiding nuclear war between the superpowers over four decades of Cold War. But absent deep, sustained cooperation between the United States, Russia and other nuclear states, such an effort is doomed to failure. In the context of the qualitatively new relationship Presidents Putin and Bush have established in the aftermath of last September 11, success in such a bold effort is within the reach of determined Russian-American leadership. Succeed we must.

Notes

1. This judgment echoes that of a Department of Energy task force on nonproliferation programs with Russia led by Howard Baker and Lloyd Cutler: "The most urgent unmet national security threat to the United States today is the danger that weapons of mass destruction or weapons-usable material in Russia could be stolen and sold to terrorists or hostile nation states and used against American troops abroad or citizens at home." *A Report Card on the Department of Energy's Nonproliferation Programs with Russia*, January 10, 2001.

2. Although biological and chemical weapons can cause huge devastation as well, "the massive, assured, instantaneous, and comprehensive destruction of life and property" of a nuclear weapon is unique. See Matthew Bunn, John P. Holdren and Anthony Wier, "Securing Nuclear Weapons and Materials: Seven Steps for Immediate Action," *Nuclear*

Threat Initiative and the Managing the Atom Project, May 20, 2002, p. 2. This report provides extensive, but not-too-technical detail on many of the points in this essay.

3. See U.S. Department of Energy, *FY 2003 Budget Request: Detailed Budget Justifications—Defense Nuclear Nonproliferation* (Washington, DC: DOE, 2002), p. 172.

4. Summarized in NIS Nuclear Trafficking Database, available at www.nti.org.

5. The Nuclear Threat Initiative maintains a database of cases and reported incidents of trafficking in nuclear and radioactive materials in and from the former Soviet Union. Available at www.nti.org.

6. U.S. Department of Defense, "Narrative Summaries of Accidents involving U.S. Nuclear Weapons: 1950-1980" (April 1981).

7. Joshua Handler, Amy Wickenheiser and William M. Arkin, "Naval Safety 1989: The Year of the Accident," *Neptune Paper No. 4* (April 1989).

8. U.S Senate Subcommittee on International Security, Proliferation, and Federal Services, "Statement of Robert Walpole before the Senate Subcommittee on International Security, Proliferation, and Federal Services," 107th Cong., 1st sess., March 11, 2002.

9. A proverbial Russian refrain: "What is to be done?"

10. Bunn, Holdren and Wier, "Securing Nuclear Weapons and Materials."

11. Speech by Senator Richard Lugar, May 27, 2002, at the Moscow Nuclear Threat Initiative Conference.

12. Sam Nunn, "Our New Security Framework," *Washington Post*, October 8, 2001.

Graham Allison is director of the Belfer Center for Science and International Affairs at Harvard's John F. Kennedy School of Government. Andrei Kokoshin is director of the Institute for International Security Studies of the Russian Academy of Sciences and a former secretary of the Security Council of Russia.

UNIT 7
Values and Visions

Unit Selections

Key Points to Consider

- Is it naive to speak of global issues in terms of ethics?

- What role can governments, international organizations, and the individual play in making the world a more ethical place?

- How is the political role of women changing, and what impacts are these changes having on conflict resolution and community building?

- The consumption of resources is the foundation of the modern economic system. What are the values underlying this economic system, and how resistant to change are they?

- What are the characteristics of leadership?

- In addition to the ideas presented here, what other new ideas are being expressed, and how likely are they to be widely accepted?

 **Links: www.dushkin.com/online/**
These sites are annotated in the World Wide Web pages.

Human Rights Web
http://www.hrweb.org
InterAction
http://www.interaction.org

The final unit of this book considers how humanity's view of itself is changing. Values, like all other elements discussed in this anthology, are dynamic. Visionary people with new ideas can have a profound impact on how a society deals with problems and adapts to changing circumstances. Therefore, to understand the forces at work in the world today, values, visions, and new ideas in many ways are every bit as important as new technology or changing demographics.

Novelist Herman Wouk, in his book *War and Remembrance*, observed that many institutions have been so embedded in the social fabric of their time that people assumed that they were part of human nature. Slavery and human sacrifice are two examples. However, forward-thinking people opposed these institutions. Many knew that they would never see the abolition of these social systems within their own lifetimes, but they pressed on in the hope that someday these institutions would be eliminated.

Wouk believes the same is true for warfare. He states, "Either we are finished with war or war will finish us." Aspects of society such as warfare, slavery, racism, and the secondary status of

women are creations of the human mind; history suggests that they can be changed by the human spirit.

The articles of this unit have been selected with the previous six units in mind. Each explores some aspect of world affairs from the perspective of values and alternative visions of the future.

New ideas are critical to meeting these challenges. The examination of well-known issues from new perspectives can yield new insights into old problems. It was feminist Susan B. Anthony who once remarked that "Social change is never made by the masses, only by educated minorities." The redefinition of human values (which, by necessity, will accompany the successful confrontation of important global issues) is a task that few people take on willingly. Nevertheless, in order to deal with the dangers of nuclear war, overpopulation, and environmental degradation, educated people must take a broad view of history. This is going to require considerable effort and much personal sacrifice.

When people first begin to consider the magnitude of contemporary global problems, many often become disheartened and depressed. Some ask: What can I do? What does it matter? Who cares? There are no easy answers to these questions, but people need only look around to see good news as well as bad. How individuals react to the world is not solely a function of so-called objective reality but a reflection of themselves.

As stated at the beginning of the first unit, the study of global issues is the study of people. The study of people, furthermore, is the study of both values and the level of commitment supporting these values and beliefs.

It is one of the goals of this book to stimulate you, the reader, to react intellectually and emotionally to the discussion and description of various global challenges. In the process of studying these issues, hopefully you have had some new insights into your own values and commitments. In the presentation of the allegory of the balloon, the fourth color added represented the "meta" component, all of those qualities that make human beings unique. It is these qualities that have brought us to this "special moment in time," and it will be these same qualities that will determine the outcome of our historically unique challenges.

Are Human Rights Universal?

Shashi Tharoor

The growing consensus in the West that human rights are universal has been fiercely opposed by critics in other parts of the world. At the very least, the idea may well pose as many questions as it answers. Beyond the more general, philosophical question of whether anything in our pluri-cultural multipolar world is truly universal, the issue of whether human rights is an essentially Western concept—ignoring the very different cultural, economic, and political realities of the other parts of the world—cannot simply be dismissed. Can the values of the consumer society be applied to societies that have nothing to consume? Isn't talking about universal rights rather like saying that the rich and the poor both have the same right to fly first class and to sleep under bridges? Don't human rights as laid out in the international convenants ignore the traditions, the religions, and the socio-cultural patterns of what used to be called the Third World? And at the risk of sounding frivolous, when you stop a man in traditional dress from beating his wife, are you upholding her human rights or violating his?

This is anything but an abstract debate. To the contrary, ours is an era in which wars have been waged in the name of human rights, and in which many of the major developments in international law have presupposed the universality of the concept. By the same token, the perception that human rights as a universal discourse is increasingly serving as a flag of convenience for other, far more questionable political agendas, accounts for the degree to which the very idea of human rights is being questioned and resisted by both intellectuals and states. These objections need to be taken very seriously.

The philosophical objection asserts essentially that nothing can be universal; that all rights and values are defined and limited by cultural perceptions. If there is no universal culture, there can be no universal human rights. In fact, some philosophers have objected that the concept of human rights is founded on an anthropocentric, that is, a human-centered, view of the world, predicated upon an individualistic view of man as an autonomous being whose greatest need is to be free from interference by the state—

free to enjoy what one Western writer summed up as the "right to private property, the right to freedom of contract, and the right to be left alone." But this view would seem to clash with the communitarian one propounded by other ideologies and cultures where society is conceived of as far more than the sum of its individual members.

Who Defines Human Rights?

Implicit in this is a series of broad, culturally grounded objections. Historically, in a number of non-Western cultures, individuals are not accorded rights in the same way as they are in the West. Critics of the universal idea of human rights contend that in the Confucian or Vedic traditions, duties are considered more important than rights, while in Africa it is the community that protects and nurtures the individual. One African writer summed up the African philosophy of existence as: "I am because we are, and because we are therefore I am." Some Africans have argued that they have a complex structure of communal entitlements and obligations grouped around what one might call four "r's": not "rights," but respect, restraint, responsibility, and reciprocity. They argue that in most African societies group rights have always taken precedence over individual rights, and political decisions have been made through group consensus, not through individual assertions of rights.

These cultural differences, to the extent that they are real, have practical implications. Many in developing countries argue that some human rights are simply not relevant to their societies—the right, for instance, to political pluralism, the right to paid vacations (always good for a laugh in the sweatshops of the Third World), and, inevitably, the rights of women. It is not just that some societies claim they are simply unable to provide certain rights to all their citizens, but rather that they see the "universal" conception of human rights as little more than an attempt to impose alien Western values on them.

Rights promoting the equality of the sexes are a contentious case in point. How, critics demand, can women's rights be universal in the face of widespread divergences of cultural practice, when in many societies, for example, marriage is not seen as a contract between two individuals but as an alliance between lineages, and when the permissible behavior of womenfolk is central to the society's perception of its honor?

And, inseparable from the issues of tradition, is the issue of religion. For religious critics of the universalist definition of human rights, nothing can be universal that is not founded on transcendent values, symbolized by God, and sanctioned by the guardians of the various faiths. They point out that the cardinal document of the contemporary human rights movement, the Universal Declaration of Human Rights, can claim no such heritage.

Recently, the fiftieth anniversary of the Universal Declaration was celebrated with much fanfare. But critics from countries that were still colonies in 1948 suggest that its provisions reflect the ethnocentric bias of the time. They go on to argue that the concept of human rights is really a cover for Western interventionism in the affairs of the developing world, and that "human rights" are merely an instrument of Western political neocolonialism. One critic in the 1970s wrote of his fear that "Human Rights might turn out to be a Trojan horse, surreptitiously introduced into other civilizations, which will then be obliged to accept those ways of living, thinking and feeling for which Human Rights is the proper solution in cases of conflict."

In practice, this argument tends to be as much about development as about civilizational integrity. Critics argue that the developing countries often cannot afford human rights, since the tasks of nation building, economic development, and the consolidation of the state structure to these ends are still unfinished. Authoritarianism, they argue, is more efficient in promoting development and economic growth. This is the premise behind the so-called Asian values case, which attributes the economic growth of Southeast Asia to the Confucian virtues of obedience, order, and respect for authority. The argument is even a little more subtle than that, because the suspension or limiting of human rights is also portrayed as the sacrifice of the few for the benefit of the many. The human rights concept is understood, applied, and argued over only, critics say, by a small Westernized minority in developing countries. Universality in these circumstances would be the universality of the privileged. Human rights is for the few who have the concerns of Westerners; it does not extend to the lowest rungs of the ladder.

The Case for the Defense

That is the case for the prosecution—the indictment of the assumption of the universality of human rights. There is, of course, a case for the defense. The philosophical objection is, perhaps surprisingly, the easiest to counter. After all, concepts of justice and law, the legitimacy of government, the dignity of the individual, protection from oppressive or arbitrary rule, and participation in the affairs of the community are found in every society on the face of this earth. Far from being difficult to identify, the number of philosophical common denominators between different cultures and political traditions makes universalism anything but a distortion of reality.

Historically, a number of developing countries—notably India, China, Chile, Cuba, Lebanon, and Panama—played an active and highly influential part in the drafting of the Universal Declaration of Human Rights. In the case of the human rights covenants, in the 1960s the developing world actually made the decisive contribution; it was the "new majority" of the Third World states emerging from colonialism—particularly Ghana and Nigeria—that broke the logjam, ending the East–West stalemate that had held up adoption of the covenants for nearly two decades. The principles of human rights have been widely adopted, imitated, and ratified by developing countries; the fact that therefore they were devised by less than a third of the states now in existence is really irrelevant.

In reality, many of the current objections to the universality of human rights reflect a false opposition between the primacy of the individual and the paramountcy of society. Many of the civil and political rights protect groups, while many of the social and economic rights protect individuals. Thus, crucially, the two sets of rights, and the two covenants that codify them, are like Siamese twins—inseparable and interdependent, sustaining and nourishing each other.

Still, while the conflict between group rights and individual rights may not be inevitable, it would be naïve to pretend that conflict would never occur. But while groups may collectively exercise rights, the individuals within them should also be permitted the exercise of their rights within the group, rights that the group may not infringe upon.

A Hidden Agenda?

Those who champion the view that human rights are not universal frequently insist that their adversaries have hidden agendas. In fairness, the same accusation can be leveled against at least some of those who cite culture as a defense against human rights. Authoritarian regimes who appeal to their own cultural traditions are cheerfully willing to crush culture domestically when it suits them to do so. Also, the "traditional culture" that is sometimes advanced to justify the nonobservance of human rights, including in Africa, in practice no longer exists in a pure form at the national level anywhere. The societies of developing countries have not remained in a pristine, pre-Western state; all have been subject to change and distortion by external influence, both as a result of colonialism in many cases and through participation in modern interstate relations.

You cannot impose the model of a "modern" nation-state cutting across tribal boundaries and conventions on your country, appoint a president and an ambassador to the United Nations, and then argue that tribal traditions should be applied to judge the human rights conduct of the resulting modern state.

In any case, there should be nothing sacrosanct about culture. Culture is constantly evolving in any living society, responding to both internal and external stimuli, and there is much in every culture that societies quite naturally outgrow and reject. Am I, as an Indian, obliged to defend, in the name of my culture, the practice of suttee, which was banned 160 years ago, of obliging widows to immolate themselves on their husbands' funeral pyres? The fact that slavery was acceptable across the world for at least 2,000 years does not make it acceptable to us now; the deep historical roots of anti-Semitism in European culture cannot justify discrimination against Jews today.

The problem with the culture argument is that it subsumes all members of a society under a cultural framework that may in fact be inimical to them. It is one thing to advocate the cultural argument with an escape clause—that is, one that does not seek to coerce the dissenters but permits individuals to opt out and to assert their individual rights. Those who freely choose to live by and to be treated according to their traditional cultures are welcome to do so, provided others who wish to be free are not oppressed in the name of a culture they prefer to disavow.

A controversial but pertinent example of an approach that seeks to strengthen both cultural integrity and individual freedom is India's Muslim Women (Protection of Rights upon Divorce) Act. This piece of legislation was enacted following the famous Shah Banu case, in which the Supreme Court upheld the right of a divorced Muslim woman to alimony, prompting howls of outrage from Muslim traditionalists who claimed this violated their religious beliefs that divorced women were only entitled to the return of the bride price paid upon marriage. The Indian parliament then passed a law to override the court's judgment, under which Muslim women married under Muslim law would be obliged to accept the return of the bride price as the only payment of alimony, but that the official Muslim charity, the Waqf Board, would assist them.

Many Muslim women and feminists were outraged by this. But the interesting point is that if a Muslim woman does not want to be subject to the provisions of the act, she can marry under the civil code; if she marries under Muslim personal law, she will be subject to its provisions. That may be the kind of balance that can be struck between the rights of Muslims as a group to protect their traditional practices and the right of a particular Muslim woman, who may not choose to be subject to that particular law, to exempt herself from it.

It needs to be emphasized that the objections that are voiced to specific (allegedly Western) rights very frequently involve the rights of women, and are usually vociferously argued by men. Even conceding, for argument's sake, that child marriage, widow inheritance, female circumcision, and the like are not found reprehensible by many societies, how do the victims of these practices feel about them? How many teenage girls who have had their genitalia mutilated would have agreed to undergo circumcision if they had the human right to refuse to permit it? For me, the standard is simple: where coercion exists, rights are violated, and these violations must be condemned whatever the traditional justification. So it is not culture that is the test, it is coercion.

Not with Faith, But with the Faithful

Nor can religion be deployed to sanction the status quo. Every religion seeks to embody certain verities that are applicable to all mankind—justice, truth, mercy, compassion—though the details of their interpretation vary according to the historical and geographical context in which the religion originated. As U.N. secretary general Kofi Annan has often said, the problem is usually not with the faith, but with the faithful. In any case, freedom is not a value found only in Western faiths: it is highly prized in Buddhism and in different aspects of Hinduism and Islam.

If religion cannot be fairly used to sanction oppression, it should be equally obvious that authoritarianism promotes repression, not development. Development is about change, but repression prevents change. The Nobel Prize–winning economist Amartya Sen has pointed out in a number of interesting pieces that there is now a generally agreed-upon list of policies that are helpful to economic development— "openness to competition, the use of international markets, a high level of literacy and school education, successful land reforms, and public provision of incentives for investment, export and industrialization"—none of which requires authoritarianism; none is incompatible with human rights. Indeed, it is the availability of political and civil rights that gives people the opportunity to draw attention to their needs and to demand action from the government. Sen's work has established, for example, that no substantial famine has ever occurred in any independent and democratic country with a relatively free press. That is striking; though there may be cases where authoritarian societies have had success in achieving economic growth, a country like Botswana, an exemplar of democracy in Africa, has grown faster than most authoritarian states.

In any case, when one hears of the unsuitability or inapplicability or ethnocentrism of human rights, it is important to ask what the unstated assumptions of this view really are. What exactly are these human rights that it is so unreasonable to promote? If one picks up the more contentious covenant—the one on civil and political rights— and looks through the list, what can one find that someone in a developing country can easily do without? Not the right to life, one trusts. Freedom from torture? The right not to be enslaved, not to be physically assaulted, not to be arbitrarily arrested, imprisoned, executed? No one actually advocates in so many words the abridgement of any of these rights. As Kofi Annan asked at a speech in Tehran University in 1997: "When have you heard a free voice demand an end to freedom? Where have you heard a slave argue for slavery? When have you heard a victim of torture endorse the ways of the torturer? Where have you heard the tolerant cry out for intolerance?"

Tolerance and mercy have always, and in all cultures, been ideals of government rule and human behavior. If we do not unequivocally assert the universality of the rights that oppressive governments abuse, and if we admit that these rights can be diluted and changed, ultimately we risk giving oppressive governments an intellectual justification for the morally indefensible. Objections to the applicability of international human rights standards have all too frequently been voiced by authoritarian rulers and power elites to rationalize their violations of human rights—violations that serve primarily, if not solely, to sustain them in power. Just as the Devil can quote scripture for his purpose, Third World communitarianism can be the slogan of a deracinated tyrant trained, as in the case of Pol Pot, at the Sorbonne. The authentic voices of the Third World know how to cry out in pain. It is time to heed them.

The "Right to Development"

At the same time, particularly in a world in which market capitalism is triumphant, it is important to stress that the right to development is also a universal human right. The very concept of development evolved in tune with the concept of human rights; decolonization and self-determination advanced side by side with a consciousness of the need to improve the standards of living of subject peoples. The idea that human rights could be ensured merely by the state not interfering with individual freedom cannot survive confrontation with a billion hungry, deprived, illiterate, and jobless human beings around the globe. Human rights, in one memorable phrase, start with breakfast.

For the sake of the deprived, the notion of human rights has to be a positive, active one: not just protection from the state but also the protection of the state, to permit these human beings to fulfill the basic aspirations of growth and development that are frustrated by poverty and scarce resources. We have to accept that social deprivation and economic exploitation are just as evil as political oppression or racial persecution. This calls for a more profound approach to both human rights and to development. Without development, human rights could not be

truly universal, since universality must be predicated upon the most underprivileged in developing countries achieving empowerment. We can not exclude the poorest of the poor from the universality of the rich.

After all, do some societies have the right to deny human beings the opportunity to fulfill their aspirations for growth and fulfillment legally and in freedom, while other societies organize themselves in such a way as to permit and encourage human beings freely to fulfill the same needs? On what basis can we accept a double standard that says that an Australian's need to develop his own potential is a right, while an Angolan's or an Albanian's is a luxury?

Universality, Not Uniformity

But it is essential to recognize that universality does not presuppose uniformity. To assert the universality of human rights is not to suggest that our views of human rights transcend all possible philosophical, cultural, or religious differences or represent a magical aggregation of the world's ethical and philosophical systems. Rather, it is enough that they do not fundamentally contradict the ideals and aspirations of any society, and that they reflect our common universal humanity, from which no human being must be excluded.

Most basically, human rights derive from the mere fact of being human; they are not the gift of a particular government or legal code. But the standards being proclaimed internationally can become reality only when applied by countries within their own legal systems. The challenge is to work towards the "indigenization" of human rights, and their assertion within each country's traditions and history. If different approaches are welcomed within the established framework—if, in other words, eclecticism can be encouraged as part of the consensus and not be seen as a threat to it—this flexibility can guarantee universality, enrich the intellectual and philosophical debate, and so complement, rather than undermine, the concept of worldwide human rights. Paradoxical as it may seem, it is a universal idea of human rights that can in fact help make the world safe for diversity.

Note

This article was adapted from the first Mahbub-ul-Haq Memorial Lecture, South Asia Forum, October 1998.

Shashi Tharoor is Director of Communications and Special Projects in the Office of the Secretary General of the United Nations. The views expressed here are the author's own and do not necessarily reflect the positions of the United Nations.

The Grameen Bank

A small experiment begun in Bangladesh has turned into a major new concept in eradicating poverty

by Muhammad Yunus

Over many years, Amena Begum had become resigned to a life of grinding poverty and physical abuse. Her family was among the poorest in Bangladesh—one of thousands that own virtually nothing, surviving as squatters on desolate tracts of land and earning a living as day laborers.

In early 1993 Amena convinced her husband to move to the village of Kholshi, 112 kilometers (70 miles) west of Dhaka. She hoped the presence of a nearby relative would reduce the number and severity of the beatings that her husband inflicted on her. The abuse continued, however—until she joined the Grameen Bank. Oloka Ghosh, a neighbor, told Amena that Grameen was forming a new group in Kholshi and encouraged her to join. Amena doubted that anyone would want her in their group. But Oloka persisted with words of encouragement. "We're all poor—or at least we all were when we joined. I'll stick up for you because I know you'll succeed in business.

Amena's group joined a Grameen Bank Center in April 1993. When she received her first loan of $60, she used it to start her own business raising chickens and ducks. When she repaid her initial loan and began preparing a proposal for a second loan of $110, her friend Oloka gave her some sage advice: "Tell your husband that Grameen does not allow borrowers who are beaten by their spouses to remain members and take loans." From that day on, Amena suffered significantly less physical abuse at the hands of her husband. Today her business continues to grow and provide for the basic needs of her family.

Unlike Amena, the majority of people in Asia, Africa and Latin America have few opportunities to escape from poverty. According to the World Bank, more than 1.3 billion people live on less than a dollar a day. Poverty has not been eradicated in the 50 years since the Universal Declaration on Human Rights asserted that each individual has a right to:

> A standard of living adequate for the health and well-being of himself and of his family, including food, clothing, housing and medical care and necessary social services, and the right to security in the event of unemployment, sickness, disability, widowhood, old age or other lack of livelihood in circumstances beyond his control.

Will poverty still be with us 50 years from now? My own experience suggests that it need not.

After completing my Ph.D. at Vanderbilt University, I returned to Bangladesh in 1972 to teach economics at Chittagong University. I was excited about the possibilities for my newly independent country. But in 1974 we were hit with a terrible famine. Faced with death and starvation outside my classroom, I began to question the very economic theories I was teaching. I started feeling there was a great distance between the actual life of poor and hungry people and the abstract world of economic theory.

I wanted to learn the real economics of the poor. Because Chittagong University is located in a rural area, it was easy for me to visit impoverished households in the neighboring village of Jobra. Over the course of many visits, I learned all about the lives of my struggling neighbors and much about economics that is never taught in the classroom. I was dismayed to see how the indigent in Jobra suffered because they could not come up with small amounts of working capital. Frequently they needed less than a dollar a person but could get that money only on extremely unfair terms. In most cases, people were required to sell their goods to moneylenders at prices fixed by the latter.

This daily tragedy moved me to action. With the help of my graduate students, I made a list of those who needed small amounts of money. We came up with 42 people. The total amount they needed was $27.

I was shocked. It was nothing for us to talk about millions of dollars in the classroom, but we were ignoring the minuscule capital needs of 42 hardworking, skilled people next door. From my own pocket, I lent $27 to those on my list.

Still, there were many others who could benefit from access to credit. I decided to approach the university's bank and try to persuade it to lend to the local poor. The branch manager said, however, that the bank could not give loans to the needy: the villagers, he argued, were not creditworthy.

I could not convince him otherwise. I met with higher officials in the banking hierarchy with similar results. Finally, I offered myself as a guarantor to get the loans.

In 1976 I took a loan from the local bank and distributed the money to poverty-stricken individuals in Jobra. Without exception, the villagers paid back their loans. Confronted with this evidence, the bank still refused to grant them loans directly. And so I tried my experiment in another village, and again it was successful. I kept expanding my work, from two to five, to 20, to 50, to 100 villages, all to convince the bankers that they should be lending to the poor. Although each time we expanded to a new village the loans were repaid, the bankers still would not change their view of those who had no collateral.

Because I could not change the banks, I decided to create a separate bank for the impoverished. After a great deal of work and negotiation with the government, the Grameen Bank ("village bank" in Bengali) was established in 1983.

From the outset, Grameen was built on principles that ran counter to the conventional wisdom of banking. We sought out the very poorest borrowers, and we required no collateral. The bank rests on the strength of its borrowers. They are required to join the bank in self-formed groups of five. The group members provide one another with peer support in the form of mutual assistance and advice. In addition, they allow for peer discipline by evaluating business viability and ensuring repayment. If one member fails to repay a loan, all members risk having their line of credit suspended or reduced.

The Power of Peers

Typically a new group submits loan proposals from two members, each requiring between $25 and $100. After these two borrowers successfully repay their first five weekly installments, the next two group members become eligible to apply for their own loans. Once they make five repayments, the final member of the group may apply. After 50 installments have been repaid, a borrower pays her interest, which is slightly above the commercial rate. The borrower is now eligible to apply for a larger loan.

The bank does not wait for borrowers to come to the bank; it brings the bank to the people. Loan payments are made in weekly meetings consisting of six to eight groups, held in the villages where the members live. Grameen staff attend these meetings and often visit individual borrowers' homes to see how the business—whether it be raising goats or growing vegetables or hawking utensils—is faring.

Today Grameen is established in nearly 39,000 villages in Bangladesh. It lends to approximately 2.4 million borrowers, 94 percent of whom are women. Grameen reached its first $1 billion in cumulative loans in March 1995, 18 years after it began in Jobra. It took only two more years to reach the $2-billion mark. After 20 years of work, Grameen's average loan size now stands at $180. The repayment rate hovers between 96 and 100 percent.

A year after joining the bank, a borrower becomes eligible to buy shares in Grameen. At present, 94 percent of the bank is owned by its borrowers. Of the 13 members of the board of directors, nine are elected from among the borrowers; the rest are government representatives, academics, myself and others.

A study carried out by Sydney R. Schuler of John Snow, Inc., a private research group, and her colleagues concluded that a Grameen loan empowers a woman by increasing her economic security and status within the family. In 1998 a study by Shahidur R. Khandker an economist with the World Bank, and others noted that participation in Grameen also has a significant positive effect on the schooling and nutrition of children—as long as women rather than men receive the loans. (Such a tendency was clear from the early days of the bank and is one reason Grameen lends primarily to women: all too often men spend the money on themselves.) In particular, a 10 percent increase in borrowing by women resulted in the arm circumference of girls—a common measure of nutritional status—expanding by 6 percent. And for every 10 percent increase in borrowing by a member the likelihood of her daughter being enrolled in school increased by almost 20 percent.

Not all the benefits derive directly from credit. When joining the bank, each member is required to memorize a list of 16 resolutions. These include commonsense items about hygiene and health—drinking clean water, growing and eating vegetables, digging and using a pit latrine, and so on—as well as social dictums such as refusing dowry and managing family size. The women usually recite the entire list at the weekly branch meetings, but the resolutions are not otherwise enforced.

Even so, Schuler's study revealed that women use contraception more consistently after joining the bank. Curiously, it appears that women who live in villages where Grameen operates, but who are not themselves members, are also more likely to adopt contraception. The population growth rate in Bangladesh has fallen dramatically in the past two decades, and it is possible that Grameen's influence has accelerated the trend.

In a typical year 5 percent of Grameen borrowers—representing 125,000 families—rise above the poverty level. Khandker concluded that among these borrowers extreme poverty (defined by consumption of less than 80 percent of the minimum requirement stipulated by the Food and Agriculture Organization of the United Nations) declined by more than 70 percent within five years of their joining the bank.

To be sure, making a microcredit program work well—so that it meets its social goals and also stays economi-

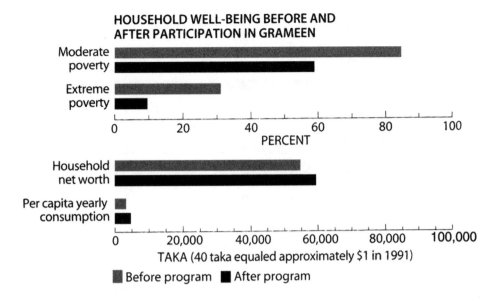

HOUSEHOLD WELL-BEING BEFORE AND
AFTER PARTICIPATION IN GRAMEEN

■ Before program   ■ After program

cally sound—is not easy. We try to ensure that the bank serves the poorest: only those living at less than half the poverty line are eligible for loans. Mixing poor participants with those who are better off would lead to the latter dominating the groups. In practice, however, it can be hard to include the most abjectly poor, who might be excluded by their peers when the borrowing groups are being formed. And despite our best efforts, it does sometimes happen that the money lent to a woman is appropriated by her husband.

Given its size and spread, the Grameen Bank has had to evolve ways to monitor the performance of its branch managers and to guarantee honesty and transparency. A manager is not allowed to remain in the same village for long, for fear that he may develop local connections that impede his performance. Moreover, a manager is never posted near his home. Because of such constraints—and because managers are required to have university degrees—very few of them are women. As a result, Grameen has been accused of adhering to a paternalistic pattern. We are sensitive to this argument and are trying to change the situation by finding new ways to recruit women.

Grameen has also often been criticized for being not a charity but a profit-making institution. Yet that status, I am convinced, is essential to its viability. Last year a disastrous flood washed away the homes, cattle and most other belongings of hundreds of thousands of Grameen borrowers. We did not forgive the loans, although we did issue new ones, and give borrowers more time to repay. Writing off loans would banish accountability, a key factor in the bank's success.

Liberating Their Potential

The Grameen model has now been applied in 40 countries. The first replication, begun in Malaysia in 1986, currently serves 40,000 poor families; their repayment rate has consistently stayed near 100 percent. In Bolivia, microcredit has allowed women to make the transition from "food for work" programs to managing their own businesses. Within two years the majority of women in the program acquire enough credit history and financial skills to qualify for loans from mainstream banks. Similar success stories are coming in from programs in poor countries everywhere. These banks all target the most impoverished, lend to groups and usually lend primarily to women.

The Grameen Bank in Bangladesh has been economically self-sufficient since 1995. Similar institutions in other countries are slowly making their way toward self-reliance. A few small programs are also running in the U.S., such as in innercity Chicago. Unfortunately, because labor costs are much higher in the U.S. than in developing countries—which often have a large pool of educated unemployed who can serve as managers or accountants—the operations are more expensive there. As a result, the U.S. programs have had to be heavily subsidized.

In all, about 22 million poor people around the world now have access to small loans. Microcredit Summit, an institution based in Washington, D.C., serves as a resource center for the various regional microcredit institutions and organizes yearly conferences. Last year the attendees pledged to provide 100 million of the world's poorest families, especially their women, with credit by the year 2005. The campaign has grown to include more than 2,000 organizations, ranging from banks to religious institutions to nongovernmental organizations to United Nations agencies.

The standard scenario for economic development in a poor country calls for industrialization via investment. In this "topdown" view, creating opportunities for employment is the only way to end poverty. But for much of the developing world, increased employment exacerbates migration from the countryside to the cities and creates low-paying jobs in miserable conditions. I firmly believe that, instead, the eradication of poverty starts with people

IMPACT OF GRAMEEN ON NUTRITIONAL MEASURES OF CHILDREN

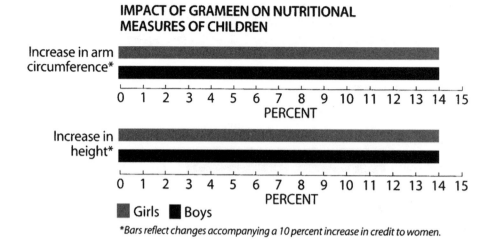

Increase in arm circumference*

PERCENT
0 1 2 3 4 5 6 7 8 9 10 11 12 13 14 15

Increase in height*

PERCENT
0 1 2 3 4 5 6 7 8 9 10 11 12 13 14 15

■ Girls ■ Boys

*Bars reflect changes accompanying a 10 percent increase in credit to women.

being able to control their own fates. It is not by creating jobs that we will save the poor but rather by providing them with the opportunity to realize their potential. Time and time again I have seen that the poor are poor not because they are lazy or untrained or illiterate but because they cannot keep the genuine returns on their labor.

Self-employment may be the only solution for such people, whom our economies refuse to hire and our taxpayers will not support. Microcredit views each person as a potential entrepreneur and turns on the tiny economic engines of a rejected portion of society. Once a large number of these engines start working, the stage can be set for enormous socioeconomic change.

Applying this philosophy, Grameen has established more than a dozen enterprises, often in partnership with other entrepreneurs. By assisting microborrowers and microsavers to take ownership of large enterprises and even infrastructure companies, we are trying to speed the process of overcoming poverty. Grameen Phone, for instance, is a cellular telephone company that aims to serve urban and rural Bangladesh. After a pilot study in 65 villages, Grameen Phone has taken a loan to extend its activities to all villages in which the bank is active. Some 50,000 women, many of whom have never seen a telephone or even an electric light, will become the providers of telephone service in their villages. Ultimately, they will become the owners of the company itself by buying its shares. Our latest innovation, Grameen Investments, allows U.S. individuals to support companies such as Grameen Phone while receiving interest on their investment. This is a significant step toward putting commercial funds to work to end poverty.

I believe it is the responsibility of any civilized society to ensure human dignity to all members and to offer each in-

dividual the best opportunity to reveal his or her creativity. Let us remember that poverty is not created by the poor but by the institutions and policies that we, the better off, have established. We can solve the problem not by means of the old concepts but by adopting radically new ones.

The Author

MUHAMMAD YUNUS, the founder and managing director of the Grameen Bank, was born in Bangladesh. He obtained a Ph.D. in economics from Vanderbilt University in 1970 and soon after returned to his home country to teach at Chittagong University. In 1976 he started the Grameen project, to which he has devoted all his time for the past decade. He has served on many advisory committees: for the government of Bangladesh, the United Nations, and other bodies concerned with poverty, women and health. He has received the World Food Prize, the Ramon Magsaysay Award, the Humanitarian Award, the Man for Peace Award and numerous other distinctions as well as six honorary degrees.

Further Reading

GRAMEEN BANK: PERFORMANCE AND SUSTAINABILITY. Shahidur R. Khandker, Baqui Khalily and Zahed Khan. World Bank Discussion Papers, No. 306. ISBN 0-8213-3463-8. World Bank, 1995.

GIVE US CREDIT. Alex Counts. Times Books (Random House), 1996.

FIGHTING POVERTY WITH MICROCREDIT: EXPERIENCE IN BANGLADESH. Shahidur R. Khandker. Oxford University Press, 1998.

Grameen Bank site is available at www.grameenfoundation.org on the World Wide Web.

From *Scientific American*, November 1999, pp. 114-119. © 1999 by Dr. Muhammad Yunus. Reprinted by permission.

America's Looming Creativity Crisis

The United States built the world's most powerful economy by producing and attracting human capital. Is America throwing that advantage away?

By Richard Florida

THE UNITED STATES OF AMERICA—for generations known around the world as the land of opportunity and innovation—is on the verge of losing its competitive edge. It is facing perhaps its greatest economic challenge since the dawn of the industrial revolution. This challenge has little to do with business costs and even less with manufacturing prowess. And, no, the main competitive threats are not from China or India.

Even though the United States led the world into the era of high-tech industry and constant innovation, it is by no means the nation's manifest destiny to stay on top. In fact, the great majority of U.S. business and political leaders, academics, and economic analysts fail to grasp the true reason behind American success in innovation, economic growth, and prosperity. It is not the country's generous endowment of natural resources, the size of its market, or some indigenous Yankee ingenuity that has powered its global competitiveness for more than a century. America's growth miracle turns on one key factor: its openness to new ideas, which has allowed it to mobilize and harness the creative energies of its people.

As Stanford University economist Paul Romer has long argued, great advances have always come from ideas. Ideas do not fall from the sky; they come from people. People write the software. People design the products. People start the new businesses. Every new thing that gives us pleasure or productivity or convenience, be it an iPod or the tweaks that make a chemical plant more efficient, is the result of human ingenuity.

True, the United States is still the world's center of ingenuity. Its GDP tops $10 trillion, and it is home to great universities, Silicon Valley, and many of the most dynamic companies in information technology, biotech, entertainment, and countless other fields. But the global talent pool and the high-end, high-margin creative industries that used to be the sole province of the U.S., and a crucial source of its prosperity, have begun to disperse around the globe. A host of countries—Ireland, Finland, Canada, Australia, New Zealand, among them-are investing in higher education, cultivating creative people, and churning out stellar products, from Nokia phones to the *Lord of the Rings* movies. Many of these countries have learned from past U.S. success and are shoring up efforts to attract foreign talent—including Americans. If even a handful of these rising nations draws away just 2% to 5% of the creative workers from the U.S., the effect on its economy will be enormous. The United States may well have been the Goliath of the twentieth century global economy, but it will take just half a dozen twenty-first-century Davids to begin to wear it down.

To stay innovative, America must continue to attract the world's sharpest minds. And to do that, it needs to invest in the further development of its creative sector. Because wherever creativity goes—and, by extension, wherever talent goes—innovation and economic growth are sure to follow.

The Dawn of the Creative Age

There's a whole new class of workers in the U.S. that's 38 million strong: the creative class. At its core are the scientists, engineers, architects, designers, educators, artists, musicians, and entertainers, whose economic function is to create new ideas, new technology, or new content. Also included are the creative professions of business and finance, law, health care, and related fields, in which knowledge workers engage in complex problem solving that involves a great deal of independent judgment. Today, the creative sector of the U.S. economy, broadly defined, employs more than 30% of the workforce (more than all of manufacturing) and accounts for nearly half of all wage and salary income (some $2 trillion)—almost as much as the manufacturing and service sectors together. Indeed, the United States has now entered what I call the Creative Age.

The roots of the Creative Age in the U.S. can be traced to the years surrounding World War II. After the war, federal funding for basic research jumped considerably, and so did the number of people pursuing higher education, thanks in part to the GI Bill. In the private sector, the newly formed venture capital industry provided an avenue for bringing research ideas to market. The social movements of the 1960s popularized the idea of openness; to be different was no longer to be an outcast but to be admired. Freedom of expression allowed new technologies and cultural forms to flourish—from biotechnology to alternative rock.

But the United States doesn't have some intrinsic advantage in the cultivation of creative people, innovative ideas, or new companies. Rather, its real advantage lies in its ability to attract these economic drivers from around the world. Of critical importance to American success in this last century has been a tremendous influx of talented immigrants. Immigrants have, of course, helped power American growth since the dawn of the Republic. But since the 1930s, the U.S. has welcomed a stream of scientific, intellectual, cultural, and entrepreneurial talent, as Europeans fled fascism and communism. This talent has helped make the U.S. university system and innovative infrastructure second to none.

The stream surged to historic levels in the 1980s and 1990s, thanks to more liberal immigration policies and a booming economy. In the 1990s alone, U.S. census figures reveal, more than 11 million people came to America. The largest wave of immigration in U.S. history, it brought with it talent from all corners of the globe. Think of high-tech luminaries Sergey Brin, the Moscow-born cofounder of Google, and Hotmail cofounder Sabeer Bhatia, who grew up in Bangalore. The foreign-born population of the United States currently numbers more than 30 million, or some 11% of the population.

The Creativity-Competitiveness Connection

But already the percentage of the population represented by immigrants is higher in Canada (18%) and Australia (22%) than in the United States. These countries understand that today's global economy centers on competition for people rather than for goods and services. As Pete Hodgson, New Zealand's minister for research, science, and technology, recently explained to me, "We no longer think of immigration as a gatekeeping function but as a talent-attraction function necessary for economic growth"

A close look at international statistics shows that the creative class represents a larger percentage of the workforce in many other countries than it does in the United States. Along with Irene Tinagli, a doctoral student at Carnegie Mellon, I set out to compare the size of the creative class in different countries by establishing the "Global Creative-Class Index" (GCCI). Using employment data and the job classifications established by the International Labour Organization (ILO), the index is a straightforward calculation of the number of people employed in creative job categories in each country divided by the country's total number of workers. In the exhibit, "The Global Creative-Class Index" we compare the percentage of workers in the creative classes in 25 nations.

The United States may well have been the Goliath of the twentieth-century global economy, but it will take just half a dozen twenty-first-century Davids to begin to wear it down.

Far from being the leader, the United States is not even in the top ten. The creative class constitutes around a third of the workforce in Ireland, Belgium, Australia, and the Netherlands; it accounts for roughly a quarter of the workforce in six other countries: New Zealand, Estonia, the United Kingdom, Canada, Finland, and Iceland. When our U.S. data are adjusted to be comparable to the ILO figures (which use a narrow definition of creative job categories that excludes "technicians"), the United States comes in, with 23.6%, at 11th, worldwide. Of course because the overall workforce in America is so large, that translates into a sizable group in absolute numbers— some 30 million people.

Still, if technicians *are* included in the international analysis, the creative class rises to more than 40% in some eight countries: the Netherlands (47%), Sweden (42.4%), Switzerland (42%), Denmark (42%), Norway (41.6%), Belgium (41.4%), Finland (41%), and Germany (40%). It constitutes more than 30% of the workforce in virtually all the remaining countries. What's more, the growth rate of the creative class in seven nations has been phenomenal over the past decade or so. Since 1991, for instance, New Zealand's creative class has jumped from 18.7% to 27.1%, and Ireland's has nearly doubled, starting from the same 18.7% and rising to 33.5%.

In today's economy, creativity and competitiveness go hand in hand. It's not surprising, then, that our GCCI rankings correlate closely with results from other studies of international competitiveness. Harvard Business School's Michael Porter, for instance, ranked the United States as the world's most competitive nation in his initial 1995 global Innovation Index. According to Porter's projections, by 2005, the U.S. will have tumbled to sixth among the 17 member countries of the Organisation for Economic Co-operation and Development (OECD)— trailing (in order) Japan, Finland, Switzerland, Denmark, and Sweden. The 2004 Globalization Index developed by A.T. Kearney and published in *Foreign Policy* ranks the United States seventh, behind Ireland, Singapore, Switzerland, the Netherlands, Finland, and Canada.

Rankings of individual companies' competitiveness yields similar results. According to *BusinessWeek*'s 2004 Information Technology 100, for instance, only six of the world's 25 most competitive high-tech companies are based in the United States, while 14 are in Asia.

In the area of patents and publications, America's formidable lead has been eroding, as well. Today, foreign owned companies and foreign-born inventors account for nearly half of all patents issued in the United States. A study by CHI Research found that inventors in Japan, Taiwan, and South Korea alone account for more than a quarter of all U.S. industrial patents awarded each year. In terms of publications, the National Science Board reports that back in 1988, U.S. scientists produced 178,000 scientific papers, or 38% of all scientific and engineering papers worldwide. But by 2001, the European Union nations were the largest producers of scientific literature. In the field of physics, the U.S. lead fell from 61% of all publications in 1983 to 29% in 2003, according to *Physical Review*.

Taken individually, none of those facts would be cause for concern about the future of the United States. It is, after all, a very rich country with diverse strengths. Cumulatively, though, the data create an unsettling picture of a nation that's allowing its creativity infrastructure to decay. Add to that greater security concerns and a highly politicized scientific climate, and it's easy to see why the nation is becoming less and less attractive to the world's brightest minds.

The Talent Gap

Today, virtually the entire public dialogue about jobs in the United States revolves around outsourcing and unemployment. But these are the short-term issues. The real long-term predicament facing the United States and the world is the looming shortage of creative talent.

Economists like Lawrence Summers, president of Harvard University and a former Treasury secretary, and Edward Montgomery, a former Labor Department deputy secretary, view the shortage of skilled and talented workers as all but inevitable. A 2003 National Association of Manufacturers report concurs, predicting that a skilled-worker gap will start to form in 2005, widening to 5.3 million workers by 2010 and 14 million by 2020. The labor shortages that plagued high-tech companies in the halcyon days of 1999 and 2000 will look like a "minor irritation" in comparison, contends labor market expert Anthony Carnevale, the report's author.

The cause of this labor squeeze is easy to see: Baby boomers now constitute about 60% of the prime-age workforce—that is, workers between the ages of 25 and 54. In the coming decades, boomers will retire in massive numbers, and there simply aren't enough younger workers to take their places. The talent shortage will hit every sector of the U.S. economy, but it will be felt most acutely at the cutting edges of science and engineering. Since 1980, the number of jobs in those segments has grown four times faster than the overall employment rate, and the Bureau of Labor Statistics expects that number to swell by nearly 50% again by 2010—adding a further 2.2

The Global Creative-Class Index

America may be the land of opportunity, but it no longer has a lock on the best and the brightest jobs—the ones that create new ideas, new technology, or new content. When we calculated the number of people engaged in such jobs as a proportion of the general workforce in scores of countries, the United States wasn't even in the top ten.

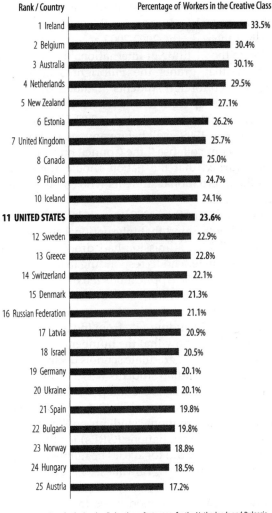

Rank / Country	Percentage of Workers in the Creative Class
1 Ireland	33.5%
2 Belgium	30.4%
3 Australia	30.1%
4 Netherlands	29.5%
5 New Zealand	27.1%
6 Estonia	26.2%
7 United Kingdom	25.7%
8 Canada	25.0%
9 Finland	24.7%
10 Iceland	24.1%
11 UNITED STATES	**23.6%**
12 Sweden	22.9%
13 Greece	22.8%
14 Switzerland	22.1%
15 Denmark	21.3%
16 Russian Federation	21.1%
17 Latvia	20.9%
18 Israel	20.5%
19 Germany	20.1%
20 Ukraine	20.1%
21 Spain	19.8%
22 Bulgaria	19.8%
23 Norway	18.8%
24 Hungary	18.5%
25 Austria	17.2%

Note: Data for the Russian Federation refer to 1999; for the Netherlands and Bulgaria, 2001; for the United States, 2003. All other figures refer to 2002, the latest year for which those data are available.

Source: Compiled by Irene Tinagli from International Labour Organization and U.S. Bureau of Labor Statistics data.

million new jobs. At the same time, the average age of the scientific and technological worker is rising. More than half are 40 or older, and many will leave the workforce in the next two decades.

You don't have to be a rocket scientist to figure out that there is only one way for the United States to fill this gap: foreign talent. Former director of the U.S. Census and Columbia University professor Kenneth Prewitt says that the United States will increasingly depend on these "replacement people" to provide vital skills and grow new industries. But that may not be as easy as it once was.

The Canaries of the Talent Mine

Students are a leading indicator of global talent flows. The countries and regions that attract them have a leg up on retaining them and also on attracting other pools of foreign talent—scientists, researchers, inventors, entrepreneurs.

For decades, international students have flocked to the United States to take advantage of the world-class education offered there. In the 2002-2003 academic year alone, according to the Institute of International Education (IIE)—the body that grants the Fulbright scholarships—roughly 585,000 foreign students attended U.S. colleges and universities, up from less than 50,000 in 1960, and international education contributed $12.9 billion to the U.S. economy. But in 1999, well before anyone had heard the phrase "dot-com collapse," the Council on Competitiveness had warned that the nation should not count on keeping the international students who come to study at elite universities.

More recently, a March 2004 report by the Council of Graduate Schools found that international student applications for fall 2004 admission had dropped sharply at 90% of the graduate schools responding to its survey. The total decline was 32%. Applications fell off most from the countries that have traditionally sent the most students: More than half of all foreign-born graduate students hailed from Asia, including 14% from India and 10% from China. The figures show that the number of Chinese students applying to U.S. graduate schools declined by 76%, and the number of Indian students was 58% lower than it was the previous year. Signs don't point to a turnaround anytime soon. The Educational Testing Service found that one-third fewer international students applied to take the Graduate Record Examinations (GREs) for the 2004 academic year than they did for 2003. The number of Chinese test takers was down 50%; Taiwanese, 43%; Indians, 37%; and Koreans, 15%.

One reason for this is good news from a global perspective. Several major economies—most notably India's and China's—have grown to the point where they can offer great opportunities for people who stay or return home. Both of those countries are investing heavily to build excellent university systems of their own. Peter Drucker said recently that India may already have the greatest engineering and medical schools in the world.

Foreign students are not only finding attractive educational opportunities in other countries, they are also facing obstacles to studying in the United States. A survey of educators at 276 U.S. campuses conducted by the IIE found a significant drop in enrollment to U.S. universities in fall 2003 from students whose home countries have large Islamic populations, especially United Arab Emirates, Saudi Arabia, and Pakistan. Fifty-nine percent of respondents cited the visa application process as a reason for the decline.

The *New York Times* reports that the rejection rate for "cultural exchange" visas, used by many medical students, rose from 5.1% in fiscal year 2001 to 7.8% in fiscal 2003. And the number of students whose visas were rejected rose from 27.6% in FY2001 to 35.2% in FY2003, according to the National Science Board's *Science & Engineering Indicators—2004*.

Having taught at several major universities—Ohio State, Harvard, MIT, and Carnegie Mellon—I've known many foreign students. They have always been quick to point out the benefits of studying and conducting research in the United States. But their impressions have changed dramatically over the past year. They complain of being hounded by immigration agencies as potential threats to security, and they feel that the war on terror is leading America to abandon its commitment to an open society. Many have told me they are thinking of leaving the U.S. for graduate education and professional positions in other nations. They also report that growing numbers of their friends and colleagues back home are no longer interested in coming to America for their education.

Over time, terrorism is less a threat to the U.S. than the possibility that creative and talented people will stop wanting to live within its borders.

James Langer, vice president of the National Academy of Sciences, spoke plainly about what the drop in foreign students could mean. At a May 2004 luncheon for the United States Senate Science and Technology Caucus, he commented: "Applications to many leading U.S. graduate schools from students in China, India, Russia, and elsewhere are already down by 30% or more, and there is evidence that these students are going elsewhere for advanced degrees. International scientific organizations, such as the International Union of Pure and Applied Physics, are refusing to hold conferences here." As one oceanographer from the University of California at San Diego recently quipped, it may be time for academics in that part of the country to "have our scientific meetings in Tijuana," because at least there international experts can get in. In short, as Langer concluded, "American science is being isolated from the rest of the world."

Sadly, restricting foreign immigration will not open up more places for homegrown talent in the top American graduate programs and research facilities. The U.S. has many brilliant young people but not nearly enough to satisfy the demand the nation's powerhouse economy has created.

Other countries are taking full advantage of America's fading allure. English-speaking Canada, the United Kingdom, and Australia are particularly well placed to capitalize on this opportunity. In June 2003, an eminent Oxford professor told me that the university had "never seen so many applications from top international students" adding that these students seem to be "looking for alternatives to top American universities" like Harvard, Chicago, MIT, and Stanford. In fact, together, the United Kingdom, Germany, France, Australia, and Japan attracted 650,000 foreign students—some 11% more than the United States—according to the *2003 Atlas of Student Mobility*, compiled by the IIE. And the stakes are growing. In 2000, UNESCO estimates, 1.7 million students worldwide were educated abroad; by 2025, it expects that number will swell to more than 8 million. The countries that attract these students will have a huge advantage in the coming war for global talent.

The Reverse Brain Drain

For the first time in its history, then, the United States is confronting the possibility of a reverse brain drain. And students are just the tip of the iceberg. The evidence suggests that the country may be losing out on the talents of a host of foreign scientists, engineers, inventors, and other professionals. Visa delays have cost U.S. businesses roughly $30 billion in two years, according to a June 2004 study commissioned by the Santangelo Group. The group is a consortium of leading U.S. industry organizations ranging from the Aerospace Industries Association to the National Foreign Trade Council to the Association for Manufacturing Technology, and its study was based on a survey of 734 of its member companies. Of the 141 companies that responded, 73% reported having had problems processing business visas since 2002, and the average financial impact per company was nearly a million dollars ($925,816). Thirty-eight percent of respondents said that visa delays caused projects to be postponed, 42% said the delays prevented them from bringing foreign employees to the United States, and 20% said training events had to be relocated outside the country.

The direct-sales giant Amway, for instance, chose to hold a convention for its 8,000 South Korean distributors in Japan this year rather than in Los Angeles or Hawaii, the *Washington Post* recently reported, because the United States would require each visitor to go through an individual interview with a consular official. Amway estimated that the attendees would have spent, on average, $1,250-translating into a $10 million loss for the potential host city.

According to a recent *New York Times* article, 6.3 million people applied for U.S. visas between October 2000 and September 2001. But in fiscal 2003, that number dropped more than 40% to 3.7 million. And fewer of those who are applying are getting in. The rejection rate for H-1B visas (also called "high-skilled visas"), which allow professionals who are not U.S. citizens to work in the country for up to six years, increased from 9.5% to 17.8% between 2001 and 2003. Almost every major American industry from high-tech to entertainment is feeling the repercussions of these decisions. A number of prominent international music groups, such as Cuba's Sierra Maestra, have canceled American tours because they were re-

fused visas. (Sierra Maestra was denied a visa when the FBI failed to complete background checks fast enough to meet INS deadlines.) These cancellations in and of themselves won't have a big impact on the U.S. economy, but think of the influence on American artists, let alone on the multibillion-dollar music business. Choking musicians and businesspeople off from those on the frontiers of this ever-evolving (and increasingly global) industry will eventually yield the same result as prohibiting scientists from carrying out potentially rewarding research. It will dull their competitive edge.

Foreign professionals already working in U.S. firms aren't having an easy go of it either. Processing times for renewing green cards and travel documents have reached glacial proportions. As the *Times* also reports, it now takes an average of 19 months to replace a lost green card. It takes seven months for legal workers in the U.S. whose green cards are pending to get travel papers- and during that period, the applicants cannot leave the country or they risk not being able to reenter. The same article claims that the number of pending green card applications has jumped by nearly 60% since 2001 because 1,000 agents who once issued documents have been reassigned to do "extensive security checks of every applicant instead"

There's no denying how important foreign-born workers are to the U.S. economy. AnnaLee Saxenian, dean of the School of Information Management and Systems at the University of California, Berkeley, conducted extensive research on immigrant-run companies in Silicon Valley. She and her team pored over census data on immigrants' education, occupations, and earnings, and they used a Dun & Bradstreet database to distill immigrant-run companies from the nearly 12,000 start-ups launched between 1980 and 1998. They found that Chinese and Indian engineers were running nearly 30% of the area's high-tech companies in the 1990s—up from 13% in the early 1980s. Saxenian estimated that in 2000, these firms collectively accounted for nearly $20 billion in sales and more than 70,000 jobs. And because Saxenian's database identified only those companies that are currently headed by a Chinese or Indian chief executive, she suspects her figures are conservative.

Trends are eye-opening, but individual cases are perhaps even more important. What if, for example, Vinod Khosla, the cofounder of Sun Microsystems and venture capital luminary who has backed so many blockbuster companies, had stayed in India? Or if An Wang, founder of Wang Laboratories, had gone to university in Europe? These are people whose creative genius has affected the trajectory of entire industries; their breakthroughs and business acumen have helped set in motion what the economist Joseph Schumpeter liked to call the "gales of creative destruction" that create new companies and industries and completely remake existing ones.

This circle-the-wagons mentality is even causing some leading American scientists and engineers to leave the country. If the status quo remains, then more people may react like Roger Pedersen, a stem cell researcher, who left the University of California, San Francisco, for Cambridge University. "I have a soft spot in my heart for America, but the UK is much better for this research. More working capital." Pedersen told Wired, "They haven't made such a political football out of stem cells." These tendencies illustrate on a small scale how the creative economy is being reshaped—both by global competitors' increasing savvy and by America's shortsightedness.

Rebuilding the Creative Infrastructure

What should the United States do? First, it must recognize that the issue is nonpartisan. Republicans, Democrats, independents—everyone has a stake in keeping the country open to foreign talent. The challenges the nation must overcome are too massive for the debate about them to become overshadowed by polarizing political bickering, culture wars, or short-term economic agendas. The United States must consider its next steps carefully and deliberately. I recommend focusing on three main areas.

Calculate the true cost of security. The United States is impeding its own progress when it makes scientific discovery pass religious tests or when it tightens visa restrictions unnecessarily. To be certain, America after September 11 does face real and vital threats to its security, and they are not going to disappear anytime soon. The departments of Defense and Homeland Security, the FBI, the Coast Guard, and the intelligence agencies naturally think in terms of security first. That is their job. But it is important for both business and political leadership to recognize the economic costs of overzealousness and to weigh carefully the serious trade-offs between current security and long-run competitiveness.

The U.S. needs to upgrade the huge number of service jobs its economy is generating. These are the port-of-entry jobs to the creative economy of today.

People around the world applaud America's efforts to improve its own security. But what the world does not like is the arbitrary and sometimes brash methods the country has adopted in its own defense. Over time, terrorism is less a threat to the United States than the possibility that creative and talented people will stop wanting to live within its borders. The nation must act in concrete ways to reassure people—both Americans and global citizens—that it values openness, diversity, and tolerance. To that end, it must focus on improving the visa process immediately.

If the government is unable or unwilling to take the lead in balancing one type of security with another, then the business and academic communities need to push for a renewed American openness. In the 1980s, Hewlett-Packard chief Jack Young spurred his colleagues to form

the U.S. Council on Competitiveness, which did much to bring the country's lagging industrial competitiveness to public attention. The private sector can similarly take the lead now by establishing a Global Creativity Commission—a coalition of world political and business leaders committed to developing strategies to ensure that global talent can move efficiently across borders.

Invest generously in research and education. Corporate R&D funding dropped by nearly $8 billion in 2002—the largest single-year decline since the 1950s, the National Science Foundation reports. And right now, the federal government is cutting key areas of defense R&D spending. Many state governments have slashed higher education funding for arts and culture while pumping millions into stadiums, convention centers, and other bricks-and-mortar projects. Never mind that the local economic benefits of such projects often dry up the minute the last construction worker drives off the site. These choices signal a profound failure to understand what's required to maintain an atmosphere of innovation.

The United States must invest generously in its creative infrastructure. Education reform must, at its core, make schools into places that cultivate creativity. Americans revel in the legendary stories of young creators like Michael Dell building new businesses in dorm rooms or in garages in their spare time. The question is: Why are they doing these things in their *spare time*? Isn't this the real stuff of education in the Creative Age?

What's needed is the equivalent of a GI Bill for creativity. The nation must spend radically more on research and development and on higher education, opening up universities and colleges to more Americans and to more of the world's best and brightest. In the same way it built the canals, railroads, and highways to power industrial growth, the United States has to build the creative infrastructure for the future.

Here again, business and academia may need to take the lead, at least in the short run. In response to the recent restrictions on federal funding for stem cell research, Lawrence Summers announced plans earlier this year to launch a multimillion-dollar Harvard Stem Cell Institute. Says George Q. Daley, an associate professor at Harvard Medical School and Children's Hospital, "Harvard has the resources, Harvard has the breadth, and, frankly, Harvard has the responsibility to take up the slack that the government is leaving."

Tap into more people's creative capabilities. If the creative class in America accounts for less than a third of the workforce, then, of course, the vast majority is not part of it. Nearly 45% of the U.S. workforce falls into the service class, for instance—janitors, low-end health care workers, office clerks, food service workers, and the like. Members of this class earn, on average, less than half of what creative-class members do—around $22,000 a year versus more than $50,000.

Employing so many citizens in noncreative ways is a terrible waste of talent and potential. So far, the U.S. has gotten away with it because few other societies do much better. But remember what happened in the 1970s and 1980s, when Japanese auto companies leaped to global prominence with manufacturing methods that tapped the intelligence of every worker on the factory floor to make continuous improvements in quality and productivity. U.S. manufacturers—stuck in the old Taylorist model, in which the engineers made the decisions and the laborers simply carried out the rote work—nearly had their doors blown off. If other nations develop better ways to harness their societies' creativity, the U.S. economy might be blown away on an inconceivable scale.

The United States needs to substantially upgrade the pay, working conditions, and status of the huge number of service jobs its economy is generating. These are the port-of-entry jobs to the creative economy of today. During the Great Depression and the New Deal, the nation succeeded in turning a large number of formerly low-skill, low-pay, blue-collar jobs into the kind of occupations that could support families and become the launch pad for upward mobility. And many of the equivalent jobs today—hairdressing, massage therapy, and aestheticians, to name only a few are virtually impervious to outsourcing.

Addressing the needs of the American creative class will be important, but it won't be enough. To prevent widespread social unrest and to benefit economically from the creative input of the maximum number of its citizens, the United States will have to find ways to bring the service and manufacturing sectors more fully into the Creative Age.

The Future of Global Creativity

Maybe I'm an eternal optimist, but I think the United States can continue to be a beacon of openness for the creative class—and, indeed, for the whole of humanity. It has a long history of resourcefulness and creativity to draw on, and it has transformed itself many times before, rebuilding after the Great Depression and bouncing back after the Asian manufacturing boom of the 1980s.

Unfortunately, America's eroding access to high-level foreign talent hasn't drawn much attention from political leaders or from the media. They have seemingly bigger and more immediate problems—from the war on terrorism to the loss of manufacturing jobs to China, India, and Mexico. But the nation is overlooking the biggest threat to its economic well-being—just as it did when its obsession with the Soviet Union in the last years of the Cold War caused it to miss the economic challenge of Japan.

The role of the United States in generating creativity and human capital is a concern not only for U.S. businesses and policy makers but for all nations. American universities and corporations have long been the educators and innovators for the world. If this engine stalls—or if political decisions about immigration, visas, and scien-

Creative Regions

Competition for talent occurs not only between nations but also between cities and regions, just as competition in many industries occurs at the business-unit, rather than the company, level. New York, for instance, is pitted against London and Hong Kong; San Francisco is up against the likes of Dublin, Vancouver, Stockholm, and Sydney. While comprehensive regional data do not exist, several studies do give a detailed picture of areas inside Canada and Australia.

According to data amassed by Kevin Stolarick, Meric Gertler, Gary Gates, and Tara Vinodrai, the percentage of workers in the creative classes in Toronto (36.4%), Montreal (35.0%), and Vancouver (35.2%) rival those in the leading American regions. Of America's ten most populous regions, only the Washington, DC (39.8%), and Boston (36.5%) areas do better. Toronto and Vancouver have the highest concentration of immigrants in North America, with 43.7% and 37.5% of their respective populations hailing from other countries. By comparison only 24.4% of New Yorkers were born outside the United States and only 30.9% of Los Angelenos. Of course, percentages don't give the full picture. The sheer number of creative-class members found in a metropolis like New York is far greater than in, say, Toronto. But the percentages do shed light on which cities are fostering creative cultures and will, therefore, be attractive to more creative types in the future.

Australia's leading regions are also well poised to compete as global creative centers, according to detailed benchmarking data compiled by the National Institute of Economic and Industry Research. Its two largest regions, Sydney and Melbourne, would rank approximately fourth or fifth if they were U.S. regions. Their creative classes are similar in size to those of Boston or Seattle. The Australian study compiled data for particular inner city neighborhoods, as well. Creative occupations make up fully half the workforce in both central Sydney (51.1%) and central Melbourne (49.5%)—far greater then in virtually any inner city in the U.S. Both of these centers have high percentages of immigrants—42.5% and 35.6% respectively—and are hotbeds of fine art, fashion, music, and street culture.

These and many other cities outside the United States boast additional attractions—stunning natural landscapes, world-class beaches, extensive recreational enticements. An added plus: They're rarely at war with anyone. These cities are fast becoming the global equivalents of Austin, Texas—transforming themselves from small, relatively obscure outposts into creative centers capable of luring talent from around the world.

tific research put sugar in its gas tank—the whole world will have to live with the repercussions.

The Creative Age requires nothing short of a change of worldview. Creativity is not a tangible asset like mineral deposits, something that can be hoarded or fought over, or even bought and sold. The U.S. must begin to think of creativity as a "common good," like liberty or security. It is something essential that belongs to everyone and must always be nourished, renewed, and maintained—or else it will slip away.

Richard Florida (florida@gmu.edu) is the Hirst Professor of Public Policy at George Mason University in Arlington, Virginia, and the author of The Rise of the Creative Class *(Basic Books, 2002). His new book,* The Flight of the Creative Class, *will be published in March 2005 by Harper Business.*

Women Waging Peace

You can't end wars simply by declaring peace. "Inclusive security" rests on the principle that fundamental social changes are necessary to prevent renewed hostilities. Women have proven time and again their unique ability to bridge seemingly insurmountable divides. So why aren't they at the negotiating table?

By Swanee Hunt and Cristina Posa

Allowing men who plan wars to plan peace is a bad habit. But international negotiators and policymakers can break that habit by including peace promoters, not just warriors, at the negotiating table. More often than not, those peace promoters are women. Certainly, some extraordinary men have changed the course of history with their peacemaking; likewise, a few belligerent women have made it to the top of the political ladder or, at the grass-roots level, have taken the roles of suicide bombers or soldiers. Exceptions aside, however, women are often the most powerful voices for moderation in times of conflict. While most men come to the negotiating table directly from the war room and battlefield, women usually arrive straight out of civil activism and—take a deep breath—family care.

Yet, traditional thinking about war and peace either ignores women or regards them as victims. This oversight costs the world dearly. The wars of the last decade have gripped the public conscience largely because civilians were not merely caught in the crossfire; they were targeted, deliberately and brutally, by military strategists. Just as warfare has become "inclusive"—with civilian deaths more common than soldiers'—so too must our approach toward ending conflict. Today, the goal is not simply the absence of war, but the creation of sustainable peace by fostering fundamental societal changes. In this respect, the United States and other countries could take a lesson from Canada, whose innovative "human security" initiative—by making human beings and their communities, rather than states, its point of reference—focuses on safety and protection, particularly of the most vulnerable segments of a population.

The concept of "inclusive security," a diverse, citizen-driven approach to global stability, emphasizes women's agency, not their vulnerability. Rather than motivated by gender fairness, this concept is driven by efficiency: Women are crucial to inclusive security since they are often at the center of nongovernmental organizations (NGOs), popular protests, electoral referendums, and other citizen-empowering movements whose influence has grown with the global spread of democracy. An inclusive security approach expands the array of tools available to police, military, and diplomatic structures by adding collaboration with local efforts to achieve peace. Every effort to bridge divides, even if unsuccessful, has value, both in lessons learned and links to be built on later. Local actors with crucial experience resolving conflicts, organizing political movements, managing relief efforts, or working with military forces bring that experience into ongoing peace processes.

International organizations are slowly recognizing the indispensable role that women play in preventing war and sustaining peace. On October 31, 2000, the United Nations Security Council issued Resolution 1325 urging the secretary-general to expand the role of women in U.N. field-based operations, especially among military observers, civilian police, human rights workers, and humanitarian personnel. The Organization for Security and Co-operation in Europe (OSCE) is working to move women off the gender sidelines and into the everyday activities of the organization—particularly in the Office for Democratic Institutions and Human Rights, which has been useful in monitoring elections and human rights throughout Europe and the former Soviet Union. Last November, the European Parliament passed a hard-hitting resolution calling on European Union members (and the European Commission and Council) to promote the equal participation of women in diplomatic conflict resolution; to ensure that women fill at least 40 percent of all reconciliation, peacekeeping, peace-enforcement, peace-building, and conflict-prevention posts; and to support the creation and strengthening of NGOs (including women's organizations) that focus on conflict prevention, peace building, and post-conflict reconstruction.

Ironically, women's status as second-class citizens is a source of empowerment, since it has made women adept at finding innovative ways to cope with problems.

But such strides by international organizations have done little to correct the deplorable extent to which local women have been relegated to the margins of police, military, and diplomatic efforts. Consider that Bosnian women were not invited to participate in the Dayton talks, which ended the war in Bosnia, even though during the conflict 40 women's associations remained organized and active across ethnic lines. Not surprisingly, this exclusion has subsequently characterized—and undermined—the implementation of the Dayton accord. During a 1997 trip to Bosnia, U.S. President Bill Clinton, Secretary of State Madeleine Albright, and National Security Advisor Samuel Berger had a miserable meeting with intransigent politicians elected under the ethnic-based requirements of Dayton. During the same period, First Lady Hillary Rodham Clinton engaged a dozen women from across the country who shared story after story of their courageous and remarkably effective work to restore their communities. At the end of the day, a grim Berger faced the press, offering no encouraging word from the meetings with the political dinosaurs. The first lady's meeting with the energetic women activists was never mentioned.

We can ignore women's work as peacemakers, or we can harness its full force across a wide range of activities relevant to the security sphere: bridging the divide between groups in conflict, influencing local security forces, collaborating with international organizations, and seeking political office.

BRIDGING THE DIVIDE

The idea of women as peacemakers is not political correctness run amok. Social science research supports the stereotype of women as generally more collaborative than men and thus more inclined toward consensus and compromise. Ironically, women's status as second-class citizens is a source of empowerment, since it has made women adept at finding innovative ways to cope with problems. Because women are not ensconced within the mainstream, those in power consider them less threatening, allowing women to work unimpeded and "below the radar screen." Since they usually have not been behind a rifle, women, in contrast to men, have less psychological distance to reach across a conflict line. (They are also more accepted on the "other side," because it is assumed that they did not do any of the actual killing.) Women often choose an identity, notably that of mothers, that cuts across international borders and ethnic enclaves. Given their roles as family nurturers,

women have a huge investment in the stability of their communities. And since women know their communities, they can predict the acceptance of peace initiatives, as well as broker agreements in their own neighborhoods.

As U.N. Secretary-General Kofi Annan remarked in October 2000 to the Security Council, "For generations, women have served as peace educators, both in their families and in their societies. They have proved instrumental in building bridges rather than walls." Women have been able to bridge the divide even in situations where leaders have deemed conflict resolution futile in the face of so-called intractable ethnic hatreds. Striking examples of women making the impossible possible come from Sudan, a country splintered by decades of civil war. In the south, women working together in the New Sudan Council of Churches conducted their own version of shuttle diplomacy—perhaps without the panache of jetting between capitals—and organized the Wunlit tribal summit in February 1999 to bring an end to bloody hostilities between the Dinka and Nuer peoples. As a result, the Wunlit Covenant guaranteed peace between the Dinka and the Nuer, who agreed to share rights to water, fishing, and grazing land, which had been key points of disagreement. The covenant also returned prisoners and guaranteed freedom of movement for members of both tribes.

On another continent, women have bridged the seemingly insurmountable differences between India and Pakistan by organizing huge rallies to unite citizens from both countries. Since 1994, the Pakistan-India People's Forum for Peace and Democracy has worked to overcome the hysterics of the nationalist media and jingoistic governing elites by holding annual conventions where Indians and Pakistanis can affirm their shared histories, forge networks, and act together on specific initiatives. In 1995, for instance, activists joined forces on behalf of fishers and their children who were languishing in each side's jails because they had strayed across maritime boundaries. As a result, the adversarial governments released the prisoners and their boats.

In addition to laying the foundation for broader accords by tackling the smaller, everyday problems that keep people apart, women have also taken the initiative in drafting principles for comprehensive settlements. The platform of Jerusalem Link, a federation of Palestinian and Israeli women's groups, served as a blueprint for negotiations over the final status of Jerusalem during the Oslo process. Former President Clinton, the week of the failed Camp David talks in July 2000, remarked simply, "If we'd had women at Camp David, we'd have an agreement."

Sometimes conflict resolution requires unshackling the media. Journalists can nourish a fair and tolerant vision of society or feed the public poisonous, one-sided, and untruthful accounts of the "news" that stimulate violent conflict. Supreme Allied Commander of Europe Wesley Clark understood as much when he ordered NATO to bomb transmitters in Kosovo to prevent the Milosevic media machine from spewing ever more inflammatory rhetoric. One of the founders of the independent Kosovo radio station RTV-21 realized that there were "many instances of male colleagues reporting with anger, which served to raise the tensions rather than lower them." As a result, RTV-

21 now runs workshops in radio, print, and TV journalism to cultivate a core of female journalists with a noninflammatory style. The OSCE and the BBC, which train promising local journalists in Kosovo and Bosnia, would do well to seek out women, who generally bring with them a reputation for moderation in unstable situations.

Nelson Mandela suggested at last summer's Arusha peace talks that if Burundian men began fighting again, their women should withhold "conjugal rights" (like cooking, he added).

INFLUENCING SECURITY FORCES

The influence of women on warriors dates back to the ancient Greek play *Lysistrata*. Borrowing from that play's story, former South African President Nelson Mandela suggested at last summer's Arusha peace talks on the conflict in Burundi that if Burundian men began fighting again, their women should withhold "conjugal rights" (like cooking, he added).

Women can also act as a valuable interface between their countries' security forces (police and military) and the public, especially in cases when rapid response is necessary to head off violence. Women in Northern Ireland, for example, have helped calm the often deadly "marching season" by facilitating mediations between Protestant unionists and Catholic nationalists. The women bring together key members of each community, many of whom are released prisoners, as mediators to calm tensions. This circle of mediators works with local police throughout the marching season, meeting quietly and maintaining contacts on a 24-hour basis. This intervention provides a powerful extension of the limited tools of the local police and security forces.

Likewise, an early goal of the Sudanese Women's Voice for Peace was to meet and talk with the military leaders of the various rebel armies. These contacts secured women's access to areas controlled by the revolutionary movements, a critical variable in the success or failure of humanitarian efforts in war zones. Women have also worked with the military to search for missing people, a common element in the cycle of violence. In Colombia, for example, women were so persistent in their demands for information regarding 150 people abducted from a church in 1999 that the army eventually gave them space on a military base for an information and strategy center. The military worked alongside the women and their families trying to track down the missing people. In short, through moral suasion, local women often have influence where outsiders, such as international human rights agencies, do not.

That influence may have allowed a female investigative reporter like Maria Cristina Caballero to go where a man could not go, venturing on horseback alone, eight hours into the jungle to tape a four-hour interview with the head of the paramilitary forces in Colombia. She also interviewed another guerilla leader and published an award-winning comparison of the transcripts, showing where the two mortal enemies shared the same vision. "This [was] bigger than a story," she later said, "this [was] hope for peace." Risking their lives to move back and forth across the divide, women like Caballero perform work that is just as important for regional stabilization as the grandest Plan Colombia.

INTERNATIONAL COLLABORATION

Given the nature of "inclusive" war, security forces are increasingly called upon to ensure the safe passage of humanitarian relief across conflict zones. Women serve as indispensable contacts between civilians, warring parties, and relief organizations. Without women's knowledge of the local scene, the mandate of the military to support NGOs would often be severely hindered, if not impossible.

In rebel-controlled areas of Sudan, women have worked closely with humanitarian organizations to prevent food from being diverted from those who need it most. According to Catherine Loria Duku Jeremano of Oxfam: "The normal pattern was to hand out relief to the men, who were then expected to take it home to be distributed to their family. However, many of the men did what they pleased with the food they received: either selling it directly, often in exchange for alcohol, or giving food to the wives they favored." Sudanese women worked closely with tribal chiefs and relief organizations to establish a system allowing women to pick up the food for their families, despite contrary cultural norms.

In Pristina, Kosovo, Vjosa Dobruna, a pediatric neurologist and human rights leader, is now the joint administrator for civil society for the U.N. Interim Administration Mission in Kosovo (UNMIK). In September 2000, at the request of NATO, she organized a multiethnic strategic planning session to integrate women throughout UNMIK. Before that gathering, women who had played very significant roles in their communities felt shunned by the international organizations that descended on Kosovo following the bombing campaign. Vjosa's conference pulled them back into the mainstream, bringing international players into the conference to hear from local women what stabilizing measures they were planning, rather than the other way around. There, as in Bosnia, the OSCE has created a quota system for elected office, mandating that women comprise one third of each party's candidate list; leaders like Vjosa helped turn that policy into reality.

In addition to helping aid organizations find better ways to distribute relief or helping the U.N. and OSCE implement their ambitious mandates, women also work closely with them to locate and exchange prisoners of war. As the peace processes in Northern Ireland, Bosnia, and the Middle East illustrate, a deadlock on the exchange and release of prisoners can be a major obstacle to achieving a final settlement. Women activists in Armenia and Azerbaijan have worked closely with the International Helsinki Citizens Assembly and the OSCE for the release

The Black and the Green

Grass-roots women's organizations in Israel come in two colors: black and green. The Women in Black, founded in 1988, and the Women in Green, founded in 1993, could not be further apart on the political spectrum, but both claim the mantle of "womanhood" and "motherhood" in the ongoing struggle to end the Israeli-Palestinian conflict.

One month after the Palestinian intifada broke out in December 1988, a small group of women decided to meet every Friday afternoon at a busy Jerusalem intersection wearing all black and holding hand-shaped signs that read: "Stop the Occupation." The weekly gatherings continued and soon spread across Israel to Europe, the United States, and then to Asia.

While the movement was originally dedicated to achieving peace in the Middle East, other groups soon protested against repression in the Balkans and India. For these activists, their status as women lends them a special authority when it comes to demanding peace. In the words of the

Asian Women's Human Rights Council: "We are the Women in Black... women, unmasking the many horrific faces of more public 'legitimate' forms of violence—state repression, communalism, ethnic cleansing, nationalism, and wars...."

Today, the Women in Black in Israel continue their nonviolent opposition to the occupation in cooperation with the umbrella group Coalition of Women for a Just Peace. They have been demonstrating against the closures of various Palestinian cities, arguing that the blockades prevent pregnant women from accessing healthcare services and keep students from attending school. The group also calls for the full participation of women in peace negotiations.

While the Women in Black stood in silent protest worldwide, a group of "grandmothers, mothers, wives, and daughters; housewives and professionals; secular and religious" formed the far-right Women in Green in 1993 out of "a shared love, devotion and concern for Israel." Known for the signature

green hats they wear at rallies, the Women in Green emerged as a protest to the Oslo accords on the grounds that Israel made too many concessions to Yasir Arafat's Palestinian Liberation Organization. The group opposes returning the Golan Heights to Syria, sharing sovereignty over Jerusalem with the Palestinians, and insists that "Israel remain a Jewish state."

The Women in Green boast some 15,000 members in Israel, and while they have not garnered the global support of the Women in Black, 15,000 Americans have joined their cause. An ardent supporter of Israeli Prime Minister Ariel Sharon, the group seeks to educate the Israeli electorate through weekly street theater and public demonstrations, as well as articles, posters, and newspaper advertisements.

White the groups' messages and methods diverge, their existence and influence demonstrate that women can mobilize support for political change—no matter what color they wear.

—FP

of hostages in the disputed region of Nagorno-Karabakh, where tens of thousands of people have been killed. In fact, these women's knowledge of the local players and the situation on the ground would make them indispensable in peace negotiations to end this 13-year-old conflict.

REACHING FOR POLITICAL OFFICE

In 1977, women organizers in Northern Ireland won the Nobel Peace Prize for their nonsectarian public demonstrations. Two decades later, Northern Irish women are showing how diligently women must still work not only to ensure a place at the negotiating table but also to sustain peace by reaching critical mass in political office. In 1996, peace activists Monica McWilliams (now a member of the Northern Ireland Assembly) and May Blood (now a member of the House of Lords) were told that only leaders of the top 10 political parties—all men—would be included in the peace talks. With only six weeks to organize, McWilliams and Blood gathered 10,000 signatures to create a new political party (the Northern Ireland Women's Coalition, or NIWC) and got themselves on the ballot. They were voted into the top 10 and earned a place at the table.

The grass-roots, get-out-the-vote work of Vox Femina convinced hesitant Yugoslav women to vote for change; those votes contributed to the margin that ousted President Slobodan Milosevic.

The NIWC's efforts paid off. The women drafted key clauses of the Good Friday Agreement regarding the importance of mixed housing, the particular difficulties of young people, and the need for resources to address these problems. The NIWC also lobbied for the early release and reintegration of political prisoners in order to combat social exclusion and pushed for a comprehensive review of the police service so that all members of society would accept it. Clearly, the women's prior work with individuals and families affected by "the Troubles" enabled them to formulate such salient contributions to the agreement. In the subsequent public referendum on the Good Friday Agreement, Mo Mowlam, then British secretary of state for Northern Ireland, attributed the overwhelming success of the YES Campaign to the NIWC's persistent canvassing and lobbying.

Women in the former Yugoslavia are also stepping forward to wrest the reins of political control from extremists (including women, such as ultranationalist Bosnian Serb President Biljana Plavsic) who destroyed their country. Last December, Zorica Trifunovic, founding member of the local Women in Black (an antiwar group formed in Belgrade in October 1991), led a meeting that united 90 women leaders of pro-democracy political campaigns across the former Yugoslavia. According to polling by the National Democratic Institute, the grass-roots, get-out-the-vote work of groups such as Vox Femina (a local NGO that participated in the December meeting) convinced hesitant women to vote for change; those votes contributed to the margin that ousted President Slobodan Milosevic.

International security forces and diplomats will find no better allies than these mobilized mothers, who are tackling the toughest, most hardened hostilities.

Argentina provides another example of women making the transition from protesters to politicians: Several leaders of the Madres de la Plaza de Mayo movement, formed in the 1970s to protest the "disappearances" of their children at the hands of the military regime, have now been elected to political office. And in Russia, the Committee of Soldiers' Mothers—a protest group founded in 1989 demanding their sons' rights amidst cruel conditions in the Russian military—has grown into a powerful organization with 300 chapters and official political status. In January, U.S. Ambassador to Moscow Jim Collins described the committee as a significant factor in countering the most aggressive voices promoting military force in Chechnya. Similar mothers' groups have sprung up across the former Soviet Union and beyond—including the Mothers of Tiananmen Square. International security forces and diplomats will find no better allies than these mobilized mothers, who are tackling the toughest, most hardened hostilities.

YOU'VE COME A LONG WAY, MAYBE

Common sense dictates that women should be central to peacemaking, where they can bring their experience in conflict resolution to bear. Yet, despite all of the instances where women have been able to play a role in peace negotiations, women remain relegated to the sidelines. Part of the problem is structural: Even though more and more women are legislators and soldiers, underrepresentation persists in the highest levels of political and military hierarchies. The presidents, prime ministers, party leaders, cabinet secretaries, and generals who typically negotiate peace settlements are overwhelmingly men. There is also a psychological barrier that precludes women from sitting in on negotiations: Waging war is still thought of as a "man's job," and as such, the task of stopping war often is delegated to men

(although if we could begin to think about the process not in terms of stopping war but promoting peace, women would emerge as the more logical choice). But the key reason behind women's marginalization may be that everyone recognizes just how good women are at forging peace. A U.N. official once stated that, in Africa, women are often excluded from negotiating teams because the war leaders "are afraid the women will compromise" and give away too much.

Some encouraging signs of change, however, are emerging. Rwandan President Paul Kagame, dismayed at his difficulty in attracting international aid to his genocide-ravaged country, recently distinguished Rwanda from the prevailing image of brutality in central Africa by appointing three women to his negotiating team for the conflict in the Democratic Republic of the Congo. In an unusually healthy tit for tat, the Ugandans responded by immediately appointing a woman to their team.

Will those women make a difference? Negotiators sometimes worry that having women participate in the discussion may change the tone of the meeting. They're right: A British participant in the Northern Ireland peace talks insightfully noted that when the parties became bogged down by abstract issues and past offenses, "the women would come and talk about their loved ones, their bereavement, their children and their hopes for the future." These deeply personal comments, rather than being a diversion, helped keep the talks focused. The women's experiences reminded the parties that security for all citizens was what really mattered.

The role of women as peacemakers can be expanded in many ways. Mediators can and should insist on gender balance among negotiators to ensure a peace plan that is workable at the community level. Cultural barriers can be overcome if high-level visitors require that a critical mass (usually one third) of the local interlocutors be women (and not simply present as wives). When drafting principles for negotiation, diplomats should determine whether women's groups have already agreed upon key conflict-bridging principles, and whether their approach can serve as a basis for general negotiations.

Moreover, to foster a larger pool of potential peacemakers, embassies in conflict areas should broaden their regular contact with local women leaders and sponsor women in training programs, both at home and abroad. Governments can also do their part by providing information technology and training to women activists through private and public partnerships. Internet communication allows women peace builders to network among themselves, as well as exchange tactics and strategies with their global counterparts.

"Women understood the cost of the war and were genuinely interested in peace," recalls retired Admiral Jonathan Howe, reflecting on his experience leading the U.N. mission in Somalia in the early 1990s. "They'd had it with their warrior husbands. They were a force willing to say enough is enough. The men were sitting around talking and chewing qat, while the women were working away. They were such a positive force.... You have to look at all elements in society and be ready to tap into those that will be constructive."

Want to Know More?

The Internet is invaluable in enabling the inclusive security approach advocated in this article. The Web offers not only a wealth of information but, just as important, relatively cheap and easy access for citizens worldwide. Most of the women's peace-building activities and strategies explored in this article can be found on the Web site of **Women Waging Peace**—a collaborative venture of Harvard University's John F. Kennedy School of Government and the nonprofit organization Hunt Alternatives, which recognize the essential role and contribution of women in preventing violent conflict, stopping war, reconstructing ravaged societies, and sustaining peace in fragile areas around the world. On the site, women active in conflict areas can communicate with each other without fear of retribution via a secure server. The women submit narratives detailing their strategies, which can then be read on the public Web site. The site also features a video archive of interviews with each of these women. You need a password to view these interviews, so contact Women Waging Peace online or call (617) 868-3910.

The Organization for Security and Co-operation in Europe (OSCE) is an outstanding resource for qualitative and quantitative studies of women's involvement in conflict prevention. Start with the final report of the *OSCE Supplementary Implementation Meeting: Gender Issues* (Vienna: UNIFEM, 1999), posted on the group's Web site. **The United Nations Development Fund for Women** (UNIFEM) also publishes reports on its colorful and easy-to-navigate site. The fund's informative book, *Women at the Peace Table: Making a Difference* (New York: UNIFEM, 2000), available online, features interviews with some of today's most prominent women peacemakers, including Hanan Ashrawi and Mo Mowlam.

For a look at how globalization is changing women's roles in governments, companies, and militaries, read Cynthia Enloe's *Bananas, Beaches and Bases: Making Feminist Sense of International Politics* (Berkeley: University of California Press, 2001). In *Maneuvers: The International Politics of Militarizing Women's Lives* (Berkeley: University of California Press, 2000), Enloe examines the military's effects on women, whether they are soldiers or soldiers' spouses. For a more general discussion of where feminism fits into academia and policymaking, see **"Searching for the Princess? Feminist Perspectives in International Relations"** (*The Harvard International Review*, Fall 1999) by J. Ann Tickner, associate professor of international relations at the University of Southern California.

The Fall 1997 issue of FOREIGN POLICY magazine features two articles that highlight how women worldwide are simultaneously gaining political clout but also bearing the brunt of poverty: **"Women in Power: From Tokenism to Critical Mass"** by Jane S. Jaquette and **"Women in Poverty: A New Global Underclass"** by Mayra Buvinic.

• For links to relevant Web sites, as well as a comprehensive index of related FOREIGN POLICY articles, access **www.foreign policy.com**.

Lasting peace must be homegrown. Inclusive security helps police forces, military leaders, and diplomats do their jobs more effectively by creating coalitions with the people most invested in stability and most adept at building peace. Women working on the ground are eager to join forces. Just let them in.

Swanee Hunt is director of the Women in Public Policy Program at Harvard University's John F. Kennedy School of Government. As the United States' ambassador to Austria (1993–97), she founded the "Vital Voices: Women in Democracy" initiative. Cristina Posa, a former judicial clerk at the United Nations International Criminal Tribunal for the former Yugoslavia, is an attorney at Cleary, Gottlieb, Steen & Hamilton in New York.

Exploring the "Singularity"

The point in time when current trends may go wildly off the charts—known as the "Singularity"—is now getting serious attention. What it suggests is that technological change will soon become so rapid that we cannot possibly envision its results.

by James John Bell

Technological change isn't just happening fast. It's happening at an exponential rate. Contrary to the commonsense, intuitive, linear view, we won't just experience 100 years of progress in the twenty-first century—it will be more like 20,000 years of progress.

The near-future results of exponential technological growth will be staggering: the merging of biological and nonbiological entities in biorobotics, plants and animals engineered to grow pharmaceutical drugs, software-based "life," smart robots, and atom-sized machines that self-replicate like living matter. Some individuals are even warning that we could lose control of this expanding techno-cornucopia and cause the total extinction of life as we know it. Others are researching how this permanent technological overdrive will affect us. They're trying to understand what this new world of ours will look like and how accelerating technology already impacts us.

A number of scientists believe machine intelligence will surpass human intelligence within a few decades, leading to what's come to be called the Singularity. Author and inventor Ray Kurzweil defines this phenomenon as "technological change so rapid and profound it could create a rupture in the very fabric of human history."

Singularity is technically a mathematical term, perhaps best described as akin to what happens on world maps in a standard atlas. Everything appears correct until we look at regions very close to the poles. In the standard Mercator projection, the poles appear not as points but as a straight line. Each line is a singularity: Everywhere along the top line contains the exact point

of the North Pole, and the bottom line is the entire South Pole.

The singularity on the edge of the map is nothing compared to the singularity at the center of a black hole. Here one finds the astrophysicist's singularity, a rift in the continuum of space and time where Einstein's rules no longer function. The approaching technological Singularity, like the singularities of black holes, marks a point of departure from reality. Explorers once wrote "Beyond here be dragons" on the edges of old maps of the known world, and the image of life as we approach these edges of change are proving to be just as mysterious, dangerous, and controversial.

There is no concise definition for the Singularity. Kurzweil and many transhumanists define it as "a future time when societal, scientific, and economic change is so fast we cannot even imagine what will happen from our present perspective." A range of dates is given for the advent of the Singularity. "I'd be surprised if it happened before 2004 or after 2030," writes author and computer science professor Vernor Vinge. A distinctive feature will be that machine intelligence will have exceeded and even merged with human intelligence. Another definition is used by extropians, who say it denotes "the singular time when technological development will be at its fastest." From an environmental perspective, the Singularity can be thought of as the point at which technology and nature become one. Whatever perspective one takes, at this juncture the world as we have known it will become extinct, and new definitions of *life, nature, and human* will take hold.

Many leading technology industries have been aware of the possibility of a Singularity for some time. There are concerns that, if the public understood its ramifications, they might panic over accepting new and untested technologies that bring us closer to Singularity. For now, the debate about the consequences of the Singularity has stayed within the halls of business and technology; the kinks are being worked out, avoiding "doomsday" hysteria. At this time, it appears to matter little if the Singularity ever truly comes to pass.

"The true believers call themselves extropians, posthumans, and transhumanists, and are actively organizing not just to bring the Singularity about, but to counter the technophobes and neo-Luddites."

What Will Singularity Look Like?

Kurzweil explains that central to the workings of the Singularity are a number of "laws," one of which is Moore's law. Intel cofounder Gordon E. Moore noted that the number of transistors that could fit on a single computer chip had doubled every year for six years from the beginnings of integrated circuits in 1959. Moore predicted that the trend would continue, and it has—although the doubling rate was later adjusted to an 18-month cycle.

Today, the smallest transistors in chips span only thousands of atoms (hundreds of nanometers). Chipmakers build such components using a process in which they apply semiconducting, metallic, and insulating layers to a semiconductor wafer to create microscopic circuitry. They accomplish the procedure using light for imprinting patterns onto the wafer. In order to keep Moore's law moving right along, researchers today have built circuits out of transistors, wires, and other components as tiny as a few atoms across that can carry out simple computations.

Kurzweil and Sun Microsystems' chief scientist Bill Joy agree that, circa 2030, the technology of the 1999 film *The Matrix* (which visualized a three-dimensional interface between humans and computers, calling conventional reality into question) will be within our grasp and that humanity will be teetering on the edge of the Singularity. (See their essays in *Taking the Red Pill: Science, Philosophy, and Religion in The Matrix*, edited by Glenn Yeffeth, 2003.) Kurzweil explains that this will become possible because Moore's law will be replaced by another computing paradigm over the next few decades. "Moore's law was not the first but the fifth paradigm to provide exponential growth of computing power," Kurzweil says. The first paradigm of computer technology was the data processing machinery used in the 1890 American census. This electromechanical computing technology was followed by the paradigms of relay-based technology, vacuum tubes, transistors, and eventually integrated circuits. "Every time a paradigm ran out of steam," states Kurzweil, "another paradigm came along and picked up where that paradigm left off." The sixth paradigm, the one that will enable technology á la *The Matrix*, will be here in 20 to 30 years. "It's obvious what the sixth paradigm will be—computing in three dimensions," says Kurzweil. "We will effectively merge with our technology."

Stewart Brand in his book *The Clock of the Long Now* discusses the Singularity and another related law, Monsanto's law, which states that the ability to identify and use genetic information doubles every 12 to 24 months. This exponential growth in biological knowledge is transforming agriculture, nutrition, and health care in the emerging life-sciences industry.

A field of research building on the exponential growth rate of biotechnology is nanotechnology—the science of building machines out of atoms. A nanometer is atomic in scale, a distance that's 0.001% of the width of human hair. One goal of this science is to change the atomic fabric of matter—to engineer machinelike atomic structures that reproduce like living matter. In this respect, it is similar to biotechnol-

ogy, except that nanotechnology needs to literally create something like an inorganic version of DNA to drive the building of its tiny machines. "We're working out the rules of biology in a realm where nature hasn't had the opportunity to work," states University of Texas biochemistry professor Angela Belcher. "What would take millions of years to evolve on its own takes about three weeks on the bench top."

Machine progress is knocking down the barriers between all the sciences. Chemists, biologists, engineers, and physicists are now finding themselves collaborating on more and more experimental research. This collaboration is best illustrated by the opening of Cornell University's Nanobiotechnology Center and other such facilities around the world. These scientists predict breakthroughs soon that will open the way to molecular-size computing and the quantum computer, creating new scientific paradigms where exponential technological progress will leap off the map. Those who have done the exponential math quickly realize the possibilities in numerous industries and scientific fields—and then they notice the anomaly of the Singularity happening within this century.

In 2005, IBM plans to introduce Blue Gene, a supercomputer that can perform at about 5% of the power of the human brain. This computer could transmit the entire contents of the Library of Congress in less than two seconds. Blue Gene/L, specifically developed to advance and serve the growing life-sciences industry, is expected to operate at about 200 teraflops (200 trillion floating-point operations per second), larger than the total computing power of the top 500 supercomputers in the world. It will be able to run extremely complex simulations, including breakthroughs in computers and information technology, creating new frontiers in biology, says IBM's Paul M. Horn. According to Moore's law, computer hardware will surpass human brainpower in the first decade of this century. Software that emulates the human mind—artificial intelligence—may take another decade to evolve.

XenoMouse, patented in 2002 by Abgenix in Fremont, California, was genetically engineered to have a human immune system. Biotechnology breaks the natural boundaries existing between species. Discussion on the consequences of such breakthroughs, at least in the United States, remains on the sidelines, says author Bell.

Nanotech Advances Promote Singularity

Physicists, mathematicians, and scientists like Vinge and Kurzweil have identified through their research the likely boundaries of the Singularity and have predicted with confidence various paths leading up to it over the next couple of decades. These scientists are currently debating what discovery could set off a chain reaction of Earth-altering technological events. They suggest that advancements in the fields of nanotechnology or the discovery of artificial intelligence could usher in the Singularity.

The majority of people closest to these theories and laws—the tech sector—can hardly wait for these technologies to arrive. The true believers call themselves extropians, posthumans, and transhumanists, and are actively organizing not just to bring the Singularity about, but to counter the technophobes and neo-Luddites who believe that unchecked technological progress will exceed our ability to reverse any destructive process that might unintentionally be set in motion.

The antithesis to neo-Luddite activists is the extropians. For example, the Progress Action Coalition, formed in 2001 by bio-artist, author, and extropian activist Natasha Vita-More, fantasizes about "the dream of true artificial intelligence…adding a new richness to the human landscape never before known." Pro-Act, AgBioworld, Biotechnology Progress, Foresight Institute, the Progress and Freedom Foundation, and other industry groups acknowledge, however, that the greatest threat to technological progress comes not just from environmental groups, but from a small faction of the scientific community.

Knowledge-Enabled Mass Destruction

In April 2000, a wrench was thrown into the arrival of the Singularity by an unlikely source: Sun Microsystems chief scientist Bill Joy. He is a neo-Luddite without being a Luddite, a technologist warning the world about technology. Joy co-founded Sun Microsystems, helped create the Unix computer operating system, and developed the Java and Jini software systems—systems that helped give the Internet "life."

In a now-infamous cover story in *Wired* magazine, "Why the Future Doesn't Need Us," Joy warned of the dangers posed by

Keys to Understanding the Singularity

Singularity is the postulated point in our future when human evolutionary development-powered by such developments as nanotechnology, neuroscience, and artificial intelligence—accelerates enormously so that nothing beyond that time can reliably be conceived. Typical developments include the merging of man and machine (cybernetic organisms-or cyborgs) and accelerated technology beyond our ability to control.

Nanotechnology is the development and use of devices that have a size of only a few nanometers, including building and manipulating complex structures on an atomic scale. As we approach the Singularity, nanodevices will be able to replicate themselves like living matter.

Biorobotics is the merging of living organisms with technologies. At a simple level, this includes implanting chips encoded with health or security information. Biorobotics also encompasses the development of cyborgs that seamlessly blend living tissue with mechanical devices.

Cloning is the growing of genetically identical cells, eliminating the natural role of human biology and bringing us closer to the Singularity.

Extropians await the Singularity, seeking to overcome human limits, live indefinitely long, and become more intelligent through technology. Related groups include transhumanists and posthumanists.

Neo-Luddites oppose the impending Singularity by raising questions about moral and ethical aspects of modern technology and the threat it may pose to humanity.

developments in genetics, nanotechnology, and robotics. Joy's warning of the impacts of exponential technological progress run amok gave new credence to the coming Singularity. Unless things change, Joy predicted, "We could be the last generation of humans." Joy warned that "knowledge alone will enable mass destruction" and termed this phenomenon "knowledge-enabled mass destruction."

The twentieth century gave rise to nuclear, biological, and chemical (NBC) technologies that, while powerful, require access to vast amounts of raw (and often rare) materials, technical information, and large-scale industries. The twenty-first-century technologies of genetics, nanotechnology, and robotics (GNR), however, will require neither large facilities nor rare raw materials.

The threat posed by GNR technologies becomes further amplified by the fact that some of these new technologies have been designed to be able to replicate—i.e., they can build new versions of themselves. Nuclear bombs did not sprout more bombs, and toxic spills did not grow more spills. If the new selfreplicating GNR technologies are released into the environment, they could be nearly impossible to recall or control.

Joy understands that the greatest dangers we face ultimately stem from a world where global corporations dominate—a future where much of the world has no voice in how the world is run. Twenty-first-century GNR technologies, he writes, "are being developed almost exclusively by corporate enterprises. We are aggressively pursuing the promises of these new technologies within the now-unchallenged system of global capitalism and its manifold financial incentives and competitive pressures."

Joy believes that the system of global capitalism, combined with our current rate of progress, gives the human race a 30% to 50% chance of going extinct around the time the Singularity is expected to happen, around 2030. "Not only are these estimates not encouraging," he adds, "but they do not include the probability of many horrid outcomes that lie short of extinction."

It is very likely that scientists and global corporations will miss key developments—or, worse, actively avoid discussion of them. A whole generation of biologists has left the field for the biotech and nanotech labs. Biologist Craig Holdredge, who has followed biotech since its beginnings in the 1970s, warns, "Biology is losing its connection with nature."

When Machines Make War

Cloning, biotechnology, nanotechnology, and robotics are blurring the lines between nature and machine. In his 1972 speech "The Android and the Human," science-fiction visionary Philip K. Dick told his audience, "Machines are becoming more human. Our environment, and I mean our man-made world of machines, is becoming alive in ways specifically and fundamentally analogous to ourselves." In the near future, Dick prophesied, a human might shoot a robot only to see it bleed from its wound. When the robot shoots back, it may be surprised to find the human gush smoke. "It would be rather a great moment of truth for both of them," Dick added.

In November 2001, Advanced Cell Technology of Massachusetts jarred the nation's focus away from recession and terrorism when it announced that it had succeeded in cloning early-stage human embryos. Debate on the topic stayed equally divided between those who support therapeutic cloning and those, like the American Medical Association, who want an outright ban.

Karel Capek coined the word robot (Czech for "forced labor") in the 1920 play *R.U.R.*, in which machines assume the drudgery of factory production, then develop feelings and proceed to wipe out humanity in a violent revolution. While the robots in *R.U.R.* could represent the "nightmare vision of the proletariat seen through middle-class eyes," as science-fiction author Thomas Disch has suggested, they also are testament to the persistent fears of man-made technology run amok.

Similar themes have manifested themselves in popular culture and folklore since at least medieval times. While some might dismiss these stories simply as popular paranoia, robots are already being deployed beyond Hollywood and are poised to take over the deadlier duties of the modern soldier. The Pentagon is replacing soldiers with sensors, vehicles, aircraft, and weapons that can be operated by remote control or are autonomous. Pilotless aircraft played an important role in the bombings of Afghanistan, and a model called the *Gnat* conducted surveillance flights in the Philippines in 2002.

Leading the Pentagon's remote-control warfare effort is the Defense Advanced Research Projects Agency (DARPA). Best known for creating the infrastructure that became the World Wide Web, DARPA is working with Boeing to develop the X-45 unmanned combat air vehicle. The 30-foot-long windowless planes will carry up to 12 bombs, each weighing 250 pounds. According to military analysts, the X-45 will be used to attack radar and antiaircraft installations as early as 2007. By 2010, it will be programmed to distinguish friends from foes without consulting humans and independently attack targets in designated areas. By 2020, robotic planes and vehicles will direct remote-controlled bombers toward targets, robotic helicopters will coordinate driverless convoys, and unmanned submarines will clear mines and launch cruise missiles.

Rising to the challenge of mixing man and machine, MIT's Institute for Soldier Nanotechnologies (backed by a five-year, $50-million U.S. Army grant) is busy innovating materials and designs to create military uniforms that rival the best science fiction. Human soldiers themselves are being transformed into modern cyborgs through robotic devices and nanotechnology.

The Biorobotic Arms Race

The 2002 International Conference on Robotics and Automation, hosted by the Institute of Electrical and Electronics Engineers, kicked off its technical session with a discussion on *biorobots*, the melding of living and artificial structures into a cybernetic organism or cyborg.

"In the past few years, the biosciences and robotics have been getting closer and closer," says Paolo Dario, founder of Italy's Advanced Robotics Technology and Systems Lab. "More and more, biological models are used for the design of biometric robots [and] robots are increasingly used by neuroscientists as clinical platforms for validating biological models." Artificial constructs are beginning to approach the scale and complexity of living systems.

Some of the scientific breakthroughs expected in the next few years promise to make cloning and robotics seem rather benign. The merging of technology and nature has already yielded some shocking progeny. Consider these examples:

- Researchers at the State University of New York Health Science Center at Brooklyn have turned a living rat into a radio-controlled automaton using three electrodes placed in the animal's brain. The animal can be remotely steered through an obstacle course, making it twist, turn, and jump on demand.
- In May 2002, eight elderly Florida residents were injected with microscopic silicon identification chips encoded with medical information. The *Los Angeles Times* reported that this made them "scannable just like a jar of peanut butter in the supermarket checkout line." Applied Digital Solutions Inc., the maker of the chip, will soon have a prototype of an implantable device able to receive GPS satellite signals and transmit a person's location.
- Human embryos have been successfully implanted and grown in artificial wombs. The experiments were halted after a few days to avoid violating in vitro fertilization regulations.
- Researchers in Israel have fashioned a "bio-computer" out of DNA that can handle a billion operations per second with 99.8% accuracy. Reuters reports that these bio-computers are so minute that "a trillion of them could fit inside a test tube."
- In England, University of Reading Professor Kevin Warwick has implanted microchips in his body to remotely monitor and control his physical motions. During Warwick's Project Cyborg experiments, computers were able to remotely monitor his movements and open doors at his approach.
- Engineers at the U.S. Sandia National Labs have built a remote-controlled spy robot equipped with a scanner, microphone, and chemical microsensor. The robot weighs one ounce and is smaller than a dime. Lab scientists predict that the microbot could prove invaluable in protecting U.S. military and economic interests.

The next arms race is not based on replicating and perfecting a single deadly technology, like the nuclear bombs of the past or some space-based weapon of the future. This new arms race is about accelerating the development and integration of advanced autonomous, biotechnological, and human-robotic systems into the military apparatus. A mishap or a massive war using these new technologies could be more catastrophic than any nuclear war.

Where the Map Exceeds the Territory

The rate at which GNR technologies are being adopted by our society—without regard to long-term safety testing or researching the political, cultural, and economic ramifications—mirrors the development and proliferation of nuclear power and weapons. The human loss caused by experimentation, production, and development is still being felt from the era of NBC technologies.

The discussion of the environmental impacts of GNR technologies, at least in the United States, has been relegated to the margins. Voices of concern and opposition have likewise been missing in discussions of the technological Singularity. The true cost of this technological progress and any coming Singularity will mean the unprecedented decline of the planet's inhabitants at an ever-increasing rate of global extinction.

The World Conservation Union, the International Botanical Congress, and a majority of the world's biologists believe that a global mass extinction already is under way. As a direct result of human activity (resource extraction, industrial agriculture, the introduction of non-native animals, and population growth), up to one-fifth of all living species are expected to disappear within 30 years. A 1998 Harris Poll of the 5,000 members of the American Institute of Biological Sciences found that 70% believed that what has been termed "The Sixth Extinction" is now under way. A simultaneous Harris Poll found that 60% of the public were totally unaware of the impending biological collapse.

At the same time that nature's ancient biological creation is on the decline, laboratory-created biotech life-forms—genetically modified soybeans, genetically engineered salmon, cloned sheep, drug-crops, biorobots—are on the rise.

Nature and technology are not just evolving; they are competing and combining with one another. Ultimately they will become one. We hear reports daily about these new technologies and new creations, while shreds of the ongoing biological collapse surface here and there. Past the edges of change, beyond the wall across the future, anything becomes possible. Beware the dragons.

James John Bell is writer/director and network administrator for the environmental communications firm Sustain (www.sustainusa.org). He recently wrote the foreword for the 2003 reprinting of John Brunner's science-fiction novel *The Sheep Look Up*, available in June. Portions of this essay were excerpted from his foreword. His address is 920 North Franklin Street, Suite 301, Chicago, Illinois 60610. E-mail jamesbell@sustain.org.

Originally published in the May/June 2003 issue of *The Futurist*. Used with permission from the World Future Society, 7910 Woodmont Avenue, Suite 450, Bethesda, MD 20814. Telephone: 301/656-8274; Fax: 301/951-0394; http://www.wfs.org. © 2003.

Why Environmental Ethics Matters to International Relations

"Environmental ethics [should] not be seen as an add-on to be approached after the important issues of security and economics have been settled. Instead, we [should] recognize that all our important social choices are inherently about the 'natural' world we create."

JOHN BARKDULL

What challenge does environmental ethics pose for international relations? International relations is usually understood as the realm of power politics, a world in which military might and the quest to survive dominate. In this world, moral concern for other human beings, much less nature, is limited or entirely lacking. Environmental ethics—a set of principles to guide human interaction with the earth—calls on us to extend moral consideration beyond humans to other living things and to natural "wholes" such as bioregions and ecosystems. Is it possible to introduce environmental ethics' far-reaching moral claims into the competitive, militarized, economically unequal world political system?

Although explorations in environmental ethics now have a long resumé, the dialogue over the human debt to the natural environment has proceeded largely without reference to international politics, to international relations theory, or even to the literature on international ethics. Practical politics is thus often removed from consideration. And scholars of international relations have barely considered the relationship between their studies and environmental ethics.

Bringing the two fields into the same conversation is possible. International political theory has profound implications for understanding how humans ought to relate to the environment. Realism and liberal institutionalism (the mainstream of international relations theory), by suggesting what political, economic, and social goals are desirable, also imply what environmental values should prevail. They indicate what kind of world humans should or can create and thus tell us how we should relate to the environment. The question then is not whether environmental ethics should matter in world politics, but in which way: which environmental ethic does in fact matter, which should, and what obstacles prevent needed changes in political practices from being made?

WHICH ENVIRONMENTAL ETHICS?

Environmental ethics can be anthropocentric, biocentric, or ecocentric. Anthropocentric ethics is about what humans owe each other. It evaluates environmental policies with regard to how they affect human well-being. For example, exploitation of natural resources such as minerals can destroy forests on which indigenous peoples depend. Moral evaluation of the environmental destruction proceeds in terms of the rights, happiness, or just treatment of all human parties, including the displaced tribes and the consumers who benefit from the minerals. Anthropocentric environmental ethics generally calls for more environmental protection than we now undertake; current unsustainable resource-use patterns and conversion of land to agricultural or urban uses mean that existing practices do more harm to humans than good, especially when future generations are considered. Still, many observers find anthropocentric environmental ethics unsatisfactory because it appears not to recognize other creatures' inherent right to share the planet and considers only their value to human beings.

Biocentric environmental ethics seeks to correct this deficiency by according moral standing to non-human creatures. Humans have moral worth but only as one species among many living things that also have moral standing. The grizzly bear's right to sufficient domain for sustaining life and reproduction has as much moral weight (if not more) as a logging company's desire to make a profit in that domain. Even if maintaining the grizzly bear's habitat means some humans must live in somewhat less spacious homes, the loss of human utility by no means cancels the animal's moral claim to the forest. In short, animals have rights. Which animals have moral standing and whether plants do as well remain matters of dispute among biocentric theorists. Nonetheless, biocentric theory expands the moral realm beyond humans and hence implies greater moral obligations than anthropocentric ethics.[1]

Ecocentric theory tackles a problem at the heart of biocentric theory. In reality, ecosystems work on the principle of eat and be eaten. We may accord the grizzly "rights" but the bear survives by consuming salmon, rodents, and so forth, thus violating other living creatures' right to life. Humans are simply part of a complex food chain or web of life. Given this, ecocentric theory asserts that moral status should attach to ecological wholes, from bioregions to the planetary ecosystem (sometimes called Gaia). Ecocentric theorists are not concerned about particular animals or even species, but with the entire evolutionary process. Evolution involves the "land" broadly understood to include all its organic and nonorganic components. To disrupt or destroy the evolutionary process, reducing the diversity of life and the stability and beauty of the natural system, is unethical. As Aldo Leopold, the environmental philosopher who first developed the land ethic, put it in his 1949 book, *A Sand County Almanac and Sketches Here and There*, "A thing is right when it tends to preserve the integrity, stability, and beauty of the biotic community. It is wrong when it tends otherwise." The emphasis here is on the word "community."

Each of these approaches suggests the need for change in the practice of international politics. Anthropocentric environmental ethics implies the least extensive reform, although these still could be far-reaching, especially with regard to current economic arrangements. Developed industrial economies rely heavily on the global commons for "free" natural services, such as areas to dispose of pollutants. For example, reliance on fossil fuels leads to increases in CO_2 in the atmosphere, and in turn to global warming. Developing countries undergoing industrialization will draw on the atmosphere's capacity to absorb greenhouse gases. The added load, along with already high levels of emissions from developed countries, could push the environment beyond a critical threshold, setting off catastrophic climate changes because of global warming. These climatic upheavals could lead to crop failures and destructive storms battering coastal cities. What is fair under these circumstances? Should developed countries make radical changes—such as decentralizing and deindustrializing—in their economic arrangements? Should they refrain from adding the potentially disastrous increment of greenhouse gases that will push the climate over the threshold of climatic catastrophe? If yes, then anthropocentric environmental ethics calls for far-reaching social and economic reform.

Biocentric environmental ethics also implies considerable economic reform. If animals have moral standing, then killing them or destroying their habitat for human benefit is unacceptable. In particular, the massive species loss resulting from deforestation is a moral failure even if humans profit. Likewise, agricultural practices that rely on pesticides and fertilizers that harm nonhuman species should be curtailed. Warfare's effects on nonhuman living things would also need to be evaluated. Just-war theory generally evaluates collateral damage's significance in the context of civilians killed or injured due to military operations. Yet collateral damage also kills and injures animals that have even less stake and less say in the conflict than civilians. Should their right to life be considered? Biocentric ethics would say yes. If so, virtually the entire practice of modern war might be held as inherently immoral.

Ecocentric ethics implies the strongest critique of current practices. Disrupting the ecological cycle or the evolutionary process is morally unacceptable. Most current economic or military practices would not pass muster. Indeed, in its strong form, ecocentric ethics would require a major reduction in the human population, since the 6 billion people now on earth are already disrupting the evolutionary process and will continue to do so as world population grows to 10 billion or more. Political institutions must be replaced, either with one-world government capable of implementing ecocentric environmental policy, or with ecologically based bioregional political units (ecocentric theorists hold differing views on whether authoritarian government or more democracy is needed to make ecocentrism effective in practice). If bioregionalism were adopted, world trade would come to a halt since each bioregion would be self-sustaining. Wasteful resource use would be curtailed. Long-term sustainability in harmony with the needs of other living things would be the desired end. For some ecocentric thinkers, the model is a hunter-gatherer society or a peasant agriculture society.

The gap between what environmental ethics calls for and what international political theory postulates may find its bridge in the land ethic.

Environmental ethics in each form carries important implications for the practice of international politics. Yet the environmental ethics literature usually pays little attention to obvious features of the international system. This is not to say that environmental ethics bears no relationship to political realities. If realism (the theory of power politics) and liberal institutionalism (the theory emphasizing interdependence and the possibilities for cooperation) both contain implicit environmental ethics, then environmental ethics contains implicit political theory. Yet without explicit attention to international political theory, environmental ethics lacks the basis to determine which of its recommendations is feasible, and which utopian.

BRIDGING THE GAP

The gap between what environmental ethics calls for and what international political theory postulates may find its bridge in the land ethic. The land ethic, as formulated by J. Baird Callicott, recognizes that environmental obligations are only part of our moral world.[2] Although the land ethic implies significant change in existing practices, it does not necessarily call for abandoning the sovereign state, relinquishing national identity or authority to a world government, or even abolishing capitalism. Rather, it asks for balance between human needs and the requisites of preserving the diversity of life flowing from the evolutionary process. The land ethic simply states that which enhances the integrity, stability, and beauty of the land is good, and that which does not is bad. Human intervention can serve good purposes by this standard. (Indeed, Aldo Leopold was himself a hunter, and found no contradiction between that pursuit and his commitment to the land ethic.) Presumably, human-induced changes to the landscape must be evaluated in context, assessing positive and negative effects.

Yet if the land ethic is to provide guidance for international politics, it needs to identify the other values that must be balanced with its requirements. In international politics, these can be determined in terms of mainstream international political theory, realism and liberal institutionalism. Realism and liberal insitutionalism capture much about how the international system works and what values shape international political practices. Thus we can observe the world for clues as to what international theory entails for the kind of world we should create. Although the verdict is not positive for existing practices, this does not foreclose the possibility of change within either paradigm. But how specifically does realism and liberal institutionalism see the relationship of humans to nature?

REALISM AND THE ENVIRONMENT

Realism is generally understood to be amoral. States do as they must to survive. Survival can justify breaking agreements, lying, deception, violence, and theft. Those who fail to play the game disappear. Those who are best at the game dominate the others. Morality, when invoked, is usually a cover for state interests. Certainly, some prominent realists have said otherwise. Hans Morgenthau recognized the moral content of foreign policy, as did E. H. Carr and Reinhold Niebuhr. Nonetheless, realists usually observe the human capacity for "evil" when the stakes are high.

But this negative perspective on morality obscures realism's highly moral claims. Realism asserts that humans naturally form groups, which experience conflicts of in-terest because resources are scarce. Maintaining the group's autonomy and freedom is the highest good. On it depends the ability of a people to work out their destiny within the borders of the state. Implicitly, this moral project justifies the extreme measures states undertake. Environmental ethics must recognize this as an extremely powerful moral claim. At the same time, international relations theory must recognize that staking this claim, which superficially appears to be a social question, implies a view of how humans should relate to the natural world.

Realism assumes that the state system, or at least some form of power politics involving contending groups, will characterize human relations as long as humans inhabit the planet. The possibilities for environmental (or any other) ethics are limited by this evidently permanent institutional arrangement. Virtually every state action must be evaluated in terms of the relative gains it offers with other states. The struggle for survival and dominance is an endless game in which any minor advantage today could have profound consequences tomorrow. Moreover, as Machiavelli observed, chance plays such a large role in human affairs that immediate advantage is all the prudent policymaker can consider. Thus, to think about long-term environmental trends, for example, is impractical, because an actor that sacrifices present advantage for future gains may not be around to enjoy the fruits.

Perhaps the most significant implication of realism lies in its emphasis on military security. Military imperatives dictate that states develop and deploy the most effective military technology available. The effects on the land of the particular choices made are little considered. No military technology could be more environmentally damaging than nuclear weapons, but these weapons confer maximum national power. Thus the environmental effects of producing, storing, deploying, and dismantling them (not to mention the effects they would have on the environment if ever used) are considered secondary. Here we see how realism as an international political theory is at the same time an implicit environmental ethic: land has little or no moral worth. This is a choice about how humans are to relate to the natural world, not only a choice about how states (or humans) are to relate to one another. A similar argument could be made about the entire range of military technologies, from cluster bombs to napalm to defoliants to biological weapons.

According to realism, the economic institutions of a society must support the most effective military establishment. Societies that attempt to structure economic relations along other lines, such as long-term sustainability, will soon find themselves overwhelmed by other states that make choices geared toward military dominance. States that wish to survive will emulate the most successful economic systems of other states and constantly seek economic innovations that will give them the edge. Capitalism as practiced in the United States and other major Western nations seems to be most compatible with military preponderance.

Realism thus implicitly endorses capitalism, albeit only because it is the most successful economic system at present for enhancing national power (as the former Soviet Union discovered). Capitalism has put the United States at the top of the international order. Others fail to emulate the United States at their peril. Moreover, realism would suggest that because economic growth facilitates military preparedness, autonomous economic growth should override other goals, including environmental protection. Saving a wetland will not contribute as much to national security as producing goods for export. Consequently, realism's emphasis on security leads to embracing the market in its most environmentally heedless form.

Realism's attitude toward the land is that it is territory, an asset of the state, a form of property. The land's status is as a mere resource with no moral standing apart from human uses. It has no life of its own. Prudent management is the most that is morally required. Hence realism shares the modern notion of nature as a spiritless "other" that humans can rightfully manipulate to serve their own ends.

We see that realism's strong moral claim—that a people's right to determine their own destiny, to define and develop their own idea of freedom and the good society, without interference from others—contradicts the institutional arrangements that realism produces. In practice, few alternatives are available. A people who decide that their destiny is to live in harmony with the land, to follow the land ethic's central precept, would quickly lose the freedom to do so through conquest and domination by other states. Aside from the evident fact that many environmental problems require cooperation across political jurisdictions, the competitiveness of the state system ensures that environmental consciousness will not long guide state policy.[3] Realism thus implies an environmental ethic; unfortunately, it is a most pernicious one. Equally unfortunate for environmental ethics is that realism is an undoubtedly incomplete, but not inaccurate, description of how the international system works.

Yet realism's historical and social argument rests on the moral claim that the group is the highest value: that it is within the group that some conception of the good society can be pursued. But surely a good society is one that fosters environmental values, a goal that suffers when nations pursue national security at all costs. As environmental crises mount, the contradiction at the heart of realist ethics becomes more obvious. Perhaps this can lead to changed conceptions of morality.

Ethical standards change. Nationalism, which underpins today's state system, has not driven human behavior for all history (nationalism, for example, had little influence in feudal Europe). Humans can change their view of how best to pursue a vision of the good society. To the extent that the land ethic becomes part of the moral dialogue, institutional change to bring about a healthier human-environment relationship is possible. Nonetheless, realism reminds us that the road to ethical change toward a land ethic likely will be long and hard.

LIBERAL INSTITUTIONALISM AND THE ENVIRONMENT

Liberal institutionalism is far more ready to accept that universal values such as respect for the land exist. Unlike realism, the liberal perspective considers human rights standards to apply across boundaries and cultures. Individuals are the moral agents and moral objects of liberal thought. Individuals have rights that exist regardless of their cultural heritage.

Furthermore, liberalism asserts that these individuals have a particular character. Partly self-interested and partly altruistic, individuals are aware of their dependence on collective action to obtain the good life. The liberal individual also acts, or should, through enlightened self-interest, which is the best way to secure the means of life and protection against bodily harm. Liberal individuals are predisposed to make certain choices. But would they choose a different way of life, namely, one more in harmony with the land?

The question becomes pertinent because it is not at all clear that liberal society is sustainable. Liberalism as manifested in practice is strongly committed to the market system. Indeed, the entire point of liberal institutional international political economy is to find the means to open the world economy to free trade and investment. From this perspective, environmental problems become unintended side-effects of otherwise desirable industrialization and economic growth. The problem for liberalism is simply managing these unfortunate consequences in ways that maintain the open economy. But as the modern market system encompasses more of the globe and penetrates deeper into social life, profound social choices occur, the result of the incremental effects of countless discrete, uncoordinated individual actions. Liberals are comfortable with this way of making social choices due to their faith in progress; the mounting ecological catastrophe might speak against this optimistic view.

Liberal institutionalism cannot escape its entanglement with and commitment to the capitalist market system. In effect, this means that liberal institutionalism can only with difficulty critique that system as it has developed in history. Hence liberal institutionalism will continue to see normal diplomacy and statecraft, the operations of multinational corporations, the growth of free trade and investment, and rising interdependence as progress toward a better world. This in turn exhibits liberal institutionalism's environmental ethic: managerial, limited to mitigation of the market's worst effects, and committed to economic growth and development. The world we should build is on display. It is embodied in the more enlightened liberal states, those that combine commitment to individual liberty, representative democracy, and free enterprise with some degree of environmental awareness. It is industrialized, or postindustrial. It is technologically advanced. It provides a wide range of goods and services to consumers. Environmental concerns enter by way of interest groups devoted to the "issue" rather than as fundamental values that determine which practices to retain and which to abandon.

Like realism, liberal institutionalism captures a large part of the truth about how contemporary international politics operates. It suggests emphasizing certain trends in the hope of dampening others; strengthening the forces for globalization to reduce the impact of military competition. But its commitment to the predominant global economic institutions leaves little room for a land ethic. Liberal institutionalism's anthropocentrism and consequent emphasis on economic growth leads to a relative lack of concern for the stability, integrity, and beauty of the land. Nonetheless, liberal institutionalism is far more open to the possibility of value change and political transformation than is realism. Liberal theory's faith in progress can imply that liberalism itself eventually will be transcended in favor of more earth-centered ethics. Yet current liberal international theory does not recognize or embrace this possibility. To the extent that theory is practice, liberal international theory contributes to the worsening environmental crisis rather than offering a way out.

A NEW DIALOGUE

Both realism and liberal institutionalism are implicitly environmental ethics. They tell us the relationship humans should have with nature, even if they largely base their claims not on ethical choice but on what we must do under existing circumstances. But humans can make conscious choices about what kind of international order to create and maintain. The realist imperative to play the game of power politics or be eliminated from the system depends on a prior choice about ethics and practice. It precludes the possibility of collaboratively engaging the "other" in democratic dialogue aimed at discovering different social practices that do not, for example, lead to environmentally heedless arms races. The "other" must always remain other in realist thought, an assumption that is far from proven. Likewise, liberalism's imperative to rely on the market if we are to achieve individual liberty and social progress is open to question. If the individual is constituted in community—that is, by social practices—then the self-definition of the community can

change. Acquisitive individualism and consumerism need not define the individuals.

How is change to come about? More authentic democracy, based on unforced, open discourse, affords the opportunity to choose consciously the kind of world we are to build. The choice need not come about indirectly, as the result of more immediate decisions on how to achieve national security, nor need it occur unintentionally as individuals make the best of circumstances not of their choosing. Engaging in this dialogue will require abandoning the notion that nature and the social are distinct. The social and the natural are inextricable. We constitute nature through our practices at the same time that we constitute the social world. Thus an ecologically informed political discourse will be one that recognizes the environmental ethics embedded in all political worldviews.

Environmental ethics will not be seen as an add-on to be approached after the important issues of security and economics have been settled. Instead, we will recognize that all our important social choices are inherently about the "natural" world we create. We will consciously raise the question of what this particular action means for that world, and we will recognize that it is our responsibility, not something external to us.

NOTES

1. For more on anthropocentric and biocentric ethics, see J. Baird Callicott, *In Defense of the Land Ethic: Essays for Environmental Philosophy* (Albany: State University of New York Press, 1989).
2. See Callicott, *In Defense of the Land Ethic.*
3. This competition also influences the abilities of states to cooperate to deal with transnational environmental problems. States are expected to attempt to free ride or otherwise exploit the global "commons." If environmental cooperation occurs, it is likely due to a hegemonic power or small group of large powers imposing an international regime. Of course, the Hobbesian use of power to make and enforce law is the antithesis of democratic decision making (which could well be a major element of the society's vision of the good life). But because states are and must be short-sighted and self-interested, no alternative to coercive imposition of regimes exists. Whether such regimes would conform to the requisites of long-term environmental sustainability—much less to the integrity, beauty, and stability of the land—is doubtful.

JOHN BARKDULL *is an associate professor of political science at Texas Tech University. His research interests include international political theory, international ethics, and environmental policy.*

Index

Index

Test Your Knowledge Form

We encourage you to photocopy and use this page as a tool to assess how the articles in *Annual Editions* expand on the information in your textbook. By reflecting on the articles you will gain enhanced text information. You can also access this useful form on a product's book support Web site at *http://www.dushkin.com/online/*.

NAME: DATE:

_____

TITLE AND NUMBER OF ARTICLE:

_____

BRIEFLY STATE THE MAIN IDEA OF THIS ARTICLE:

_____

LIST THREE IMPORTANT FACTS THAT THE AUTHOR USES TO SUPPORT THE MAIN IDEA:

_____

WHAT INFORMATION OR IDEAS DISCUSSED IN THIS ARTICLE ARE ALSO DISCUSSED IN YOUR TEXTBOOK OR OTHER READINGS THAT YOU HAVE DONE? LIST THE TEXTBOOK CHAPTERS AND PAGE NUMBERS:

_____

LIST ANY EXAMPLES OF BIAS OR FAULTY REASONING THAT YOU FOUND IN THE ARTICLE:

_____

LIST ANY NEW TERMS/CONCEPTS THAT WERE DISCUSSED IN THE ARTICLE, AND WRITE A SHORT DEFINITION:

We Want Your Advice

ANNUAL EDITIONS revisions depend on two major opinion sources: one is our Advisory Board, listed in the front of this volume, which works with us in scanning the thousands of articles published in the public press each year; the other is you—the person actually using the book. Please help us and the users of the next edition by completing the prepaid article rating form on this page and returning it to us. Thank you for your help!

ANNUAL EDITIONS: Global Issues 05/06

ARTICLE RATING FORM

Here is an opportunity for you to have direct input into the next revision of this volume.
We would like you to rate each of the articles listed below, using the following scale:

1. **Excellent: should definitely be retained**
2. **Above average: should probably be retained**
3. **Below average: should probably be deleted**
4. **Poor: should definitely be deleted**

Your ratings will play a vital part in the next revision.
Please mail this prepaid form to us as soon as possible.
Thanks for your help!

RATING	ARTICLE	RATING	ARTICLE
	1. A Special Moment in History		34. The Ultimate Crop Insurance
	2. America's Sticky Power		35. Medicine Without Doctors
	3. Five Meta-Trends Changing the World		36. Countdown to Eradication
	4. Holy Orders: Religious Opposition to Modern States		37. The New Containment: An Alliance Against Nuclear Terrorism
	5. The Big Crunch		38. Are Human Rights Universal?
	6. Scary Strains		39. The Grameen Bank
	7. Bittersweet Harvest: The Debate Over Genetically Modified Crops		40. America's Looming Creativity Crisis
	8. Deflating the World's Bubble Economy		41. Women Waging Peace
	9. Shifting the Pain: World's Resources Feed California's Growing Appetite		42. Exploring the "Singularity"
	10. Water Scarcity Could Overwhelm the Next Generation		43. Why Environmental Ethics Matters to International Relations
	11. Vanishing Alaska		
	12. The Complexities and Contradictions of Globalization		
	13. Three Cheers for Global Capitalism		
	14. The Five Wars of Globalization		
	15. Will Globalization Go Bankrupt?		
	16. Soccer vs. McWorld		
	17. Croesus and Caesar		
	18. Where the Money Went		
	19. Render Unto Caesar: Putin and the Oligarchs		
	20. Is Chile a Neoliberal Success?		
	21. The Fall of the House of Saud		
	22. Thirty Years of Petro-Politics		
	23. India's Hype, Hope, and Hazards		
	24. How Nike Figured Out China		
	25. What's Wrong With This Picture?		
	26. The Transformation of National Security		
	27. Nuclear Nightmares		
	28. Lifting the Veil: Understanding the Roots of Islamic Militancy		
	29. The Great War on Militant Islam		
	30. Changing Course on China		
	31. The Korea Crisis		
	32. Strategies for World Peace: The View of the UN Secretary-General		
	33. Peace in Our Time		

(Continued on next page)

BUSINESS REPLY MAIL
FIRST CLASS MAIL PERMIT NO. 551 DUBUQUE IA

POSTAGE WILL BE PAID BY ADDRESEE

McGraw-Hill/Dushkin
2460 KERPER BLVD
DUBUQUE, IA 52001-9902

NO POSTAGE NECESSARY IF MAILED IN THE UNITED STATES

ABOUT YOU

Name

Date

Are you a teacher? ☐   A student? ☐
Your school's name

Department

Address   City   State   Zip

School telephone #

YOUR COMMENTS ARE IMPORTANT TO US!

Please fill in the following information:
For which course did you use this book?

Did you use a text with this ANNUAL EDITION?   ☐ yes ☐ no
What was the title of the text?

What are your general reactions to the *Annual Editions* concept?

Have you read any pertinent articles recently that you think should be included in the next edition? Explain.

Are there any articles that you feel should be replaced in the next edition? Why?

Are there any World Wide Web sites that you feel should be included in the next edition? Please annotate.

May we contact you for editorial input?   ☐ yes ☐ no
May we quote your comments? ☐ yes ☐ no